Piling Engineering

Piling Engineering

Third Edition

Ken Fleming, Austin Weltman,
Mark Randolph and Keith Elson

Taylor & Francis
Taylor & Francis Group

LONDON AND NEW YORK

First published 1985 by Kluwer Academic Publishers

Second edition published 1992 by Blackie & Son Ltd

This edition published 2009 by Taylor & Francis
2 Park Square, Milton Park, Abingdon, Oxon OX14 4RN

Simultaneously published in the USA and Canada
by Taylor & Francis
52 Vanderbilt Avenue, New York, NY 10017

First issued in paperback 2020

*Taylor & Francis is an imprint of the Taylor & Francis Group,
an informa business*

Typeset in Sabon by Keyword Group Ltd

British Library Cataloguing in Publication Data
A catalogue record for this book is available
from the British Library

Library of Congress Cataloging-in-Publication Data
Piling engineering / Ken Fleming ... [et al.]. – 3rd ed.
   p. cm.
  Includes bibliographical references and index.
  1. Piling (Civil engineering) I. Fleming, W. G. K.

TA780.P494 2008
624.1′54–dc22   2007044205

ISBN13: 978-0-367-65938-7 (pbk)
ISBN13: 978-0-415-26646-8 (hbk)

Contents

Chapter 1

Introduction

Since man first sought to establish secure dwellings and to cross streams and rivers where fluctuating water levels gave an element of uncertainty to travel arrangements, the driving of robust stakes or piles in the ground has provided a means whereby the hazards of living could be reduced.

Various peoples in different parts of the world found it convenient to dwell by lake shores where food, water and easy transport were readily to hand and where water levels remained reasonably constant. Evidence of piled settlements has been found on the borders of lakes in, for example, Switzerland, Italy, Scotland and Ireland.

It is believed that some of these settlements were in use about 4000 years ago and they were sometimes of considerable size, as on the shores of Lake Geneva opposite Morges. In another settlement at Robenhausen, it has been estimated that over 100 000 piles had been used. In this case the piles were beneath a covering of peat moss which it is reckoned would have taken at least 2000 years to form. Recent archaeological excavations elsewhere in Switzerland and in France have confirmed that piles were used widely for housing along marshy lake shores.

At Lough Drumkeery in Co. Cavan, Ireland, about 30 000 ancient piles including primitive sheet piles were found in 1863. These consisted of birch and oak, and the oak piles had been carefully pointed and driven to a depth of about 3 m.

The first historical reference to piling appears to be by Herodotus, the Greek writer and traveller who is sometimes referred to as the 'father of history' and who lived in the fourth century B.C. He records how a Thracian tribe, the Paeonions, lived in dwellings erected on lofty piles driven into a lake bed. The piles were driven under some kind of communal arrangement but after a time a law had been made that when a man wished to marry, he had first to drive three piles. Since the tribe was polygamous, the number of piles installed was considerable. This system provided a unique and practical method of founding an expanding settlement.

One of the earliest uses of piling appears to have been by the Phoenicians who used sheet piles for dock and shore constructions in connection with their expanding sea trade. In effect the sheet piles appear to have been derived from the skill of boat builders in the planking of ships.

The cedars of Lebanon were exported to the Egyptians who were great sailors and builders but were without suitable timber – soon cedar wood sheet piles were used to enable them to sink wells. Indeed the sustained demand for cedar was such that in time the great cedar forests of the area were reduced to only small remnants.

Greek and Roman engineers used piles for shore works at many places along the Mediterranean coast, but in this situation they encountered difficulties with the destructive 'teredo navalis' so that few of their structures have survived in this environment to give archaeological evidence. In Britain, a Roman bridge spanned the Tyne at Corbridge, about 20 m west of Newcastle on Tyne, using piles to support the construction. These were discovered when an old bridge was taken down in 1771, and it was found that they were of black oak and about 3 m in length. Much evidence exists elsewhere, testifying to the skill of the Romans in solving difficult foundation problems. The Emperor Trajan built a bridge over the Danube, the foundations of which were exposed during the eighteenth century. When examined it was said that the surface of the timber was petrified to a depth of about 20 mm, and that the timber beneath this surface layer was completely sound. About A.D. 0 Vitruvius, a Roman architect, wrote a treatise *De Architectura*, and described in it a method of sheet piling for use in creating dams and other water-retaining structures, again testifying to the skills of his generation.

The city of Venice came into being, it is believed, as refugees began to seek security on the islands of the Lagoon of Venice from the barbarian invaders who pillaged the remnants of the Roman Empire. The position of the city, in an easily defended location, was determined in the early eighth century and piles were driven for the support of all the buildings which were gradually constructed as the city became more and more prosperous.

Amsterdam, founded about 1000 years ago, was built almost entirely on piled foundations of 15 to 20 metres length; the piles in this case were sawn off level and capped with thick planks. The Romans by contrast often capped their piles with a mixture of stone rubble and concrete.

It is of interest to note that Creasy (*An Encyclopedia of Civil Engineering*, 1861) says that in Holland piling and capping by planking was still in use, with rough stones rammed between the planks, but that timber and stone used together did not produce 'solidity and duration'. He comments that pile driving has for its object the consolidation of a soil which is not sufficiently compact and he recommends the test loading of piles, with load being kept in place for long enough to remove any doubts about the security of the piles.

Little progress in the art of driving piles seems to have been made between Roman times and the beginning of the nineteenth century. The 'Pile Engine' of the early 1800s can have varied little from the Roman type. For a small engine a timber frame was used to suspend the 'ram' which was actuated by a single rope, divided at its end into as many small ropes or 'tails' as required. At each 'tail' a man worked and the method was known as the 'whip ram' method. Small rams would have a mass of about 120 kg and often consisted of oak strongly hooped by iron. Coulomb stated that for this kind of work a man's effort was normally the equivalent of raising a 75 kg weight through a height of 1000 m in one day. For man operated piling frames in about A.D. 1700 it was said that 20 men could operate a 360 kg ram, the procedure being to raise the ram 25 or 30 times in succession over a period of about 1 min. There was then an equal rest period, and the 'volley' was repeated, but it was said (somewhat sadly) that a further minute was often wasted. Larger engines raised their heavier cast-iron rams, which could weigh up to $1\frac{1}{2}$ tonnes, by a winch, which might be powered by men or horses. When the ram reached a prearranged height a scissor catch at the ram head compressed and disengaged it, so that it slid down its guide to produce the necessary impact.

At this time, as no doubt for hundreds of years previously, it was fully appreciated that timber piles subject to wetting and drying cycles were much less durable than piles which were permanently submerged. Where permanent submergence was not possible, the piles above water were often coated with tar. Iron shoes were also fitted to piles where driving involved penetration into stony layers.

It was in the nineteenth century that, as in many other engineering fields, big changes began to take place both in terms of materials and motive power. Metal piles, mostly in the form of cast-iron pipes, became available in the mid-1830s and these were used for important structures on account of their durability. Screw piles were first used for the foundations of the Maplin Sands Lighthouse in the Thames Estuary in 1838, the inventor being Alexander Mitchell. In 1824 Joseph Aspdin patented his hydraulic cement, later known as Portland cement on the grounds that its appearance bore a resemblance to Portland stone. While in his initial patent he apparently withheld some manufacturing details, by the middle of the century hydraulic cement had become widely available and before the end of the century Coignet and Hennebique, French competitors, had introduced the successful reinforced concrete which was later to be brought to Britain by L.G. Mouchel.

Steam power had first been applied to the driving of piles in Britain by John Rennie in 1801 or 1802 at the Bell Dock at the entrance to the London Docks. The ram was hoisted by an 8-hp engine, constructed by Boulton and Watt, and this engine, or one similar to it, was used again at Hull docks in 1804. In 1843, Nasmyth produced a radical departure from this type of machine, introducing a revolutionary type of hammer. In this new hammer the ram was attached to the lower end of the piston rod and the whole weight of the steam cylinder acted on the pile. It achieved striking rates of 80 blows per min and was used, for example, at the site of the Newcastle High-Level Bridge to drive 70-ton piles, the initial supervision of the works being carried out by Robert Stephenson. In 1846, a compressed-air hammer was proposed by Clarke, Freeman and Varley, and in 1849, Clarke and Motley produced a vacuum hammer. This was indeed an era of invention and engineering excitement.

In about 1870, a pile driver powered by gunpowder was launched upon the world by a Mr Shaw of Philadelphia. The machine was claimed to have several advantages over any other type available at the time, not least being that it required only a man and a boy to work it. A cast steel cap or 'mortar' rested on the pile head, in which was formed a recess sufficient to contain a few ounces of gunpowder. Mounted on a frame above this anvil or mortar was a steel dropweight with a projecting plunger, designed on falling to strike the gunpowder in the recess and cause an explosion. This action then threw the dropweight back up into its retainers ready for the next strike. The function of the man was to pull a lever which released the dropweight, while the boy hung on a ladder adjacent to the pile head and threw in the charges of gunpowder as required. The operating rate achieved was 15 to 20 blows per minute, but if it became any faster, the heat of the mortar became sufficient to cause premature explosions, which apparently were not good for the morale of the boy.

A machine of this type was brought to England to the Severn Tunnel works, but it proved unsuccessful in the sandy conditions which had previously baffled other piling machines. It was then brought to London to St. Katherine's Dock to work for the

East London Railway but here its use was again short-lived. After only three charges of gunpowder had exploded, it caused such a state of alarm among the dock authorities that they ordered its removal forthwith. The same, or a similar machine, was then used to drive piles for the third bridge over the River Elbe at Dresden and was highly praised for efficiency, economy and speed. Piles of 1.8 m to 2.5 m length were driven at a rate of up to 30 per day using 60 gunpowder cartridges per pile. The cost per pile for driving was reported as '8 shillings and 6 pence'.

In 1897, A.A. Raymond patented the Raymond pile system and was, it appears, the first person to develop a practical economical cast-in-place concrete pile. In 1903, R.J. Beale also developed a method of driving a steel pipe, plugged at its lower end, and of subsequently filling it with concrete and then withdrawing the tube, while in 1908 E. Frankignoul, a Belgian, invented an early version of the Franki driven-tube pile which was later developed to provide expanded bases and which became widely known and successful throughout the world. At about the same time the first pre-cast concrete driven piles appeared and although the originator is not known, the earliest drawings of precast concrete piles in the United Kingdom appear to have come from the Hennebique Company.

The use of steel I-beam piles originated in the United States before 1900, when fabricated sections were used for highway bridges in Nebraska, but after 1908 the Bethlehem Steel Co. produced rolled H-sections which quickly captured the market for this type of pile.

The use of steam-operated hammers continued throughout the first half of this century for driving piles of all types but declined in the post-war period after 1946, and in-place diesel-operated hammers became popular. Drop hammers, operated from diesel winches and crane drums, have nevertheless remained to the present time because of the basic simplicity of the method. Now diesel hammers have lost some of their former popularity for environmental reasons, and it would appear that hydraulically-powered hammers are finding increased usage because of relative efficiency and easy variable control.

Bored piles, formed by percussion boring tools, have certainly been known and used in the United Kingdom since the early 1930s, but it would be wrong to think of bored piling as a development of the twentieth century. Long before the term 'bored pile' was coined, and before hydraulic cements came into common use, the 'well foundation' was used in many countries for the support of major structures. Such foundations have been dug in India for example for hundreds of years, with the stone foundation being carried up from the base of each excavation or boring. The Taj Mahal, which was built in the period 1632 to 1650, made extensive use of this type of foundation, and throughout the Mogul period from 1526 this type of support was used for many bridges across deep river beds where scour occurred.

The same technique proved particularly effective for railway bridges when the railways were first introduced into India and was equally successful in England where three railway bridges across the Thames in London at Charing Cross, Cannon Street and Victoria Station were founded on well-type foundations in the mid-nineteenth century, casting what were in effect large-diameter piles within hand-dug caissons.

With the advent of hydraulic cements, however, it was no longer necessary to restrict the dimensions of well foundations to those which would allow workers to descend and construct a pier from the bottom of the excavation upwards, since concrete could be mixed at the surface and poured in. This meant that for relatively light loads, small-diameter piles could be constructed by percussion boring equipment as used for well sinking, and so the art of bored piling developed.

Bored piling offered certain advantages over driven types for at least part of the market, since a small tripod machine could be used to suspend the boring tool and could at the same time provide enough effort to enable relatively short lengths of steel tube, which were necessary to retain soft and unstable upper ground strata, to be removed after concrete placing was complete. When compared with the relatively large machines of the same period needed to insert driven piles, the equipment was small and could readily operate on sites to which the larger machines could not gain easy access.

In its early days, the method was probably a good deal less reliable than the methods used to install driven piles, but as companies built up experience, reliability improved. Nevertheless it was still difficult to cope with water inflows at depth, and in order to overcome this problem the 'pressure pile' was invented. In this system an airlock was screwed on to the top of the steel lining tube which was used to stabilize the hole. This was done at the end of the excavation stage and, using appropriate air pressures within the tube, water was expelled from the boring. Concrete was then placed in batches through the airlock and thus the segregation which might have occurred had concrete been poured directly into water, or by means of an unreliable bottom-opening skip, was avoided. The method proved rather cumbersome and has now fallen into disuse. Instead concrete is now poured into water-filled boreholes using tremie pipes.

Today, although small tripod or similar percussion boring equipment remains, the majority of bored piles are formed using rotary augering machines. This results from rapid development of this type of plant since about 1950. Machines are now available to bore a wide range of pile diameters from about 150 mm to over 2 m, while yet other equipment is designed to enable concrete or grout to be injected to form a pile through the hollow stem of a continuously flighted boring auger as the laden auger is withdrawn from the soil.

At the same time driven piles have developed considerably and two main forms exist for land-based work, namely the pre-cast jointed pile and the cast-in-place pile. The former type of pile replaces to a large extent the long precast piles which were common in the early part of this century and requires much smaller and lighter handling and driving plant, while the latter offers a cheap and reliable pile where the pile penetration required is not more than about 18 m.

Steel H piles and tubular steel piles are used for limited applications. The H section pile causes minimal soil displacement and can stand up to fairly heavy driving conditions while the tubular pile finds application in many shoreline works where both direct load and bending capacity are often needed. Very large and long tubular piles are used extensively in major offshore works such as the founding of oil platforms at sea. The cost of steel is however such that these types of pile find only rather occasional application for inland jobs.

The design of piles and pile groups has advanced steadily in recent times with much of this work being carried out by engineers who specialize particularly in foundation engineering. Effective stress methods are being developed for individual pile designs, while computer-based techniques for assessing the settlement behaviour of large and complex pile groups are finding increasing application.

The construction of piles today is also carried out for the most part by specialist contractors who exercise considerable skill in adapting methods of working to enable sound piles to be formed even in the most forbidding ground conditions, although it should still be borne in mind that every technique used can encounter difficulties and unforeseen problems at times. The skill of a given contractor often depends on their particular experience and this is only gained over many years, perhaps using a range of methods.

In modern piling practice site investigation has become an essential precursor to making sensible decisions both regarding pile design and the choice of appropriate construction methods, and is all the more important because of the wide range of methods and equipment now available. Financial savings made at the investigation stage are often followed by the unwelcome discovery of adverse conditions which can eventually impose an extra burden of cost far exceeding the earlier savings which may have been made.

Clearly it is essential that, with ever-increasing pile stresses which are consequent upon the strong incentives which exist to use materials economically, piles must be made both sound and durable. The integrity of different types of pile in a wide range of ground conditions has been investigated and presented in various Construction Industry Research and Information Association reports (Healy and Weltman, 1980; Sliwinski and Fleming, 1984). The reports PG2 and PG8 give some insight into the potential problems which can arise in certain circumstances. At the same time the development of integrity testing methods now provides a means for detection of major pile construction problems, and this in turn implies that pile construction can reach higher standards than ever before (Thorburn and Thorburn, 1977; Weltman, 1977).

Good specification of piling works is also important, and with reference to several of the common types of piling, information on this is provided in the Institution of Civil Engineers publication *Specification for Piling and Embedded Retaining Walls* (2007). This should at least be a standard reference work for those preparing such documents for any land-based contract. So much in piling depends on an exercise of judgment by an experienced engineer that those who undertake this type of work must remain both well informed and close to the realities of dealing with one of nature's most variable materials, the ground. Performance in the field equally requires alert supervision and sound understanding of the processes.

While the evaluation of methods and the advancement of engineering knowledge have changed the piling industry in important ways throughout its history and while the pace of change has been particularly rapid in recent years, it should still be realized that whatever the equipment or methods employed, a great deal of reliance must be placed on individuals whose skill, experience and honest endeavour eventually turn engineering concepts into dependable reality. It has ever been so.

Chapter 2

Site investigation for piling

2.1 Planning

The planning of a site investigation for piling works is essentially similar to that for shallow foundations, apart from the obvious requirement for greater depth of exploration. The main additional factors that should be considered are the practical aspects of pile installation, including the potential need for a piling working platform, the design for which would be based on the properties of the shallow soils.

The first stage in planning a site investigation comprises a desk study of the available information. This subject is comprehensively discussed by Dumbleton and West (1976), Weltman and Head (1983) and BS 5930 (BSI, 1999). It should also be noted that the Eurocode 7 drafts which relate to site investigation are in preparation. Three of these, concerning the description of soil and also rock, are completed (BSI, 2002, 2003, 2004). In some remote overseas locations little published information may be available and in this instance, more emphasis must be placed on the exploratory work. The second stage should include a site inspection in order to confirm, as far as is possible, the data collected during the desk study, and to make as many additional preliminary observations as possible.

The investigation should take into account details of the foundation design if known, including the tolerance of the structure to settlement, since this can dictate the extent of the investigation. Where these details are unknown it is important that the exploratory work provides sufficient information to permit decisions concerning the type of foundation, as well as the actual foundation design, to be made. The major requirement of the investigation in terms of the design is to provide comprehensive information over the full depth of the proposed foundations and well below any possible pile toe level. This is necessary to permit flexibility in the pile design and to be certain that the anticipated strata thicknesses exist. Of equal importance is a thorough appraisal of the groundwater conditions, including its occurrence and response during boring operations. Advice on the depth and number of exploratory holes is given in Eurocode 7.

For the case of straightforward pile design in well-documented homogeneous deposits, it is generally possible to make firm proposals for the number and depth of boreholes at the planning stage, although some aspects of the programme of exploratory work, including the field and laboratory testing, can usually only be outlined. As the complexity of the proposed piling works increases, however, a more flexible approach to the investigation has to be made to allow adequate feedback and analysis of information as it is obtained. This is particularly true where there are large variations in design loads for heavily loaded piles or for piles into rock-head which varies in level or which has a weathered profile. For complex projects, a two-stage approach may be the most cost effective. The second stage is then based on the preliminary findings of the first. Good communications must exist between the site investigation contractor and client in order to tailor the exploration to the requirements of the foundation design and the practical aspects of pile installation *vis-à-vis* the prevailing soil conditions. Where a piled foundation is proposed for an extensive project, piling trials including pile testing can be cost effective and should be included as part of the investigation. Reference may be made to the ICE publication *Specification for Ground Investigation with Bills of Quantities* (1989) for additional guidance. Detailed information of the ground conditions reduces the possibility of encountering unexpected problems due to adverse soil or groundwater conditions. In this respect, it is essential that all the available factual information is passed on to the contractor at the tender stage since the offer to carry out the work will be made on the basis of the information provided.

Not uncommonly, soil deformation at and around a piled foundation is a critical aspect of a foundation design. Soil stiffness (Young's modulus, E, or shear modulus, G) is seldom measured by normal ground investigation methods and does not usually form part of an investigation. However, there has been a growing use of seismic cone penetration testing in recent years, where the small strain shear modulus, G_0, may be estimated from the shear wave velocity in the soil. While G_0 will need to be reduced for application to foundation movements, it represents a relatively simple method for assessing the soil stiffness in situ. When deformation is an aspect of a design appraisal, special measures to assess this aspect of soil performance may need to be considered. Alternatively, stiffness may be estimated from strength measurement, from accumulated data of foundation behaviour related to the specific or near identical circumstances, or from the back-analysis of well-executed pile tests, for example using the (Fleming, 1992) method.

Other specific aspects of the investigation may include matters such as the seismic risk of the area, aggressive soil conditions (especially in fills) and the possibility of aquifer pollution if piles are to penetrate contaminated ground. Consideration should also be given to acceptable levels of noise for the area, the extent of existing foundations adjoining the site and the sensitivity of neighbouring structures and buried services to vibration and soil displacement. For piled foundations the investigation should include comprehensive details of trafficability (which can vary with seasonal changes in the weather) and access (including headroom) for heavy plant. Water supplies and the facilities for waste disposal and site drainage would also be relevant to the investigation, since the installation of deep foundations can produce large quantities of spoil and slurries.

In the following sections, only those aspects of site investigation that relate to piling works are discussed.

2.2 Depth of exploration

The extent of the exploration is not limited solely by the zone of influence caused by the loaded piles, but is dependent on other factors such as considerations of general stability, the necessity to understand the overall groundwater regime, highly heterogeneous soil conditions or the effects of piling on adjacent structures. The zone of

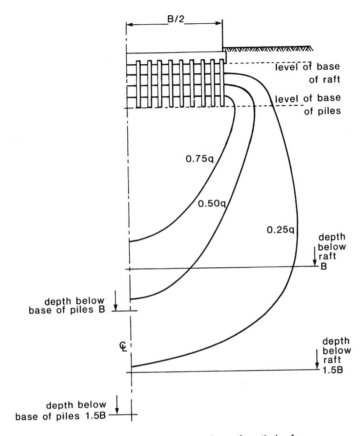

Figure 2.1 Vertical stress envelopes for piled raft.

influence is a good starting point, since it quantifies to some degree the lateral and vertical extent of the stressed zone. In the case of a piled raft, significant stresses are transferred to a considerable depth (see Figure 2.1), and it is recommended that the depth of exploration should be at least one and a half times the width of the loaded area. In the special circumstance where the geological conditions or rockhead level are well known, such depths may be reduced. As a general guide the distance is measured below the base of the pile group, and the loaded area is calculated allowing a spread of one in the horizontal direction to four in the vertical direction over the estimated length of the pile. For isolated pile groups (under a single column for instance), the stressed zone based on the spread of load concept is relatively much shallower (see Figure 2.2). Therefore a single proving borehole should be taken to a depth in excess of $2B$ in order to ensure that 'punching' through a relatively strong stratum into a weaker stratum cannot occur. In addition, the intensity of loading at the tip of end-bearing piles may be high, and a rock-bearing stratum should be adequately investigated particularly where heavy column loads are to be transmitted to single or small groups of highly loaded piles.

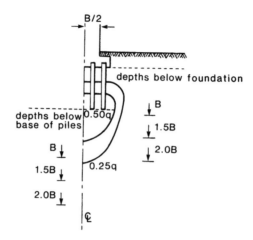

Figure 2.2 Vertical stress envelopes for isolated pile group.

Where piles are to be founded in a bearing stratum of limited thickness (such as a gravel layer overlying clay) it is essential that a large number of bores or probings are employed, possibly spaced on a close grid. These will determine whether the bearing stratum thickness and level is maintained over the proposed area of loading. In these cases, detailed exploration may be restricted to a limited number of locations, permitting the calibration of a lower cost penetration method that may be used to provide the necessary coverage. It is important in this situation that the competence of the underlying stratum to sustain the transmitted stresses is established. Where piles are to be founded in a rock-bearing stratum, the likely variability of rock-head level should be assessed from a geological point of view and the investigation should be sufficient to establish rock-head contours in adequate detail and to obtain profiles of strength and weathering.

It should be noted that the strict geological definition of bedrock (and therefore rock-head level) may not concur with the views of a piling specialist, and this point should be made clear in the dissemination of information. In ground conditions that are known to be of a heterogeneous nature, a qualified geologist or engineering geologist should be involved with the work from an early stage. Even if piles are to be taken through heterogeneous soils to found in underlying bedrock, variations in groundwater conditions, or the occurrence of lenses of silt, sand, large boulders or open work gravel may jeopardize the successful installation of displacement or non-displacement piles and these conditions should be fully investigated.

2.3 Groundwater conditions

The investigation of groundwater conditions is particularly pertinent to piling works. The findings may influence the choice of pile type and it is essential that the piling contractor has the data necessary to assess the rate of water inflow that may occur during the installation of conventional bored piles.

Observations of the ingress of water including the levels at which water seeps and collects should be made and noted on the site investigation logs in such a way that they may be related to the casing and borehole depths. These water level observations will be unlikely to reflect the true hydrostatic conditions, but the information is of use to piling contractors tendering for bored or drilled piles.

Some fine soils of low permeability permit fairly rapid boring and excavation without seepage of water for the period of time that the bore is open, although the groundwater level may be near the ground surface. This is typical of silty clays, such as London Clay. However, the possibility of the occurrence of silt or fine sand lenses within these soils should be recognized and they should be detected if present, since such layers may result in rapid accumulation of water in the bore. Where these conditions are anticipated, appropriate measures can be taken, but the contractor will require the information at the tender stage.

In many situations pile bores in homogeneous clay may be similar in length to the clay layer thickness, and with natural variation in bed levels, there can be a risk of a pile boring extending into underlying strata. The underlying beds may be permeable, and will probably be under a considerable head of water. The site investigation must delineate the interface between the clay and underlying beds with some accuracy, to minimize the risk of penetrating such aquifers during pile construction.

Where piles are bored through clays to penetrate more permeable strata with the rig situated at a reduced ground level such as a basement, a hitherto shallow water table may effectively become artesian at the lower level. If this condition is not recognized in the planning stage it could lead to an unprogrammed and costly revision to the method of working especially concerning the casting of piles by tremie methods.

The heads of water recorded by observation in boreholes, or measurement of levels in piezometers may be subject to seasonal variation, and ideally measurements should be collected over a period of months where the levels are critical. The period should of course include the 'wetter' parts of the year. This is not always possible, but the topography and geology of the surrounding area should enable an assessment of any potential problem to be made. A special case concerns tidal variation. Accurate estimates of maximum groundwater levels may be essential if, for instance, piles are to be constructed using polymer or bentonite mud, when a positive head of support fluid has to be maintained during construction. Any tidal variation in groundwater level that can occur at the proposed location of the piles has therefore to be monitored during spring tides in order to record the maximum potential level.

Whilst measures can be taken to combat erosion of the concrete in bored piles by flowing groundwater immediately after casting, the possibility of this occurrence must be known beforehand, and openwork, highly permeable gravels under a seepage head, should be detected. Their presence may be suspected in glaciated valleys for instance, where coarse granular material may be found in outwash fans. The situation could also arise in alluvial gravels if there is a marked absence of finer material. Simple permeability tests can be useful in the assessment of this problem.

Chemical attack of steel or concrete piles may occur due to dissolved salts in acidic groundwater. Groundwater samples should be taken from boreholes in such a way that they are uncontaminated or diluted by water added during boring or drilling. Special samplers may be required to obtain water samples at given levels. In made ground or backfill such as quarry or mine waste, extremely corrosive conditions may be present and this situation should be adequately investigated. Industrial waste may also give rise to corrosive ground conditions and in built-up areas, the previous use of the ground should be established. Chemical tests on soil samples may be carried out to give additional information with respect to the leachability of contaminants.

2.4 Sampling, in-situ testing and laboratory testing

In certain respects, decisions regarding the optimum programme of sampling and testing may only be made after some knowledge of the ground conditions at the site has been gained, and some conclusions regarding the possible piling alternatives have been reached. This situation emphasizes the need for a flexible approach and the value of preliminary studies to facilitate the initial assessment. In many cases, the selection of a suitable pile type is controlled by external factors such as access conditions, cost, vibration or noise level, so that there is in general no simple way to relate a pile type to the prevailing soil and groundwater conditions.

Guidelines are given in B5930, section 4.3, for sample spacing, the size of bulk samples, types of sampling equipment and the intervals between in-situ tests, and these are applicable to the recommendations given in the following notes.

Some situations may arise where special tests may be required or the intensity of sampling or testing may have to be increased to provide sufficient data on particular aspects. An example would be the estimation of rock modulus in a weak fissured or blocky rock, where, for instance, plate-bearing tests combined with intensive large-diameter rotary diamond coring and possibly other in-situ tests may be necessary. Such an approach would be very dependent on the costs involved in the additional exploration compared with the costs of deeper, or larger (possibly over-conservative) pile foundations.

The usual forms of cable tool boring equipment and rotary drilling equipment are appropriate, but in view of the additional depth of boring likely to be necessary, larger than usual starting diameters for boreholes and drillholes may be prudent. This permits 'stepping down' at an intermediate depth in order to maintain a reasonable rate of progress. Normal sampling and testing methods suffice, but there is a bias towards in-situ static and dynamic penetration tests, especially the standard penetration tests (SPT) and (static) cone penetration test (CPT), since pile design by empirical methods can be based directly on the results. Provided the ground does not contain cemented material or large boulders, cone penetration testing may be preferred to the SPT, particularly as many pile design methods are linked increasingly to the cone resistance. Table 2.1 summarizes typical in-situ tests that may be carried out, and their application to pile design.

2.5 Methods of exploration for various soil types

In the following sections, specific soil types are described with appropriate methods of exploration. Whilst the ground conditions may well closely approximate those described, intermediate or mixed conditions are quite likely, and a combination of methods would then be employed. Pile design is related to the soil conditions, and if established methods of design are to be employed, the investigation should provide appropriate parameters.

Table 2.1 Typical in-situ tests and their application to pile design

In-situ test	Application to pile design
Vane test	For measurement of in-situ undrained strength in soft to firm clays. Results may be applied to the estimation of downdrag on a pile shaft. The re-moulded strength of sensitive clays may be relevant to overall stability problems associated with pile driving in soft clays.
Standard penetration test (SPT)	Investigation of thickness of bearing strata. Direct application of N value in empirical formulae for pile load capacity. Estimation of the angle of friction ϕ', in granular deposits, for use in pile design. Crude estimates of cohesion in stiff clays or matrix dominant fills. For estimation of approximate compressive strength of very weak rock. A tentative relationship between N value and modulus for use in pile/soil systems has been proposed.
Static cone test	Direct application of sleeve friction and point resistance to the design of driven piles. Estimation of the shear strength of clays and production of detailed soil profiles.
Pressuremeter (including self drilling types)	Estimates of modulus of soil for possible application in pile design, but results may not be strictly appropriate. Estimates of shear strength in weak rock.
Plate-bearing test (over a range of depths)	Gives shear strength and modulus in all soil types. The shear strength parameters are highly relevant to pile design, as similar volumes of soil are stressed. The modulus values may not be quite so relevant as the pile installation can markedly influence this parameter. Important that this test is done carefully.
Simple permeability tests	For estimation of flow in permeable gravels or fissured rock. This may be a factor relevant to the selection of pile type. The tests can also be used in weak rock, to indicate highly fissured or fractured zones.

2.5.1 Stiff fissured clays

In these soils, bored cast-in-place piles or low-displacement driven piles are usually employed. At the present time pile design is generally based on values of undrained strength obtained from triaxial compression tests on 'undisturbed' open-drive samples, using the 'α' adhesion factor method. Scattered shear strength values are usual as a result of the fissures and of sample disturbance, and a sufficient number of tests should be carried out to permit selection of a mean or upper and lower bound strength profiles without too much guesswork.

In fissured clays, sample size and orientation influence the results of strength tests. If the 'α' method is employed, using the α values recommended by Skempton (1959), undrained triaxial tests should be carried out on 38 mm diameter specimens cut from open-drive U100 samples. Due care should be exercised to relate the strength obtained from undrained tests to the mass behaviour of the soil below a pile base where the operational strength will generally be lower than that obtained from laboratory tests. Estimates of modulus of deformation from laboratory tests are very approximate, and even the results obtained from plate-bearing tests may not be wholly representative of the installed pile.

In very stiff fissured clays with high silt contents, open-drive sampling becomes impracticable and in-situ dynamic or static penetration tests such as the SPT or CPT may be employed. Pile design from the results of these tests is highly empirical although it is generally adequate for a preliminary test pile. A number of correlations have been proposed between the SPT 'N' value and the undrained shear strength of silt and hard clays. These are given below.

$$c_u = 20N \text{ kN/m}^2 \text{ Meyerhof (1956)}$$

$$c_u = 13N \text{ kN/m}^2 \text{ Terzaghi and Peck (1967)}$$

$$c_u = 7N \text{ kN/m}^2 \text{ Reese, Touma and O'Neill (1976)}$$

$$c_u = 4 \text{ to } 6N \text{ kN/m}^2 \text{ Stroud and Butler (1975)}$$

[Note that modern nomenclature is to use s_u for undrained shear strength rather than c_u, emphasizing the nature of shear strength in clays; however, for historical reasons c_u has been retained in this text].

Clearly the range of values is large, and the correlations should not be employed in isolation, but checked against any available field and laboratory test results for the particular clay being investigated. Corroboration is particularly necessary for the higher c_u derivations. It should be noted that UK practice tends towards the use of lower ratios compared with typical US practice. This could have come about as a result of a number of factors, such as typical SPT field methods and may not relate to differences in the soil. The relationship between 'N' and c_u proposed by Stroud and Butler (1975) is frequently adopted for use in UK non-sensitive clays where other measurements are not available. The Stroud and Butler N–c_u relationship is dependent upon the plasticity index of the clay (PI) as shown in Figure 2.4.

Where the static cone penetrometer is used to estimate soil strength, the undrained shear strength, c_u, may be predicted from the relationship

$$q_c = N_k c_u + p_0$$

where q_c denotes the loading on unit area of the cone (kN/m²).
           N_k denotes the cone factor.
           c_u denotes the undrained shear strength (kN/m²).
           p_0 denotes the total overburden pressure (kN/m²).

Estimation of shear strength by this method has been used extensively in North Sea glacial clays, where electrical cones are used (see Figure 2.3). Cone factors, N_k, of 25 are recommended by Marsland (1976), who also noted that the ultimate bearing capacities measured by the cone were greater by factors of between 1.2 and 3.0 than those obtained from plate-bearing tests. Further information on cone testing may be found in Meigh (1987) and Lunne et al. (1997).

Of the latest methods available for the determination of deformation and strength parameters from in-situ measurements, there is the self-boring pressuremeter such as

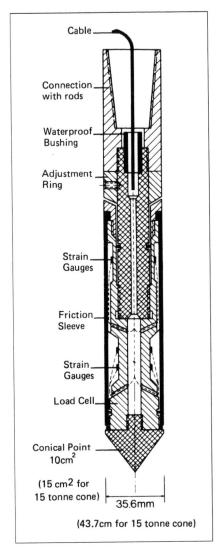

Cable

Connection
with rods

Waterproof
Bushing

Adjustment
Ring

Strain
Gauges

Friction
Sleeve

Strain
Gauges

Load Cell

Conical Point
10cm²

(15 cm² for
15 tonne cone)

35.6mm

(43.7cm for 15 tonne cone)

Figure 2.3 Cross-section of electric static cone. (Illustration courtesy of Fugro Ltd., Hemel
Hempstead.)

the 'Camkometer' or the Marchetti spade dilatometer Powell and Uglow (1988).
Further information on pressuremeter testing may be found in Mair and Woods
(1987).

   Where pile design is to be based on effective stress methods, the sampling programme should provide material for consolidated undrained tests, with measurement
of pore-water pressure in order to estimate the effective strength parameters, in particular the friction angle ϕ'. For driven piles, ring shear tests can provide useful
information on the interface friction angle between pile and soil (Ramsey *et al.*,
1998).

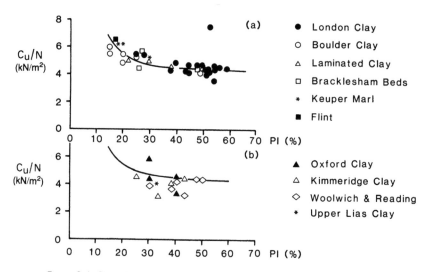

Figure 2.4 Correlation of shear strength with SPT value (Stroud, 1974).

2.5.2 Glacial till

Tills vary in properties considerably and range from heavily over-consolidated clays to coarse grained granular deposits, where both types may contain extremely large boulders. There should therefore be an increased awareness of the probability of significant soil variations, which could be of considerable importance in pile selection and installation. An understanding of the morphology of the terrain under investigation is of considerable assistance in providing an awareness of the likely range of glacial deposits that may be encountered. Whilst progress in boring exploratory holes through glacial soils may be slow, it is imperative that these deposits are investigated fully in depth, a common error being to mistake a large boulder for bedrock. To provide some assurance that bedrock rather than a boulder has encountered, hard deposits should be proved for at least 3 m, and preferably deeper, and bedrock should not be assumed unless there is some corroborative evidence. When it is not possible to penetrate boulders by cable tool methods, rotary coring can be used although this may preclude further soft ground sampling and some types of in-situ test.

Representative sampling of glacial tills is often difficult, and much guidance in this respect can be found in 'Engineering in glacial tills' (Trenter, 1999), but earlier work by McKinlay *et al.* (1974) presents some detailed information on sample recovery and testing, both in the laboratory and in-situ. Open-drive sampling and in-situ testing become more difficult as the proportion of coarse elastic material in a till increases, i.e. as the till becomes Clast-dominant rather than Matrix-dominant, properties that can be reflected in the actual pile installation. In these situations, it is important to establish a grading for the till, and particle size distribution tests on suitably sized bulk samples should be considered. It is difficult to secure samples that are free from sampling bias in a glacial soil of this type, unless the deposits are at shallow depth in which case an excavated trial pit may be considered. The depth of piled foundations usually precludes this approach and the advice of geologists or engineers familiar with the local ground conditions should be sought.

The site investigation borehole logs should record chiselling time and the rate of borehole and casing advance, since this may give valuable information regarding the difficulties of installing small-diameter temporary casings for bored cast-in-situ piles. Resort may be made to rotary diamond coring using a suitable flushing medium, which may include air or drilling mud (Hepton, 1995), although these methods may be extremely costly and core recovery may be poor. The static cone penetrometer has proved a useful method of predicting shear strength in glacial tills in the special environment of the North Sea. As discussed in section 2.5.1, a cone factor, N_k, of 25 has been suggested by Marsland (1976). Other authors recommend factors of between 15 and 30 (Toolan and Fox, 1977) and 15 and 20 (Kjekstad and Stub, 1978).

Less satisfactory, but more economical, is the SPT using a relationship between the 'N' value and undrained strength as proposed by Stroud and Butler (1975) as described in connection with stiff fissured clays. However, wide variations in this relationship (when applied to Matrix-dominant tills) are known to exist, with correlation coefficients of between 2.5 and 10 being cited in published literature. The lower values can be attributed to the presence of granular material in the till, but it is often difficult to avoid this occurrence and few SPT drives are likely to be within the till matrix material alone. Where there is good knowledge of the undrained strength of a till from laboratory tests or back-analysis of the results of pile tests in the locality (Weltman and Healy, 1978; Fleming, 1992) a site-specific correlation may provide a more certain means of determining strength (or modulus) from the SPT.

Whilst in-situ tests overcome the difficulty of sampling in cohesive tills containing granular material, the presence of the clastic components interferes with some of the tests, and every effort should be made to secure undisturbed samples for laboratory tests. If the recovered material is too mechanically disturbed for triaxial testing, moisture content, PL and LL determinations should be made. Note too that there is a minimum shear strength for re-moulded clays based on the Skempton–Northey plot (Skempton and Northey, 1952) of liquidity index against undrained shear strength (see Figure 2.5), which gives a lower bound to the strength of the material. Points representing test results that fall below this line are probably from excessively disturbed samples, or from material that has dried out, the plot being for saturated clays. Envelopes for the undrained cohesion of matrix dominant tills for the 'CL' and 'CH' Casagrande sub-group classification according to moisture content are presented in Figure 2.6. Estimates of the shear strength from various sources can be combined to give upper and lower bounds, and factors of safety selected accordingly.

The notes made in connection with the observation and recording of groundwater conditions are especially pertinent in glacial soils, and great care should be taken to relate the observed conditions to the stratigraphy. To this end, a number of suitably disposed piezometers should be installed to record the hydrostatic head in selected strata.

2.5.3 Fine-grained granular deposits

For the purpose of establishing parameters suitable for pile design, deposits of silts and sands are generally investigated by means of in-situ tests. Dynamic tests such as the SPT

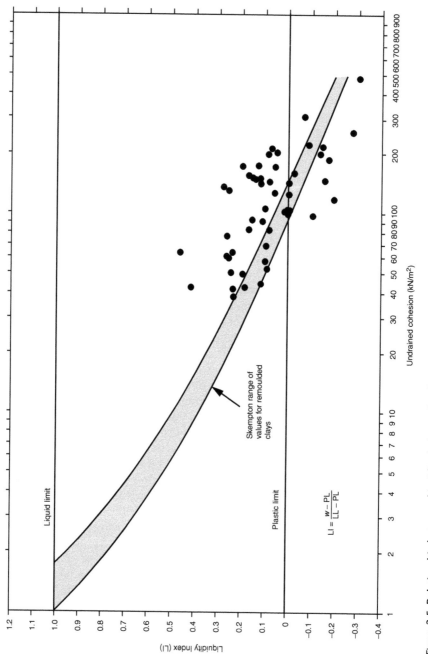

Figure 2.5 Relationship between Liquidity Index and undrained cohesion showing typical range of values in glacial till. CIRIA Survey (1977). (Illustration courtesy of CIRIA, London.)

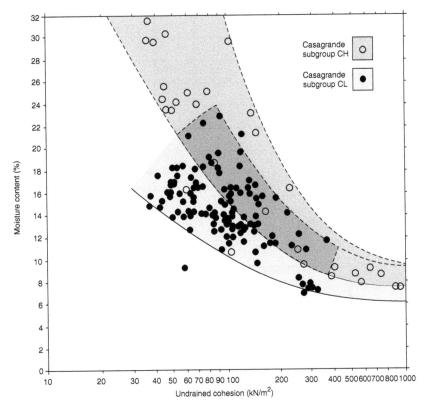

Figure 2.6 Relationship between undrained cohesion and moisture content for matrix-dominant till. (Illustration courtesy of CIRIA, London.)

are sometimes used and the resulting N values employed in estimating pile capacity. However the use of the static cone penetrometer test (CPT) is generally advantageous in thick deposits of silts or fine sands to give a more reliable ground profile. The method also overcomes difficulties which are commonly met in these soils where water pressures cause piping into the base of open boreholes giving rise to low N values, although the SPT drive can be extended in such soils and the penetration recorded for each 75 mm. Values of N derived in this way are not however strictly comparable to those obtained in the normal way. For a given relative density, the N value is dependent on the effective overburden pressure, the over-consolidation ratio of the deposit, and, for derived parameters such as angles of internal resistance, the grading and shape of the particles. Of these potential influences, correction to the N value for the effective overburden pressure has been given most attention by a number of researchers. This is discussed further in section 2.5.4 in connection with medium- and coarse-grained granular deposits.

The modern electric cone penetrometer has overcome the problem of rod friction, which was present with the mechanical type, and no correction is required for cumulative rod weight. The cone provides a measure of end resistance, q_c, and sleeve friction, f_s, which can be used for the design of driven piles. Where the static cone is

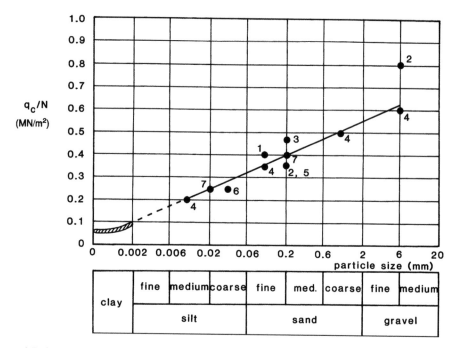

Figure 2.7 Correlation of cone resistance with SPT value. I, Meigh and Nixon; 2, Meyerhof; 3, Rodin; 4, Schmertman; 5, Shulze and Knausenberger; 6, Sutherland; 7, Thorburn and McVicar.

employed, and a correlation is required with SPT N values (or vice versa), approximate conversion between the two parameters is possible using the relationship between cone resistance, q_c, and N for soil according to particle size as proposed by a number of authors. These are summarized in Figure 2.7.

In fine-grained soils, the use of the CPT is particularly useful for assessing the appropriate founding levels for driven piles and the reasonable length to which they may be installed. However in dense silts a high penetration resistance to driven piles during driving can occur, without the realization of correspondingly high bearing capacities and their presence should be adequately established. Accurate gradings of such deposits are not generally required for assessment of pile performance, and representative samples are in any event difficult to secure. Water pressures should be known in such strata especially where bored piles might be employed, as severe loosening of these materials can occur during boring where there are seepage pressures. Loosening during continuous-flight auger (CFA) boring can be troublesome and special care may be necessary to prevent excessive loss of ground. Stand-pipes should be installed, or a piezometer used to monitor the water pressures, in specific strata.

2.5.4 Medium and coarse-grained granular deposits

Exploration for pile foundations in this type of material is often carried out using the SPT, but the heavier capacity static cone tests are equally suitable. In-situ permeability tests, and bulk sampling for grading analyses, provide useful and sometimes essential additional information. It is general practice to employ empirical methods of pile design based on the results of penetration tests. Where the size range is large, and includes a finer fraction (as may be the case in glacially derived materials) the results of penetration tests may be misleading, and care should be exercised in the selection of representative values. In such soils, groundwater seepage pressures may cause loosening in the base of boreholes, and the penetration values may be affected if suitable precautions are not taken. These would include topping up boreholes with water, the use of small-diameter shells to clean out the base of the borehole and reducing the rate of retrieval of the shell from the base of the borehole. Notwithstanding these devices, some reduction in N value may still occur. Whilst low values may lead to over-conservative design, it is also possible that unexpectedly heavy driving may be experienced with displacement piles and full penetration may not be possible.

Correction to the N value for effective vertical stress at the depth of the test may be carried out. This should not be applied to correct for poor field methods. Where tests have been carried out in loosened ground for instance, it is recommended that the correction is only applied where consistent values have been obtained over a series of tests. A comprehensive review of the SPT by Skempton (1986) gives corrections that may be applied to N for the effective vertical overburden pressure. The test as currently carried out in the UK is also compared with the original test when Terzaghi and Peck (1948) proposed the relationship between N and relative density, D_r. The overburden pressure relates the N value of the test to a reference value at a stress level of 100 kN/m^2 (C_{N1}), as shown in Figure 2.8. A revision to the relationship between the N value and relative density, using a standardized N_{60} value as given by the modern test with an automatic trip hammer is provided in Figure 2.9, for $(N_1)_{60}$ values, these being N_{60} values corrected for overburden pressure.

In terms of the parameters that may be derived from the N value, Stroud (1989) presents charts relating the angle of internal friction with $(N_1)_{60}$. Figure 2.10 shows a relationship between the effective angle of friction (ϕ') less the constant volume angle of friction (ϕ'_{cv}) for a range of relative density and Figure 2.11 gives a relationship between the constant volume angle of friction and particle shape. Stroud additionally considers the effects of over-consolidation ratio on the effective angle of friction.

These relationships enable a value of ϕ' to be derived from 'N' via the relative density, D_r.

Further causes of inaccurate N values occur in tests that are carried out in material below the base of a borehole that is mechanically disturbed to an excessive depth by overshelling, or testing partially or totally within the borehole casing. Double range tests, i.e. driving the tool an additional 300 mm, may reveal this occurrence, exhibited either as a marked increase in resistance over the final range, or a sudden drop, respectively. In dense granular soils, it is often impracticable to use the SPT shoe, as it is too easily damaged. In these particular soils, the SPT 60° cone can be substituted without significantly influencing the results.

Chiselling depths and times should be recorded on the borehole logs (with the weight of the chisel) since this may be the only evidence available to provide an indication of the maximum boulder size present. If rotary diamond coring is used to penetrate a boulder, there may be some difficulty drilling below the boulder in purely granular material.

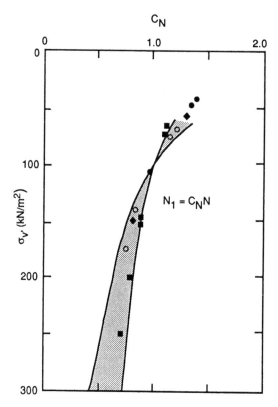

Figure 2.8 Correction for overburden pressure: (O) fine sands fill; (●) fine sands NC; (♦) fine sands OC; (■) coarse sands NC. Shaded region indicates laboratory tests (Marcuson and Bieganousky, 1977; Gibbs and Holtz, 1957); unshaded region indicates field test data. (After Skempton, 1986.)

A rock roller 'tricone' bit may permit progress, but at the expense of anything other than minimal geotechnical data. Every effort must be made to penetrate dense layers and to carry the investigation to the full depth to ensure that weaker or less dense layers at depth are fully explored. In order to install bored piles through, or into these strata, the groundwater conditions should be thoroughly investigated as discussed in section 2.3. The permeability of the deposits should be established by field permeability tests and the values should be related to the particle size distribution as seepage conditions are particularly relevant if non-displacement piles are selected for the foundation.

2.5.5 Normally consolidated clays

These clays are not employed as bearing strata but influence pile design as a result of 'downdrag' or by the effects they may have on the construction of bored cast-in-situ piles. The magnitude of the downdrag may be estimated approximately from the undrained shear strength of the clay (using for example the results of field

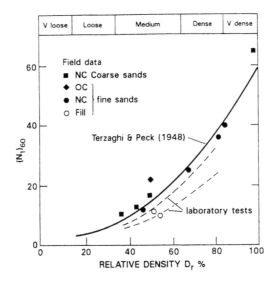

Figure 2.9 Effect of relative density, based on field data. (After Skempton, 1986.)

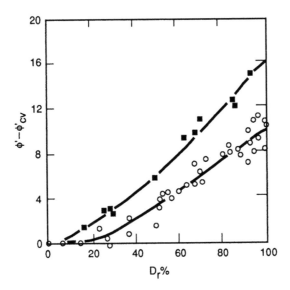

Figure 2.10 Variation of $\phi' - \phi_{cv}$; with relative density: (O) triaxial; (■) plane strain; mean effective stress (p') 150 to 600 kN/m^2; ϕ' measured as secant value. (After Bolton, 1986.)

vane shear tests, or from laboratory triaxial compression tests on carefully taken undisturbed samples). It is generally necessary to employ thin-walled piston samples for this purpose. Such samples may also be used for laboratory determination of the shear strength of the clay in terms of effective stress to permit the preferred calculation of downdrag by effective stress methods.

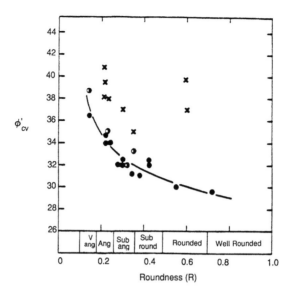

Figure 2.11 Relationship between particle shape and ϕ' (after Youd, 1972) based on triaxial tests: (●) uniformly-graded quartz; (◐) uniformly-graded feldspar or feldspar quartz mixtures; (×) well-graded.

Where the undrained strength of these clays is low, there is a possibility of producing ground failure and deformed pile shafts under the head of concrete employed in the construction of bored cast-in-situ piles. Typically, ground failure occurs in the depth range 2 to 4 m below ground level in very soft alluvial clays, especially at the time of extracting any temporary casing where $2\sigma'_v > c_u > \sigma'_v$. A sufficient number of in situ tests or laboratory tests should be carried out to establish a strength profile so that this possibility may be examined.

2.5.6 Weak rocks

Some information relating to the strength/density of weak or weathered rock may be obtained from static or dynamic penetration tests. The profiles may show broad changes in rock strength and may be useful in estimating the penetration of driven piles. Pile design from the results of dynamic penetration tests is generally not straightforward, especially in chalk and weak or weathered Triassic strata. In chalk, guidance on the use of the SPT is given by Hobbs and Healy (1979) and also Lord *et al.* (2003). It is now generally recognized that the use of the 60° cone, rather than the shoe, can significantly affect the N values obtained, and the shoe should be used for correlation with other work and to provide N values for use in empirical formulae. The sample obtained from the SPT tool is not representative.

In many cases penetration results are insufficient on their own, and it will be necessary to carry out rotary diamond coring in a diameter large enough to ensure as close to 100% core recovery as possible, even in weak friable rocks. Open-drive samples (U100), even if possible to obtain, do not tend to yield representative material.

Weathered sedimentary rocks frequently exhibit erratic strength profiles and weak zones below less weathered beds must be detected if they are present.

Percussive drilling combined with penetration resistance measurement may give some guide to the driving resistance of driven piles, but significant zones may be missed and there is generally little substitute for adequately detailed logs prepared from drill cores in order to establish a rock grade. In variably weak rock core, laboratory tests such as the point load test or unconfined compression tests tend to favour the stronger zones as the weakest material is often lost and therefore the results need to be carefully interpreted to avoid drawing the wrong (and unsafe) conclusions about the rock strength.

The results of tests carried out on rock cores in the field (e.g. point load tests) should be interpreted in a similar way. Useful information on the in-situ rock mass may however be obtained from permeability tests (packer tests for instance) which reflect the fracturing, fissuring and jointing of the rock structure. Where detailed information of in-situ rock condition or rock grades is required, especially in formations such as chalk or weak Triassic strata, a purpose-excavated shaft is recommended. Additional information can be obtained by carrying out plate-bearing tests at selected horizons as the shaft is excavated.

Settlement of piled foundations on rock may be estimated from in-situ modulus measurements using the pressuremeter or by plate-bearing tests carried out over a range of depths, though with adequate geological information, settlement may be shown to be too small to justify any extensive investigation. Pressuremeter testing is comparatively cheap, but suffers the disadvantages of possible disturbance of the rock when drilling the pocket for the probe, and 'sample' size and orientation effects. The plate-bearing test is costly, but is likely to give superior information.

It is however of critical necessity to recognize the presence of clay layers in rocks that are vertically or near vertically fractured or jointed (i.e. 'blocky'). Piles end bearing in such rocks may give a satisfactory blow count during driving, or may, if drilled or augered, appear to be within apparently satisfactory, massive strata. When loaded, poor load-settlement behaviour is obtained as a result of the toe load being transferred via a column of blocks to the clay layers which compress, or may undergo lateral yield. It is essential that the local geology is understood at the time of the investigation if this condition is to be identified correctly and emphasis needs to be placed on good drilling to achieve high core recovery, combined with accurate and suitably detailed core logging.

Adequate information relating to the groundwater conditions is particularly important for bored or drilled piles, and detailed notes should be made of water encountered during drilling operations. Some qualitative information may be obtained from the quantity of water/mud returns if a fluid flushing medium is being used, and water levels in drilled holes should be recorded if the borehole is left standing for any length of time. In continuous aquifers, simple standpipes may be adequate to establish the hydrostatic head, but in more complex ground, piezometers will be required to isolate water pressures associated with a particular structure.

Limestone is a special case as it may contain large cavities that can transfer considerable quantities of water with strong cross flows. Pile installation can be compromised in such strata and grouting may not be a successful precursor to the piling. As with the problem posed by clay layers in blocky rock, an understanding of the geology is a necessary adjunct to good quality fieldwork if these circumstances are to be properly recognized.

2.6 Backfilling site investigation boreholes

On sites likely to be piled, the backfilling of site investigation boreholes should be carried out with special care. This is particularly important in the case of bored piles where the loss of a pile shaft due to accidental connection to waterbearing strata via a badly backfilled borehole may occur. This may be accompanied by loss of ground in granular deposits. In an under-reamed pile base, the likelihood of striking a site investigation borehole is increased and the consequences of a collapse can be more serious.

Where the risks associated with striking an open loosely backfilled borehole are high it may be necessary to employ a grouted weak cement backfill or, if suitable material is available, heavily compacted clay plugs may be formed at frequent intervals in a bore otherwise filled with arisings. For future reference, an accurate survey of the boreholes should be carried out on completion of the work.

Special requirements relate to the boring/drilling of boreholes where contamination of underlying aquifers can occur. In the United Kingdom, the Environment Agency will need to be aware of any work where such an occurrence could take place, and special precautions may well be necessary, such as boring through a bentonite seal and backfilling with cement/bentonite.

2.7 Piling working platforms

In the United Kingdom it is generally necessary that piling rigs (and indeed other tracked plant) operate from a granular platform of over-site fill that is compacted into place in order to provide a safe working area. There are exceptions when a rig is operating from an existing hard-standing or in certain situations 'navvi' mats can be employed i.e. thick timber decking units that can be moved around the site. Such mats are not as effective as a properly designed and installed working platform and tend to slow down plant movement. They are only employed where a working platform is impracticable. The platform design is carried out using the recommendations of the Federation of Piling Specialists (FPS) and Building Research Establishment (BRE), Working platforms for tracked plant (2004).

In order to design a platform, it is necessary to know in detail the properties of the site soils at shallow depth. Therefore considerably more effort needs to be expended at the site investigation stage to provide an adequate coverage of shallow exploratory locations (e.g. trial pits) and to extend the sampling and testing (including as appropriate, in situ testing) to provide proper and representative information, particularly the undrained shear strength or relative density profile and any variation across a site. As the upper site surface is quite likely to include made ground this may not be a straight-forward task but every effort needs to be made to satisfy these requirements. It is of course essential to thoroughly compact the backfill to any trial pits such they do not create a new hazard.

It should be appreciated that together with the exploratory work, a thorough desk-study needs to be carried out to establish the location of potential backfilled excavations for services or voids from basements, buried tanks, etc.

Chapter 3

Basic piling methods

In this chapter, the main pile types and methods of installing them are described, with details of currently available plant, pile types and proprietary methods. However, to meet ever stricter environmental requirements and commercial pressures, piling plant and methods of pile installation are constantly changing so that the text should be regarded as an overview of the equipment and techniques in use. As with many fields of engineering, descriptions of the latest equipment and methods offered by particular manufactures are available on appropriate web-sites and scrutiny of the relevant sites is advised for the latest information.

The two basic methods of installing piles are well known, namely driving them into the ground, or excavation of the ground, usually by boring, and filling the void with concrete. Other methods have evolved to cope with certain ground or groundwater conditions, or to be more economic over particular (usually restricted) load capacities. Over the years, methods have also evolved to overcome the need for temporary casing during boring. Continuous flight auger (CFA) rigs developed to achieve this are now well established. In this process the bore is supported by the spoil around the auger right to the base of the hole until fluid concrete is pumped down the stem of the auger to fill the bore as the auger is withdrawn. Some methods involve a combination of driven and bored techniques, where a casing is driven or rotated into the ground followed by in-situ concreting. There are also hybrid types of pile which bridge the gap between vibro-stone columns and concrete columns, a stone-cement mix being vibro-driven into the ground. Where the ground is amenable to boring by water jetting, grouted columns can be produced which can be considered as a form of pile. Within each category, variations exist associated with the particular proprietary method in use.

A major advance is the degree to which piling rigs have become instrumented. The piling operation is now, in general, highly monitored and some operations such as concreting even carried out automatically where highly developed CFA rigs are employed. This level of instrumentation raises the possibility of deciding on the pile length as it is bored, i.e., the monitoring could eventually be sufficient to determine soil profile to a point where the ultimate capacity of the pile is assessed at the drilling stage.

In another development, rigs can be located with high precision over a pile position using GPS methods. The positioning of a rig with a high mast in this way requires a different approach to the guidance of other mechanical plant, as a considerable distance separates the receiving antenna on top of the (unstable) mast from the intended pile position. In practice, the operator 'sees' a stable target pile position on the in-cab display derived from the brief instants of time when the tall oscillating mast is momentarily vertical. Hence the pile is located in the correct position without attempting to steady the rig to an unattainable degree* (Stent Piling – 'SAPPAR' system). In a further advance of this system, the final positioning of the rig can be carried out automatically. Telemetry of one kind or another can also ensure that much of the monitoring of the piling operation is available on-line in 'real time' whether in the site office or back at a Company office.

3.1 Pile types

It used to be possible to categorize the various types of pile and their method of installation, using a simple division into 'driven' or 'bored' piles. This is adequate in many situations, but does not satisfactorily cope with the many different forms of pile now in use. A more rigorous division into 'displacement' or 'non-displacement' piles overcomes this difficulty to some extent, but some piles are installed by a combination of these methods and their description may require qualification.

In the displacement (generally driven) pile, soil is displaced radially as the pile shaft penetrates the ground. There may also be a component of movement of the soil in the vertical direction. Granular soils tend to become compacted by the displacement process, and clay soils may heave, with little immediate volume change as the clay is displaced.

Piles of relatively small cross-sectional area, such as steel 'H' section piles or open pipe piles, are termed 'low displacement piles', and the effects of compaction or soil heave are reduced. This can be advantageous if long lengths of pile are to be driven through granular deposits, if the piles are at close centres, or if clay heave is a problem.

In the non-displacement (generally bored) pile, lateral stresses in the ground are reduced during excavation and only partly reinstated by concreting. Problems resulting from soil displacement are therefore eliminated, but the benefit of compaction in granular soils is lost and in all soils spoil is produced which may be costly to remove from a site, especially if it is contaminated.

The displacement of the soil by a pile during installation is therefore a fundamental property, and its recognition in any classification of pile type is clearly advantageous. Little-used types such as pre-formed screw piles can also be covered by the (low) displacement classification, whereas they could not be correctly termed 'driven piles'. In a further development of the screw pile that is becoming more frequently employed, especially on contaminated sites where it reduces or eliminates the production of spoil, a hollow screw-form auger is rotated into the ground and the bore filled with concrete as it is back-rotated out or retracted without rotation.

The two main categories of pile types may be classified further according to whether pre-formed units are used, and whether the pre-formed unit is used as temporary support for the ground and withdrawn during concreting or left in place. For non-displacement piles, factors such as pile diameter and underreaming are introduced to the classification, as they have a bearing on the method of installation, and particularly

*Original Patent by A Weltman 1995; transferred to BICC 1996.

in the type of plant employed. Forming the pile bore using a continuous flight auger to mix the soil into a supporting slurry and concreting the pile from the base-up as the auger is withdrawn is a widely used method of pile installation. This type of pile (the 'CFA' pile) may be termed a 'replacement' pile.

A scheme of classification of bearing pile types is presented in Figure 3.1. A further distinction is added in this scheme which relates to casting of concrete against the soil, which can occur with both displacement and non-displacement piles. This factor may be significant when ground or groundwater conditions are aggressive to concrete, or where water flows through highly permeable ground and there is a possibility of cement and fines removal from the fluid concrete.

3.1.1 Displacement piles in contaminated ground

It has been mentioned that in the use of non-displacement piles, the spoil from the pile bores needs special consideration if it is contaminated since its disposal may be costly. By using displacement piles, this problem is eliminated, and where ground is contaminated it is generally considered that displacement piles are preferable. However, small amounts of overlying materials can be dragged down around the shaft as the toe of a driven pile emerges from contaminated strata to enter deeper uncontaminated strata. It is also possible that a very small plug of contaminated material could be driven ahead of a flat-bottomed displacement pile into previously uncontaminated soil. In general, these mechanisms are not considered to be capable of causing significant mixing, but should always be taken into account. The risk imposed by an unusual contaminant which could behave in an atypical manner will require special evaluation, (e.g. metallic mercury inevitably finds its way to the base of any formed bore). Large diameter displacement piles will produce greater compaction of the surrounding soil and could be considered to be superior in the elimination of contaminant pathways along a pile shaft than low displacement piles. Screw piles are of intermediate displacement, and some spoil may be produced, but with cast-in-place shaft formation, there should be good contact with the soil and the flow path is long thereby reducing the rate of any contaminant transfer.

3.2 Displacement piles

Figure 3.1 includes the main types of displacement piles to comprise either totally pre-formed piles or piles in which a displacement method is used to form a void which is then filled with concrete.

3.2.1 Totally pre-formed displacement piles

These piles comprise either tubular or solid sections which are driven (or sometimes screwed) into the ground. The hollow tubular types may be formed from steel or concrete, and the solid sections from steel, timber or concrete; the latter are pre-cast and may also be pre-stressed. The use of reinforced plastics in the form of glass or carbon fibre composites has entered use in soil-nailing applications, and this material is potentially of use as a pile former. For pre-cast concrete piles, the ease of transportation, handling on site and length adjustment has led to popularity of the jointed type.

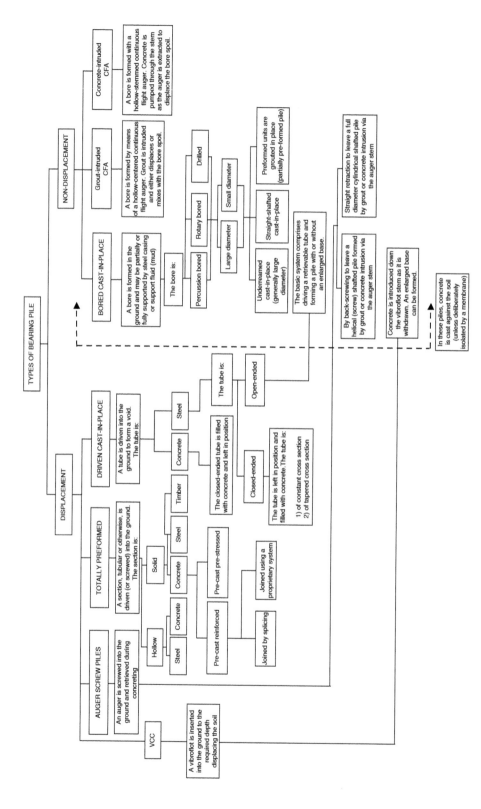

Figure 3.1 Classification of bearing pile types.

Some micro-piles are also included in this category. A summary of the main types of pre-formed displacement pile is given below.

Totally pre-formed displacement piles

(a)   Pre-cast concrete : full length reinforced (pre-stressed)
                         : jointed (reinforced)
                         : hollow (tubular) section
(b)   steel   : 'H' section
                         : tubular section
                         : other (e.g. screw-form piles)

3.2.2 Pre-cast reinforced concrete piles

Pre-cast concrete piles are now usually of the jointed type, unless a large contract with a more or less constant depth of piling makes it economical to pre-cast the piles on site, thus overcoming a potential difficulty in transport. Pre-cast non-jointed piles are generally of square section and may be up to 600 × 600 mm to work at loads up to approximetely 3000 kN in suitable ground. Typical sizes and capacities are given in Table 3.1. Extending pre-cast piles that do not have pre-formed joints is a lengthy process, involving breaking down the projecting pile head to provide a suitable lap for the steel and casting concrete to form a joining surface. The pile sections are then butted together in a steel sleeve using an epoxy cement, or joined by inserting steel dowel bars into drilled holes and using an epoxy cement to fix them in place. Good alignment of the pile sections is required to prevent excessive bending stresses developing on subsequent re-driving.

There are some benefits from pre-stressing concrete piles. Tensile stresses which can be set up in a pile during driving are better resisted, and the pile is less likely to be damaged during handling in the casting yard and when being pitched. Bending stresses which can occur during driving are also less likely to produce cracking. However, the ultimate strength in axial compression is decreased as the level of pre-stress is increased, and pre-stressed piles are therefore more vulnerable to damage from striking obstructions during driving. They are also difficult to shorten and special techniques have to be employed. As a result they are most suitable for a constant-length application.

Table 3.1 Typical load capacities for pre-cast concrete piles for normal driving conditions

Square pile section (mm)	Typical working capacity (kN)
200 × 200	400
250 × 250	500
275 × 275	750
300 × 300	900
350 × 350	1200

Note
Characteristic strength of concrete $50\,N/mm^2$ at 28 days. Piles driven with a 4 to 5 tonne hammer.
Courtesy of Teranto De Pol.

Figure 3.2 Detail of the 'Stent' pile joint. (Photographs courtesy of Stent Piling and Foundations.)

3.2.3 Jointed piles

Jointed piles are not pre-stressed, but as each unit is a shorter length, generally between 5 and 14 m long, handling stresses are much lower, reducing the chance of damage at this stage. The piles are joined by various means; bayonet fittings or wedges being typical but can include the use of epoxy resin. The Stent Piling and Foundations pinned type of pile joint with a strength equivalent to that of the reinforced concrete part of the pile is illustrated above in Figure 3.2.

With a robust method of jointing, the joint is as strong as the pile, and is of similar moment resistance. These criteria may not be met with the plain epoxy jointed pile, certainly not until the resin has hardened, and combined with the use of a single reinforcement bar, this type of jointed pile needs to be employed with care.

In any jointed pile the fit of joints should be good, as energy is lost through them, and increased hammering is then required to drive the pile. Angularity produced by the joints can lead to high bending stresses in the pile as it is loaded together with a high chance of damage during driving. Therefore any joints that could produce this effect should be rejected, with an angular misalignment of 1 to 300 considered a permitted maximum.

Jointed piles are usually of square section, but other forms, such as hexagonal or even triangular sections have been produced. Typical capacities range from 700 to 2500 kN with pile cross-sections usually in the range 250 × 250 to 400 × 400 mm although sizes above and below this range are sometimes employed. In suitable ground conditions, piles have been driven to depths of up to 100 m, whilst the vast majority of typical applications do not exceed 30 m or so.

The units are cast from a high-cement-content mix with a water/cement ratio usually around 0.4. The minimum compressive strength of cubes is generally around 60 N/mm at 28 days. Pile units with square cross-sections will usually have a main reinforcement bar at each corner; a 12 to 20 mm dia. deformed bar with a yield stress of 600 N/mm being typical. Transverse reinforcement in the form of 5 mm dia. wire 'spiral' controls longitudinal cracking, sometimes with additional bars near the pile ends. The joints are normally formed from a steel plate with integral starter bars projecting into the pile. A fixture in the casting formwork sets the plate square to the pile, but strict quality control of the squareness of the plate and concrete strength are required to produce successful jointed piles. Concrete cover should be the minimum to provide protection from corrosion, as large thicknesses of cover can lead to spalling. Accurate positioning of an adequate number of spacers is therefore an important aspect of casting the pile units. Hard steel points can be fixed at the toe of jointed piles for limited penetration into weak rock. The piles are generally driven with a drop hammer of 3 to 4 tonnes weight, although hydraulic hammers, vibrators or diesel hammers are sometimes used.

3.2.4 Hollow tubular-section concrete piles (cylinder piles)

Hollow concrete tube piles are generally of large diameters, typically in excess of 600 mm and often nearer 1500 mm. Pile working loads may be in excess of 2000 kN. The concrete tubular sections are pre-tensioned and make efficient use of reinforced concrete, having twice the moment of resistance of a solid concrete pile of equivalent weight. Large-diameter tubular concrete piles are used in marine applications to take heavy axial loads combined with significant bending loads or alternatively in thick deposits of loose sand where great depth of driving is required but with good shaft and end bearing required in the final bearing stratum.

During driving, a point may be used to close the end of the pile for penetrating hard strata, but the pile may also be driven open-ended. In either condition, large soil displacements occur, and the pile is not suitable for high-density application unless in a loose soil. In marine applications the corrosion resistance of the high-quality concrete from which the piles are constructed is generally adequate, but in freeze-thaw conditions, the long-term resistance may suffer. The piles have been driven to depths of around 80 m, but because of difficulties in cutting off any excess lengths of pile, or the costs of extending them, they are best suited for driving to a pre-determined level. Tubular concrete piles are not in common use at the present time except in the Far-East where they are favoured for heavy load applications. To drive very long piles, pile extension is typically achieved by welding together the steel plates that restrain the tensile reinforcement at the end of the pile sections. Tensile loads that can occur during easy driving conditions in soft soils could be troublesome if these joints are not sufficiently robust to accommodate stress reversals and under hard driving conditions, distress can occur in the concrete immediately behind the steel end plate.

3.2.5 Pre-formed steel piles

Totally pre-formed steel piles comprise 'H' and 'I' sections, screw sections and also some tubular sections which can be filled with concrete after driving. In addition, an 'X' section pile which can be jointed via plated end pieces has been available in Scandinavia. Because of fears of rapid corrosion, there has been reluctance by some engineers to employ steel piles, but in natural, undisturbed ground, free from contaminants, the corrosion rates of steel are low. The advantages of steel piles are (a) ease of handling, (b) little likelihood of over-stressing during lifting or pitching, and (c) ease of extension or reduction in length to cope exactly with variations in driven length.

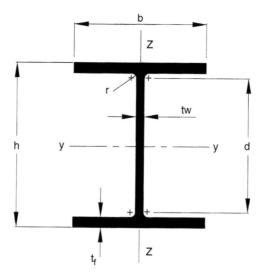

Figure 3.3 Typical cross-section of an 'H' section pile. (Courtesy of Arcelor Commercial RPS UK Ltd.)

They are more suitable than concrete piles for driving through thin layers of harder strata, or natural obstructions, but against this, slender steel piles such as 'H' sections (see Figure 3.3) may wander off line and this is exacerbated if the piles are joined by welding. The tendency for slender piles of this type to wander off line during driving has been investigated by Hanna (1968). The uneven pressure distribution across the toe of a pile as the soil is displaced from the pile base is implicated as the cause. Other reasons, such as a form of dynamic instability, have also been cited. Length for length, steel piles tend to be more costly than concrete piles, but good load-carrying capacity for a given weight of pile can reduce driving costs. Slender 'H' section piles have a low volume displacement which is a useful property for high-density piling in soils which tend to heave or become over-compacted. In some ground conditions, such as chalk or weak siltstones or sandstones, there may be a problem in achieving a suitable resistance with 'H' section piles. In chalk, this is believed to be the result of the shaft whipping as it is driven (Hobbs and Healy, 1979). In the other rocks, George *et al.* (1977) recommend adding 'wings' to 'H' piles, i.e. welded-on steel plate 'T' sections to increase the surface area of the 'H' pile. Typical properties of 'H' section Universal Bearing piles are given in Table 3.2.

Screw piles in which steel screw sections are wound into the ground can provide economical foundations in suitable circumstances but a balance has to be found between the torque that can be safely applied during installation and the bearing resistance that can be developed. The shafts tend to be of small diameter limiting lateral capacity but specialist concerns such as the GTL Partnership in the United Kingdom have developed a screw pile with a relatively large diameter shaft to enhance the lateral resistance of the pile and the torque that can be used for installation. It is usual for screw section piles to be fabricated. Early versions of the screw pile were cast iron and often used in marine application to support, for example, seaside piers in Victorian times.

Table 3.2 Dimensions and properties of H section bearing piles (Arcelor HP piles)

Section	h (mm)	b (mm)	t_w (mm)	t_f (mm)	Sectional area (cm²)	Mass* (kg/m)	Perimeter (m)	Moment of inertia y-y (cm⁴)	Moment of inertia z-z (cm⁴)	Elastic section modulus y-y (cm³)	Elastic section modulus z-z (cm³)
HP 220 × 57.2	210	225	11	11	72.85	57.2	1.265	5729	2079	545.6	185.2
HP 260 × 75	249	265	12	12	95.54	75	1.493	10650	3733	855.1	281.7
HP 260 × 87.3	253	267	14	14	111.2	87.3	1.505	12590	4455	994.9	333.7
HP 305 × 88	302	307	12.3	12.3	111.6	88	1.782	18380	5949	1218	387.3
HP 305 × 95	304	308	13.4	13.4	121.7	95	1.788	20170	6552	1328	425.1
HP 305 × 110	308	310	15.4	15.4	140.2	110	1.800	23550	7680	1530	495
HP 305 × 126	312	313	17.7	17.7	161.6	126	1.813	27540	9019	1763	577.2
HP 305 × 149	319	316	20.7	20.7	190	149	1.832	33050	10870	2075	688.8
HP 305 × 180	327	320	24.8	24.8	229.3	180	1.857	40970	13550	2508	847.4
HP 305 × 186	328	321	25.6	25.6	237	186	1.861	42580	14090	2594	879.3
HP 305 × 223	338	325	30.5	30.5	285	232	1.891	52840	17590	3127	1081
HP 320 × 88.5	303	304	12	12	112.7	88.5	1.752	18740	5634	1237	370.6
HP 320 × 103	307	306	14	14	131	103	1.764	22050	6704	1437	438.2
HP 320 × 117	311	308	16	16	149.5	117	1.776	25480	7815	1638	507.5
HP 320 × 147	319	312	20	20	186.9	147	1.800	32670	10160	2048	651.3
HP 320 × 184	329	317	25	25	234.5	184	1.830	42340	13330	2574	841.2
HP 360 × 84.3	340	367	10	10	107.3	84.3	2.102	23190	8243	1364	449.2
HP 360 × 109	346	371	12.9	12.9	138.9	109	2.123	30620	10940	1768	590.7
HP 360 × 133	352	373	15.6	15.6	168.5	133	2.140	37730	13540	2144	725.3
HP 360 × 152	356	376	17.9	17.9	193.8	152	2.153	43950	15810	2466	842.3
HP 360 × 174	362	378	20.4	20.4	221.7	174	2.169	51020	18400	2823	973.5
HP 360 × 180	363	379	21.1	21.1	229.5	180	2.173	53040	19140	2923	1011
HP 400 × 122	348	390	14	14	155.9	122	2.202	34770	13850	1998	710.3
HP 400 × 140	352	392	16	16	178.6	140	2.214	40270	16080	2288	820.2
HP 400 × 158	356	394	18	18	201.4	158	2.226	45940	18370	2581	932.4
HP 400 × 176	360	396	20	20	224.3	176	2.238	51770	20720	2876	1047
HP 400 × 194	364	398	22	22	247.5	194	2.250	57760	23150	3174	1163
HP 400 × 213	368	400	24	24	270.7	213	2.262	63920	25640	3474	1282
HP 400 × 231	372	402	26	26	294.2	231	2.274	70260	28200	3777	1403

3.2.5.1 Hollow steel piles

Hollow steel piles are of cylindrical or box type sections. They overcome some of the problems encountered with the flexibility of slender 'H' sections, although if the open end becomes plugged, i.e. soil partially enters the pile section but then moves down with the pile (relatively rare during driving) they effectively become large displacement piles. In that situation, heave of the ground and the possibility of over-compacting granular deposits occurs so that piling becomes progressively more difficult. If plugging occurs, it may prove necessary to remove the internal soil plug and possibly pre-bore to some depth, provided there is no risk of instability. Whilst there has been much debate on whether or not open-ended tubular steel piles plug under driving, the view of the authors is that in most circumstances driving such piles will occur with minimal plugging because of the high inertial resistance of the soil plug. By contrast, under most conditions piles will tend to plug during static loading.

Hollow piles may be driven closed-ended through cohesive soil containing cobbles and small boulders, but some ground heave may then occur as for any large displacement pile. This type of steel pile performs well in resisting impact and bending loads, and the availability of large-diameter sections enables considerable loads to be carried. Such a combination of properties has led to their extensive use in marine structures, where long freestanding lengths of pile are common in, for example, jetties, piers and mooring dolphins.

Box section piles are typically formed by welding two or four standard 'Z' section or 'U' section sheet piles, edge-to-edge, to form a box. The properties of piles formed in this way, using Arcelor sheet piles are given in Table 3.3, with typical dimensions identified in Figure 3.4. Permitted maximum pile loads will be dependent on the grade of steel employed, provided suitable ground reaction is present. Greater load-bearing capacity may be obtained by using different combinations of standard sections. In a practical application, with extended lengths of pile above a sea bed for example, the loads may be limited by the slenderness ratio of the effective length of pile above the supporting soil.

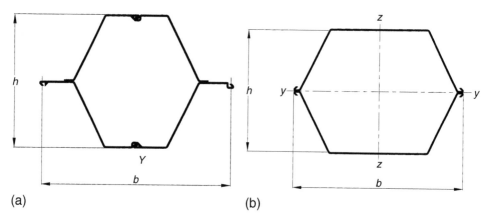

(a)   (b)

Figure 3.4 Typical cross-section of an Arcelor 'CAZ' box pile (a) and CAU box pile (b). (Courtesy of Arcelor Commercial RPS UK Ltd.)

Table 3.3a Dimensions and properties of CAZ box piles

Section	b (mm)	h (mm)	Peri-meter (cm)	Steel section (cm²)	Total section (cm²)	Mass* (kg/m)	Moment of inertia		Elastic section modulus		Min. radius of gyration (cm)	Coating area** (m²/m)
							y-y (cm⁴)	z-z (cm⁴)	y-y (cm³)	z-z (cm³)		
CAZ 12	1340	604	348	293	4166	230	125610	369510	4135	5295	20.7	3.29
CAZ 13	1340	606	349	320	4191	251	136850	402270	4490	5765	20.7	3.29
CAZ 14	1340	608	349	348	4217	273	148770	436260	4865	6255	20.7	3.29
CAZ 17	1260	758	360	305	4900	239	205040	335880	5385	5105	25.9	3.41
CAZ 18	1260	760	361	333	4925	261	222930	365500	5840	5560	25.9	3.41
CAZ 19	1260	762	361	362	4951	284	242210	396600	6330	6035	25.9	3.41
CAZ 25	1260	852	376	411	5540	323	343000	450240	8020	6925	28.9	3.57
CAZ 26	1260	854	377	440	5566	346	366820	480410	8555	7385	28.9	3.57
CAZ 28	1260	856	377	471	5592	370	392170	513050	9125	7820	28.9	3.57
CAZ 34	1260	918	392	516	5999	405	507890	552570	11020	8520	31.4	3.73
CAZ 36	1260	920	393	547	6026	430	537860	585200	11645	9030	31.4	3.73
CAZ 38	1260	922	393	579	6053	455	568840	618770	12290	9550	31.4	3.73
CAZ 36-700	1400	998	430	528	7209	414	627090	701250	12520	10015	31.4	4.10
CAZ 38-700	1400	1000	431	563	7239	442	667260	747360	13295	10675	31.4	4.10
CAZ 40-700	1400	1002	432	599	7269	470	707630	793470	14070	11335	34.4	4.10
CAZ 46	1160	962	401	595	5831	467	645940	527590	13380	8825	32.9	3.81
CAZ 48	1160	964	402	628	5858	493	681190	556070	14080	9300	32.9	3.81
CAZ 50	1160	966	402	661	5884	519	716620	584560	14780	9780	32.9	3.81

* The mass of the welds is not taken into account.
** Outside surface, excluding inside of interlocks.

Table 3.3b Dimensions and properties of CAU, CU and LP box piles

Section	b (mm)	h (mm)	Perimeter (cm)	Steel section (cm²)	Total section (cm²)	Mass* (kg/m)	Moment of inertia		Elastic section modulus		Min. radius of gyration (cm)	Coating area** (m²/m)
							y-y (cm⁴)	z-z (cm⁴)	y-y (cm³)	z-z (cm³)		
CAU 14–2	785	449	230	199	2584	**155.8**	53850	121300	**2400**	**3095**	16.5	2.04
CAU 16–2	785	454	231	220	2620	**172.5**	62240	130380	**2745**	**3325**	16.8	2.04
CAU 17–2	785	455	231	227	2626	**178.1**	64840	133330	**2855**	**3400**	16.9	2.04
CAU 18–2	786	486	239	225	2888	**177.0**	73770	142380	**3035**	**3625**	18.1	2.14
CAU 20–2	786	489	240	247	2910	**193.8**	83370	151220	**3405**	**3850**	18.4	2.14
CAU 21–2	786	490	240	254	2916	**199.3**	86540	153990	**3530**	**3920**	18.5	2.14
CAU 23–2	786	492	244	260	3013	**204.2**	94540	157900	**3845**	**4020**	19.1	2.19
CAU 25–2	786	495	245	281	3034	**220.8**	104810	166600	**4235**	**4240**	19.3	2.19
CAU 26–2	786	496	245	288	3041	**226.4**	108260	169510	**4365**	**4315**	19.4	2.19
CU 6–2	632	264	180	116	1315	**91.2**	11600	48300	**875**	**1530**	10.0	1.55
CU 8–2	633	321	189	139	1569	**109.1**	19200	60000	**1195**	**1895**	11.8	1.63
CU 12–2	635	403	198	168	1850	**132.2**	34000	70000	**1685**	**2205**	14.2	1.72
CU 12 10/10–2	635	403	198	177	1850	**139.2**	35580	73460	**1765**	**2315**	14.2	1.72
CU 18–2	635	473	212	196	2184	**153.8**	58020	78300	**2455**	**2470**	17.2	1.86
CU 22–2	636	494	220	219	2347	**172.3**	73740	88960	**2965**	**2800**	18.3	1.94
CU 25–2	636	497	222	239	2450	**187.2**	84300	97000	**3395**	**3050**	18.6	1.97
CU 32–2	636	499	223	291	2461	**226.3**	108800	109200	**4360**	**3435**	19.3	1.97
LP 3 S	537	447	195	201	1748	**157.8**	51800	57100	**2320**	**2130**	16.1	1.67
LP 4 S	537	487	204	220	1927	**172.4**	69600	61800	**2860**	**2300**	17.8	1.76

* The mass of the welds is not taken into account.
** Outside surface, excluding inside of interlocks.

Typical ratios of diameter (D) to wall thickness (t) are in the range 30 to 50 and such piles may be top-driven in the conventional way or bottom driven using an internal hammer and some means of closing the pile base. Short sections of pile may be welded together or alternatively if they comprise thicker-walled cylindrical piles they can be screwed together to achieve longer pile lengths. Once installed any internal soil plug can be removed and the pile filled with concrete although it is more usual to leave the soil plug in place. Where thin-walled piles are bottom driven they are unlikely to be used as load-bearing piles without concrete infill as discussed under driven cast-in-place piles. Spirally welded tubular steel piles in diameters from around 200 to 3000 mm are available with wall thicknesses from 10 mm up to 25 mm. At the other end of the scale, small diameter steel piles (100 to 200 mm) with thin walls (typically 3 mm) can be driven with a down-hole air hammer ('Grundomat piles'). These may have working capacities of 20 to about 50 kN and are described in the section dealing with micro-piles.

3.2.5.2 Steel piles as retaining walls

In retaining wall applications, various combinations of box and plain 'U' or 'Z' section piles may be employed to greatly increase the bending resistance of the stand-alone sections. For additional stiffness, 'H' sections can be combined with sheet pile sections. The connections can be achieved with 'ball and socket' adaptors that can be welded to the edge of the 'H' section flanges that will link up with the clutch of the sheet pile. At the same time, within limits, a degree of rotation can be introduced into the horizontal wall alignment. Special interlock connectors permit corners to be formed as required, not necessarily at 90 degrees. Up to around 5 degrees of angular change can be achieved with standard pile sections allowing circular cofferdams to be formed to diameters of around 15 m without special sections. When sheet steel piles are used to form cofferdams in marine applications, water penetration through the pile joints may be significantly reduced or eliminated using sealants applied to the pile interlocks preferably prior to delivery to site. For sheet piled basements, permanent sealing against groundwater may be achieved by welding, using either with a bead of weld along the interlock or with a separate steel plate over the exposed edge of the pile interlock. Horizontal sealing into a basement may be achieved by welding a horizontal steel rib onto the piles following installation which is then cast-into the basement floor slab.

3.2.6 Timber piles

For modest loads and piling depths of about 12 m, timber piles are quite suitable, provided certain precautions are taken. They are infrequently used in the United Kingdom, although they are still in common use in Scandinavian countries and the United States. In Australia, treated hardwood piles are produced and installed by Koppers Timber Piles P/L who specialize in this type of pile and have developed a strong market for it. In general, working loads are unlikely to exceed 500 kN, partly because of difficulties in driving piles to give sets appropriate to greater working loads and partly because of the generally smaller cross-sections of timber piles and lower compressive strengths. Stresses should be checked according to CP112 for compressive loads parallel to the grain. Pile cross-sections are usually square although circular sections can

Table 3.4 Lengths of timber piles

Timber		Square section side (mm)	Max. length approx. (m)
Douglas fir	(sw)	400	15
Pitch pine	(sw)	500	15
Greenheart	(hw)	475+	18+
Jarrah	(hw)	250	9
Okan	(hw)	250	9
Opepe	(hw)	250	9

Note
sw softwood.
hw hardwood.

be used. Section sizes depend on the type of timber; BS 8004:1986 (BSI, 1986) quotes the cross-sections and lengths given in Table 3.4. Note that pile lengths as pitched may need to generously exceed the expected installed lengths to allow sufficient up-stand for trimming after driving. Timber piles may be successfully subjected to CAPWAP testing as employed for piles formed in concrete or steel.

3.2.6.1 Driving timber piles

Drop hammers are usual for driving timber piles or alternatively low-capacity air/hydraulic hammers may be employed. Damage through over-driving manifests itself in 'brooming' of the pile head, cracking of the shaft and unseen damage to the pile tip. The driving process crushes and separates the fibres at the head of the pile – hence the 'broom' effect. A steel band around the pile head may restrict this damage to manageable proportions, but timber piles are not particularly suitable for driving through dense strata, or strata with obstructions where they can disintegrate at their toe if driving is protracted and give the false impression of penetration. To mitigate this potential problem a steel or cast-iron shoe is generally attached to the tip of the pile. Hammer weight should be approximately equal to the weight of the pile or as is some-times recommended, 50% more especially if heavy driving is expected. Particular care should be taken to make sure that the pile heads are square, that the hammer blows are axial to the pile, that the pile is properly set up in the leaders and not strained to bring it vertical after a bad start to the drive. If damage occurs, and the hammer weight is suitable for the driving conditions, it may be necessary to de-rate the pile (i.e. reduce the design load of the pile), so that smaller drops and fewer blows can be used for its installation.

3.2.6.2 Preservative treatment

Both softwoods and hardwoods should be treated. Although some hardwoods such as greenheart have slight inherent resistance to decay, it is not usual to rely on this for long-term protection. The procedures set out in British Standards 4072:1999, (BSI, 1999) or in the United States, the American Wood Preservers' Association (who regularly publish standards in this connection) should be followed. Note that if crack-ing of the pile shaft occurs, or there is damage at the head that extends below the cut-off level, the preservative treatment will not be effective and the trimmed end of the pile should be re-treated. Where softwoods are used as bearing piles, it is normal to trim these below the groundwater level (for the driest season) and extend the shaft as necessary with a concrete or possibly a suitably corrosion-protected steel section.

3.2.6.3 Quality and tolerances

Timber for piles is specially selected for straightness of grain and should also be free from defects such as shakes which would weaken it and render it liable to fail during driving, or impair preservative treatment. Piles machined to a section should not be out of straightness by more than 25 mm over their length. Round piles may be out of straightness by up to 25 mm over a 6 m chord.

3.2.6.4 Marine conditions

Timber piles have long been favoured for construction of small jetties on account of their energy-absorbing qualities. They are, however, open to attack above the mud line from a variety of marine organisms, and the marine borers *Limnoria* and *Teredo* are usually mentioned in this context, for which arsenates and creosotes (if permitted) are recommended preservatives respectively. Greenheart has some natural resistance to attack. In shallow, warm waters, pile attack is more rapid than in cold, rough water conditions.

3.2.7 Driven cast-in-place displacement piles

Driven cast-in-place displacement piles are suitable for a wide range of loads have been available for a great many years in the form of the Raymond pile. This type of pile may be of two kinds, one employing a driven temporary tube to form the void which is then filled with concrete as the tube is withdrawn, and the other employing a driven tube to form a permanent casing (as in the Raymond pile). The former variation is most commonly used in the lower load range, whilst the latter can be used to support considerable loads in suitable ground conditions. As the load-supporting medium is generally the concrete infill, methods of driving have been evolved which do not over-stress the non-load-bearing forming tube. Thus this type of pile may be bottom driven using an internal hammer to reduce driving stresses on the tube which can then be relatively thin.

Typical working loads, pile sizes and hammer sizes for spirally welded tubular steel piles are given in Table 3.5. Reinforcement for these piles is limited to starter bars at the top of the pile, unless large column stresses are to be resisted. Other forms including the Raymond pile employ a mandrel, which distributes the driving stresses to a tapered and spirally corrugated casing over a large area, and provides internal support to the tube. A previously popular version of this pile was the Westpile 'Shell' pile in which hollow concrete 'shell' sections were top driven via a mandrel, but this is no longer available.

In those piles in which the tube is retrieved, it is necessary either to bottom drive the tube with a plug that may be driven out, or to top drive the tube and close off the bottom end of the pile with an expendable blanking plate or 'shoe'. When the tube

Table 3.5 Sizes, drop hammer weights and normal design loads for cased piles driven with internal drop hammer (From BSP data; now Tex Steel Tubes Ltd.)

Internal dia. of casing (mm)	Weight of internal drop hammer (tonnes)	Design load (tonnes)		Casing thickness (mm)	Welded plate shoe thickness (mm)
		Ordinary soil*	Rock etc.†		
254	0.75	15	20	3.3	9.5
305	1.25	30	35	3.3	9.5
	1.75	(40)§	45		
356	2.0	40	50	3.3	12.5
	2.75	(50)§	65		
406	2.5	50	60	3.7	12.5
	3.5	(60)§	85		
457	3.0	65	80	4.0	16
	4.0	(75)§	100		
508	4.0	80	100	4.4	16
	5.5	(90)§	130		
559	5.0	100	125	4.8	19
610	6.0	120	150	5.3	19

*Ordinary soil; sand, gravel or very stiff clay.
†Rock etc.; rock, very dense sand or gravel, very hard marl or hard shale.
§ Subject to individual consideration.
The above table is normally applicable for pile lengths up to about 15/18 m. If the piles are driven in two lengths, the upper length may be of lesser thickness. For longer piles an increased casing thickness will be required for the lower portion of the piles.

is withdrawn, the shoe remains at the base of the pile. In the former system, a plug of 'earth' dry concrete mix or gravel is compacted in the base of the tube using an internal drop hammer. The hammer is then used to drive the tube into the ground via the compacted plug. At the required depth the tube is restrained at ground surface, and the plug driven out to form an enlarged base ('basing out'). Alternatively, a heavier gauge tube can be used and the pile driven from the top. In this situation an expendable shoe closes the lower end of the pile.

The 'Franki' pile (Skanska/Cementation) typifies this type of 'driven cast-in-place' where the temporary pile forming tube with an end plate is driven via an external hydraulic hammer (see Figure 3.5).

Reinforcement is installed in the pile shaft before concreting. Where the pile is to resist compressive forces only, the reinforcement may be restricted to the upper section. A highly workable pumped concrete mix is employed with a suitable head of fluid concrete being maintained in the temporary casing as it is vibrated out of the ground. Current rigs used for this type of work (e.g. Franki) are highly instrumented with pile penetration logging and concrete volume monitoring.

Alternative methods of forming the shaft are to employ a pre-cast member, or a permanent casing, seated in an enlarged plug. These methods can be helpful when downdrag forces or aggressive ground conditions are present. The plug would then consist of an inert aggregate, and the shaft suitably coated or wrapped to resist downdrag forces or the adverse ground chemistry as appropriate.

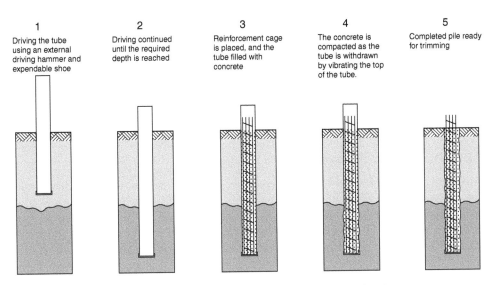

1	2	3	4	5
Driving the tube using an external driving hammer and expendable shoe	Driving continued until the required depth is reached	Reinforcement cage is placed, and the tube filled with concrete	The concrete is compacted as the tube is withdrawn by vibrating the top of the tube.	Completed pile ready for trimming

Figure 3.5 Sequence of installation of a top-driven Franki pile.

3.2.7.1 Concrete mixes for driven cast-in-place piles

The mixes used for driven cast-in-place piles vary with the type of compaction applied, as discussed in the previous section. The mixes should not however contain less than 300 kg/m³ of cement. It is suggested in BS 8004:1986 that the mean compressive stress over the total cross-section of the pile under working load should not exceed 25% of the specified 28-day works cube strength. Some increase in working compressive stresses is possible with permanent steel casing.

3.2.7.2 Damage to adjacent piles during driving

Driven cast-in-place piles can be damaged by driving adjacent piles too close or before the concrete has reached a suitable strength. The piles may be damaged by lateral forces or by tensile forces, as the ground heaves. When it is suspected that pile damage of this type has occurred it may be decided to carry out a pile load test or integrity test as a check. On a pile which is cracked by this means, a load test may yield an apparently satisfactory result but the long-term performance of the pile may be impaired if the steel reinforcement is exposed via the cracks. To lessen the risk of cracking caused by soil movements, a minimum spacing of 5D, centre to centre is often employed when driving adjacent piles when the concrete is less than 7 days old. The use of integrity tests may be considered to provide sufficient information to modify this rule if necessary.

During installation of cast-in-place piles with relatively thin bottom-driven permanent steel casing, collapse of the tube can occur from lateral soil displacement if the piles are driven at centres that are too close. This has sometimes resulted in the loss of the hammer at the base of the pile, when the collapse occurs above the hammer as the pile is driven. The occurrence is more likely, however, when driving piles inside a coffer-dam. Where this problem is encountered, and there is no way to reduce the piling density, pre-boring may be considered as a method of reducing the effect over the upper part of the pile. At the design stage, if high-density piling is unavoidable in soils prone to heave such as stiff clays, a low displacement 'H' section pile may be selected as more suitable. Alternatively, the multi-tube technique described by Cole (1972) can be employed. All piles within 12 diameters of each other are considered to form a part of a group, and are driven (and if necessary, re-driven) to final level before basing out and concreting.

3.2.7.3 Driven expanded piles

Although currently not commercially available, a form of low displacement driven pile, the Burland 'wedge pile', was developed in the 1990s. It comprises a cruciform cross-section fabricated from four steel angle sections, tack welded back-to-back at their edges, to form a closed sleeve. The sleeve is fitted with a shoe and driven (or jacked) into the ground to the required depth. A tapered cruciform expander mandrel is then forced down the entire length of the sleeve, splitting the tack welds. An expansion of about 10% of the width of the pile occurs, which increases the shaft friction to a significant degree. The mandrel is then withdrawn, leaving a set of spacers behind that maintain the pile in the expanded condition in order that it may be grouted for permanence.

3.2.8 Screw cast-in-place displacement piles

Whilst the installation of this type of pile is effected by means of a type of auger, the process involves compaction rather than removal of the soil and, in this respect; the piles are of a displacement type. In forming the pile, a heavy-duty single-start auger head with a short flight is screwed into the ground to the required depth. The auger head is carried on a hollow stem which transmits the considerable torque and compressive forces required, and through which the reinforcement cage is inserted after completion of the installation process. The end of the hollow stem is sealed with a disposable tip. Following placement of the reinforcement, concrete is placed through this tube from a hopper at its head. As concrete filling takes place, the auger is unscrewed and removed, leaving behind a screw-threaded cast-in-place pile. By virtue of the combined rotation and controlled lifting applied at the extraction stage the 'threads' are of robust dimensions. The sequence of pile construction is shown in Figure 3.6.

This method of forming a pile is known as the Atlas Piling System, and is marketed by Cementation Foundations: Skanska Limited in the United Kingdom, in association with N.V. Franki S.A of Belgium. A purpose-designed, track-mounted rig provides hydraulic power for auger rotation and the application of downward force and is fitted with a crane boom for handling reinforcement and concrete skips. For a given pile size and volume of concrete, pile capacities are greater than for traditionally con-structed bored piles, although the restricted diameter of the reinforcement cage may be a disadvantage if the pile is required to resist high bending stresses. The system does however combine many of the advantages of a displacement pile with the low noise and vibration characteristics of a bored pile. It will operate in most cohesive

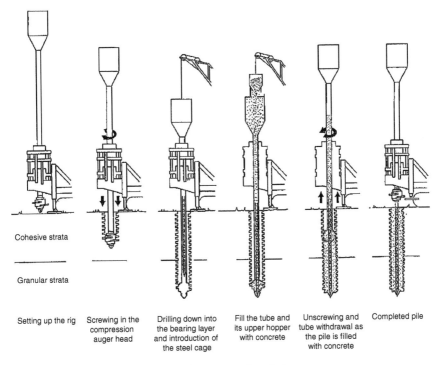

Cohesive strata

Granular strata

| Setting up the rig | Screwing in the compression auger head | Drilling down into the bearing layer and introduction of the steel cage | Fill the tube and its upper hopper with concrete | Unscrewing and tube withdrawal as the pile is filled with concrete | Completed pile |

Figure 3.6 Atlas piling system: installation process.

and granular strata to a maximum depth of 22 m, providing piles ranging in diameter from 360 to 560 mm. To achieve the torque of perhaps 250 to 350 kNm required at the auger, power requirements are relatively high. Whilst the pile is environmentally friendly in that it reduces the amount of potentially contaminated spoil that needs disposal, set against this there may be an increase in atmospheric pollution from the more powerful plant employed compared with a traditional auger pile machine. It has been found that when screw displacement augers are used in dense granular deposits the rate of auger wear is high and in extreme cases this can make the pile uneconomical.

The Omega pile introduces a variation at the concreting and auger removal stage. After screwing the auger into the ground and displacing the soil to the full depth of the pile, the auger is rotated in the same direction on extraction as for installation. This then leaves a straight-shafted pile of the diameter of the outer auger diameter, but producing a small amount of spoil. Similarly, in the Fundex pile, a flighted casing is pulled vertically out of the bore at the time of concreting, leaving a straight shafted pile. In this pile, however, if required, the casing can be left in place to give protection to the concrete shaft in aggressive ground conditions. In the patented 'Screwsol' piling system by Bachy Soletanche, a fin extends from the main hollow-stem auger to create a helically threaded pile that typically extends the central 350 mm diameter shaft to 500 mm diameter across the threads with minimum spoil production.

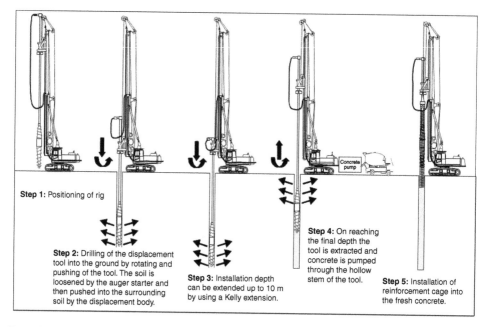

Figure 3.7 Installation sequence for a soil-displacement pile. (Illustration courtesy of Cementation Foundations.)

Concreting of soil-displacement piles is carried out with computerized control of the rate of extraction, rotational speed and concrete feed. The installation sequence for the Cementation Foundations' type of soil-displacement pile is shown in Figure 3.7 with the production log shown in Figure 3.8.

In a further variation, the continuous helical displacement (CHD) pile produced by Roger Bullivant employs a short flighted 'bullet' head with a disposable tip which is screwed into the ground via a hollow auger stem to displace the soil laterally without the production of spoil at the surface (Figure 3.9). The pile shaft is formed by pumping concrete down the hollow stem as the bullet is back-screwed to produce a simple screw-form with the reinforcement cage added afterwards. Especially developed for contaminated sites the system has been found to be cost-effective on other sites where the disruption caused by removal of spoil and noise and vibration is to be minimized.

Yet further development of this type of pile has led to a 'bored displacement pile' by Rock and Alluvium Ltd in which a short tapered auger is drilled into the ground. Being of smaller diameter, the auger stem that follows the auger reduces the friction against the soil and the auger can be more easily advanced as a result. At full depth the auger may be rotated out or raised without rotation for a straight shafted pile. In either case, concrete is pumped into the bore via the hollow auger stem to support and form the shaft.

In all these types of pile, comprehensive rig instrumentation is essential to ensure that the pile shafts are suitably formed and where helical flanges are cast on the shaft of the pile, these need to be robust enough to resist pre-mature failure under load.

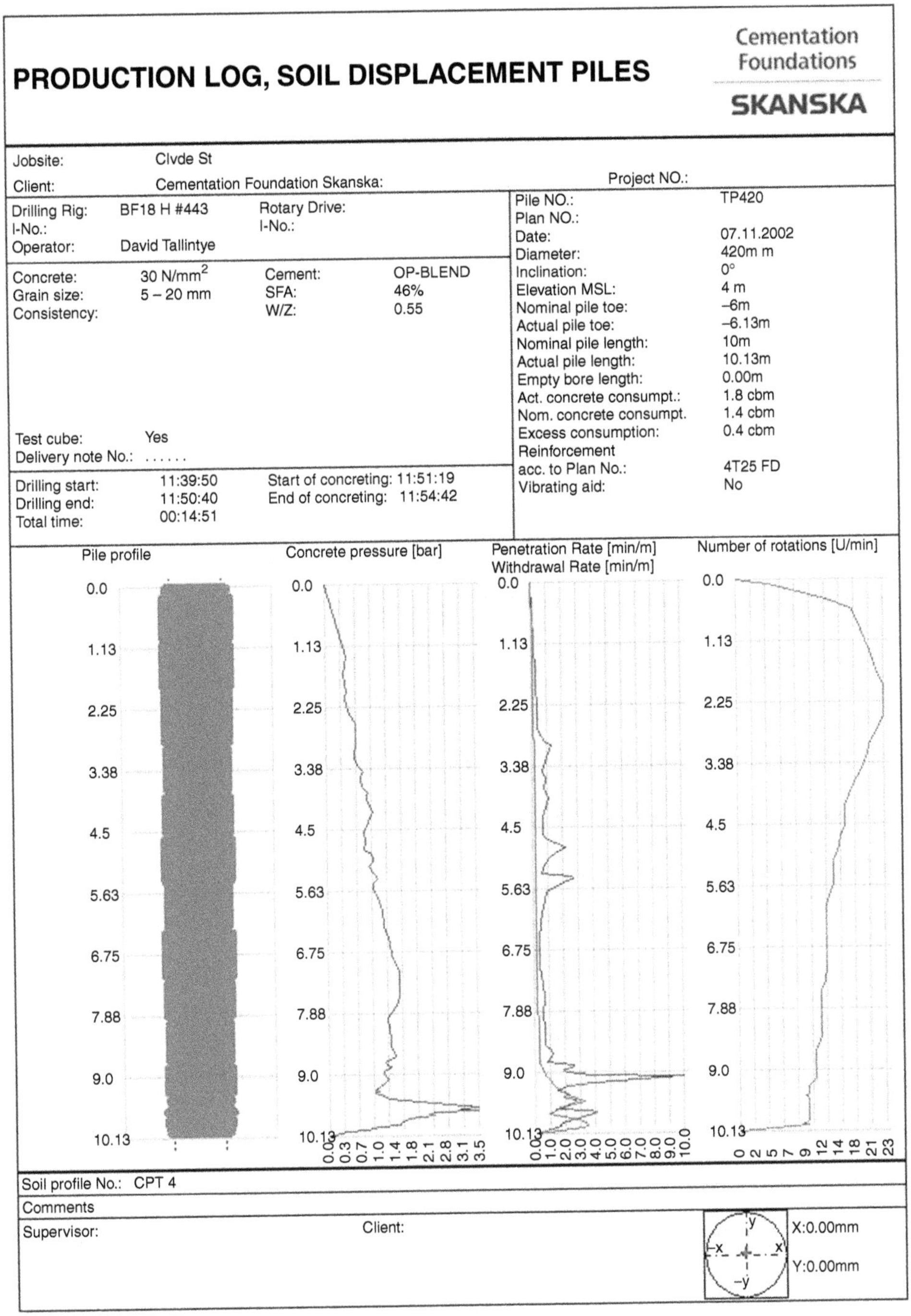

PRODUCTION LOG, SOIL DISPLACEMENT PILES

Cementation
Foundations

SKANSKA

Jobsite:	Clvde St			Project NO.:	
Client:	Cementation Foundation Skanska:		Pile NO.:		TP420
Drilling Rig:	BF18 H #443	Rotary Drive:	Plan NO.:		
I-No.:		I-No.:	Date:		07.11.2002
Operator:	David Tallintye		Diameter:		420m m
Concrete:	30 N/mm^2	Cement: OP-BLEND	Inclination:		0°
Grain size:	5 – 20 mm	SFA: 46%	Elevation MSL:		4 m
Consistency:		W/Z: 0.55	Nominal pile toe:		–6m
			Actual pile toe:		–6.13m
			Nominal pile length:		10m
			Actual pile length:		10.13m
			Empty bore length:		0.00m
			Act. concrete consumpt.:		1.8 cbm
Test cube:	Yes		Nom. concrete consumpt.		1.4 cbm
Delivery note No.:	......		Excess consumption:		0.4 cbm
			Reinforcement acc. to Plan No.:		4T25 FD
Drilling start:	11:39:50	Start of concreting: 11:51:19	Vibrating aid:		No
Drilling end:	11:50:40	End of concreting: 11:54:42			
Total time:	00:14:51				

Pile profile | Concrete pressure [bar] | Penetration Rate [min/m] Withdrawal Rate [min/m] | Number of rotations [U/min]

Soil profile No.: CPT 4

Comments

Supervisor: Client: X:0.00mm   Y:0.00mm

Figure 3.8 Production log from a soil-displacement pile installation. (Courtesy Cementation Foundations.)

Figure 3.9 CHD pile auger head. (Photograph courtesy of Roger Bullivant Limited.)

3.2.9 Vibrated Concrete Columns (VCCs)

A new form of displacement piling methods is the vibrated concrete column (VCC). This type of pile has similarities with vibro-displacement stone columns, but goes further than simply adding cement to the stone feed, as concrete is injected down the vibroflot stem once the design depth is reached. The installation is thus a displacement process; no spoil appears at the surface, and the final product is a pile with a concrete shaft and an enlarged base. A reinforcement cage can be installed into the wet concrete. Rigs for the installation of VCCs are typically instrumented with on-board computer-aided monitoring systems. The VCC pile is best suited to ground conditions where weaker deposits overlie a dense gravel or possibly a weak rock. The vibroflot will not penetrate a stiff clay to any great extent and therefore the VCC pile has become associated with the strengthening of alluvial deposits under embankments, for example, or the provision of support for geotextile reinforced granular rafts. For this type of application, it is a simple matter to provide a 'mushroom' head to the VCC to give some spread of load reaction. In suitable ground, the pile can however be used in the traditional way for column bases, strip footings or slabs. Nominal shaft diameters range from about 400 to 750 mm, and with load capacities similar to those for traditional piles.

3.3 Methods of pile driving

The methods of driving piles fall into the following categories:

(i)   dropping weight
(ii) explosion
(iii) vibration
(iv) jacking against a reaction.

With the exception of the installation of displacement piles by jacking methods, driven piling is an intrinsically noisy process. Today, as environmental issues become more important, the need to develop silenced hammers and the use of noise shrouds forms an important aspect of the technology associated with this form of pile installation. In the following sections the various means of installation of displacement pile is discussed, but in any particular application, it should be borne in mind that the suitability of a particular driving method can be strongly influenced by the environmental sensitivity of the site. In keeping with the trend towards instrumentation, modern piling rigs allow the operator to control the hammer input and such parameters as hammer velocity, energy input and drop height can be monitored in the driver's cab.

3.3.1 Drop hammers

The dropping weight, or drop hammer, is the traditional method of pile driving and is still employed. A weight approximately equal to that of the pile is raised a suitable distance in a guide and released to strike the pile head. For single tubular piles a free-standing timber trestle support is sometimes employed to support the pile above ground, provided that an internal drop hammer can be used. Alternatively, tubular guides (so called 'chimney pots') can serve to align piles during driving.

For most driven piling, purpose-made tracked rigs will hoist the pile into position, support it during driving and incorporate a guide for the drop hammer (Figure 3.10).

In the past, it was usual for a crane chassis to be employed as the base machine for use with a drop hammer, with a pile guide frame, or 'hanging leaders' supported from the crane jib and strutted from the base of the machine (Figure 3.11). This type of rig is still used today especially for large driven piles and is almost as adaptable as a purpose-designed machine so that raking piles or piles of extreme length can be handled with relative ease. The rig may be winched on skids or rollers, or mounted on tracks.

In guiding the pile, a balance has to be struck between providing suitable directional control, and allowing some freedom for slight pile movement within the guides, particularly at the base of the frame. Twist or slight lateral shifts often occur as a pile is driven. If this is prevented by a tight-fitting gate at the base of the pile frame,

Figure 3.10 Banut 850 piling rig, crawler-mounted, hydraulically operated machine with forward, backward and side-to-side raking facilities. Overall height of rig 25 m. (Photograph courtesy of Banut.)

or worse, by inclining the guides after driving has started, the pile may be over-stressed in torsion or bending. The resultant pile damage may not be apparent at the surface. Slight out-of-verticality rarely affects the pile performance, and a small deviation from a specified pile location can often be tolerated. A positional tolerance of 75 mm and a maximum of 1 in 75 on axial line for vertical piles are typically specified figures. A deviation from line of 1 in 25 is generally allowed for raking piles. These tolerances accord with those given in the ICE *Specification for Piling and Embedded Retaining Walls* (2007). A check calculation should

Figure 3.11 Pile driving rig with hanging leaders. (Photograph courtesy of RDL Munro Piling.)

be made, or proof supplied by the piling contractor, that, at the maximum tolerable inclination and eccentricity, the pile is not overstressed in bending when under load.

Drop hammer weights generally vary from 0.5 to 2 times the pile weight, with drops usually in the range of 0.2 to 2 m. Since the peak stresses at the pile head can be greatly increased if the hammer strikes the pile eccentrically, a long narrow hammer is preferable to a short large diameter one, as there is more chance of the blow being axial with the slender hammer and it has better impact characteristics. Some increase in hammer drop is usually required for raking piles, the driving energy being reduced by approximately 10% for a 1 in 3 rake. Distress at the pile head is more likely to be the result of using a hammer that is too light and hence needing an excessive drop, than using a hammer that is too heavy. The effect of increasing the ratio of hammer weight to the pile weight on pile driving efficiency is shown in

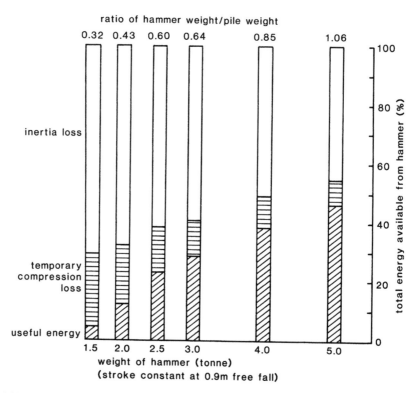

Figure 3.12 Energy distribution during driving of a pre-cast reinforced concrete pile for various hammer weights. (After Packshaw, 1951.)

Figure 3.12, for a 15.2 m long, 355 mm square-section concrete pile. In this example the pile weight is 4.7 tonne. The energy available from the hammer varies from a few percent at a hammer/pile weight ratio of 0.32 to almost 50% at a ratio of 1.06. Inertia losses and the temporary compression losses are reduced as the ratio increases to provide the increase in available energy. The maximum stresses at the top of a pile may be calculated from the relationships given by Broms (1978), as follows.

For a pre-cast concrete pile $\sigma_h = 3\sqrt{h_e}$ MN/m²
For a steel pile (with wooden packing) $\sigma_h = 12\sqrt{h_e}$ MN/m²
For a steel pile (without wooden packing) $\sigma_h = 18\sqrt{h_e}$ MN/m²
For a timber pile $\sigma_h = 1.2\sqrt{h_e}$ MN/m²

where σ_h denotes the maximum stress, and h_e (cm) denotes the equivalent height of free fall of the hammer. Note that the velocity of the hammer striking the pile head, v_0, is given by

$$v_0 = \sqrt{2h_e g}$$

The effect of the mass of the hammer is to alter the length of the stress wave produced in the pile. The length L_w of the stress wave is given approximately by the relationship

$$L_w = \frac{3W_h}{W_p/L}$$

where W_h denotes the total weight of the hammer and W_p/L denotes the weight of the pile per unit length.

During driving, the ground conditions will dictate the way in which the compressive stresses produced at the pile head by the hammer blow are reflected from the pile tip. Under easy driving conditions, the compressive stress is reflected from the toe as a tensile wave. When the pile length is somewhat greater than half the length of the stress wave, there is a likelihood of tensile stresses appearing towards the top of a pile. These can be damaging to pre-cast concrete piles, and when driving such piles in soft ground the hammer drop should be reduced, especially if using a light hammer to drive a long pile.

When heavy driving conditions are encountered, the initial compressive wave will be reflected from the pile toe as a compressive wave. Depending on the dynamics of pile/hammer contact, this compressive wave can be further reflected as a tension wave at the pile head if the hammer is not in contact with the pile head at the time the returning wave arrives. Tensile forces can therefore also occur under heavy driving conditions, which can be particularly damaging to concrete piles and may lead to a reduction in overall concrete modulus as a series of fine transverse cracks can be generated. Large compressive stresses occur at the pile toe and head when obstructions or dense strata are encountered, which may be sufficient to cause steel piles to deform or, if driving is protracted, fatigue can occur with similar results.

Variants of the simple winch-raised drop hammer include rams raised by steam or compressed air or as is most usual today, hydraulically. Hammers raised by fluid pressure may be free falling from the top of the stroke (single-acting hammers), but by applying fluid pressure on the down-stroke, double-acting or 'differential-acting' hammers will apply more impact energy for a given hammer weight. Double-acting air hammers are not generally suitable for driving concrete piles. However, large single-acting hammers have been developed with maximum net hammer weights approaching 300 tonnes, producing rated energies of up to 2.5 MNm using steam pressure (Vulcan 6300). By using a fluid cushion, the impact of the hammer is controlled in the larger pile hammers of the Hydroblok type, and a sustained push rather than a sharp peak of stress is transmitted to the pile at each blow. This type of hammer blow is better for driving through soils where the tip resistance of the pile is low as in clays.

Hydraulic hammers such as those manufactured by BSP, Junttan and Menck tend to be more efficient and quieter in operation than other types, and may be preferred to simple drop hammers and diesel hammers in many applications. The rated energies (as ram weight × stroke) are as high as 3 MNm. They are versatile and may be used on a wide variety of bearing piles, including raking piles and usually with only slight modification, may be used under water. They are also suitable for pile extraction. A typical hydraulic hammer is shown in Figure 3.13.

Figure 3.13 A Junttan hydraulic hammer driving steel piles. (Photograph courtesy of Junttan Oy.)

Approximate drop hammer and single-acting hammer sizes to drive various sizes of piles based on the design load of the pile are listed in Table 3.6 with the typical characteristics of selected BSP hydraulic hammers given in Table 3.7.

3.3.2 Pile driving by diesel hammer

Early experiments in driving by the explosive energy of gunpowder are related in Chapter 1. The modern equivalent is the use of the diesel hammer to produce rapid, controlled explosions. In operation, the diesel hammer employs a ram that is raised by explosion at the base of a closed cylinder. Alternatively, in double-acting diesel hammer, a vacuum is created in a separate annular chamber as the ram moves upward. This assists in the return of the ram, almost doubling the output rate of the hammer over the single-acting type. The ram weight is less than that of a drop hammer, but

Table 3.6 Approximate minimum hammer sizes for driving bearing piles based on design load of pile for drop hammers and single-acting hammers (BSP data)

Design load of pile (kN)	Approximate minimum hammer mass	
	Steel piles (tonne)	Concrete piles (tonne)
400	–	2
600	2	3
800[1] to 900[2]	3	4
1500	5	–
2250	8	–
3000	10	–

[1] Concrete pile.
[2] Steel pile.

Table 3.7 Typical characteristics of selected BSP hydraulic hammers

Hammer Designation	Impact Energy (kNm)	Length (m)	Weight (kg)
CX 40	40	4.022	5200
CX 50	51	4.022	6100
CX 60	60	4.975	7600
CX 75	71	4.975	8700
CX 85	83	4.975	9800
CX 110	106	5.475	12050
CG 180	180	7.00	17240
CG 210	210	7.45	20430
CG 240	240	7.75	22760
CG 270	270	8.05	25110
CG 300	300	8.35	27460

Note
CX range operates at 36 to 100 blows/minute. Generally suitable for steel sheet piles in pairs, tubular steel-bearing piles, steel 'H' piles inc. combination piles, reinforced concrete piles, cast-in-place piles and timber piles.

CG range operates at 29 to 100 blows/minute. Generally suitable for tubular steel-bearing piles, steel 'H' piles, reinforced concrete piles, cast-in-place piles and timber piles.

the more rapid action can make up for this loss in efficiency. The action of a single acting diesel hammer (e.g. Delmag, Mitsubishi, Vulcan or Kobelco type) is best suited to driving piles which derive the greater part of their resistance at the pile tip, such as occurs when driving through granular deposits. The diesel hammer is more versatile than the drop hammer and for driving vertical piles it can be set on the top of a pile without guide rails (leaders) attached to the rig. This does not apply to the driving of raking piles where the pile and hammer still require to be supported. A summary of selected Delmag hammer specifications is given in Table 3.8.

Using pile-driving analyzers, it has been found that unless very carefully maintained, the energy delivered by a diesel hammer starting in good condition rapidly diminishes

Table 3.8 Summary specification for selected Delmag diesel hammers

Hammer type	D6-32	D8-22	D12-42	D16-32	D19-42	D36-32	D62-22	D100-13	D200-42
Ram wt. (kg)	600	800	1280	1600	1820	3600	6200	10000	20000
Blows (min/max)	38/52	36/52	35/52	36/52	35/52	36/53	35/50	35/45	36/52
Max. Energy per blow (Nm)	18500	28000	46000	54000	60000	123000	224000	370000	670000
Hammer wt (kg)	2470	2670	3455	4045	4320	9770	12900	19900	57120
Hammer length (mm)	3810	4699	4724	4724	4724	5283	5918	7366	8230

in use. It should be appreciated however that site conditions are not always conducive to such levels of maintenance and lower than optimum efficiency frequently exists.

3.3.3 Vibratory methods pile driving

Vibratory hammers are electrically or hydraulically powered and consist of contra-rotating eccentric masses within a housing attached to the pile head. Power requirements are up to 250 kVA, for the electrically powered units and up to about 1000 HP for the hydraulically powered units, although usually in the range of 100 to 500 HP. One of the problems that can be encountered with large vibratory hammers is that of providing secure clamping to the pile so that the available pile clamp force is an important specification parameter. The American Pile Driving Corporation (APE) use a patented 'Quad-Clamp' system with four separate clamps of 125 tons clamp force each for driving tubular steel caissons of up to 5 m diameter using the APE 'Super Kong' vibro-driver with a 1000 HP engine. The power for vibratory pile drivers is usually supplied from a mobile generator or hydraulic power pack. The majority of pile vibrators run at low frequencies, typically 20 to 40 Hz. At these frequencies, neither the exposed length of pile nor the soil will be in resonance. Typical amplitudes (in terms of total undamped travel) are between 5 and 30 mm. Sound propagation is low, and in cohesionless soils good rates of progress can be realized. During the driving progress, the granular soil immediately adjacent to the pile is effectively fluidized, and friction on the shaft is considerably reduced.

In cohesive soils, fluidization will not occur, and vibratory pile driving methods are not generally as effective. Recent developments in vibratory piling have been summarized by Holeyman et al. (2002).

3.3.3.1 Resonance pile driving

If the frequency of vibration is increased up to perhaps 100 Hz, a pile will resonate longitudinally, and penetration rates can approach 20 m/min in loose to moderately dense granular soils, falling to 5 m/min or less in dense granular soils. At these frequencies non-cohesive soils are fluidized to the point where the frictional resistance on the pile shaft is reduced to close to zero, and the pile driving energy is delivered to the pile toe.

This method of pile installation lacks thorough investigation. It is potentially very effective, but there is a tendency for the driving equipment to suffer from the damaging effect of the severe vibration associated with the large energy inputs required to drive bearing piles and the results are to some extent unpredictable. Noise and vibration propagation can be high, leading to settlement in nearby structures. The method can be used to good effect for pile extraction.

3.3.4 Jacking methods of pile installation

By jacking (usually hydraulically) against a reaction, lengths of pile, either in short units or in continuous lengths, may be forced into the ground. The method is well suited to micro-piling (see section 3.6), as the reaction loads can then provided by the structure being underpinned. Other, portable reaction systems may be used in situations other than underpinning. An early application was the 'Taywood Pilemaster' system, in which sheet piles were forced into the ground against the reaction provided by adjacent piles. The Taywood Pilemaster is no longer available but is superseded by more modern systems typified by the Giken (Arcelor) (Figure 3.14) and the 'Still Worker' (Stent Piling). The system can in certain circumstances be extended to install tubular piles by using adjacent sheet piling which is likely be present within a cofferdam for reaction. Jacking methods are exceptionally quiet and vibration-free in use and by monitoring the pressure in the hydraulic system an indication of the

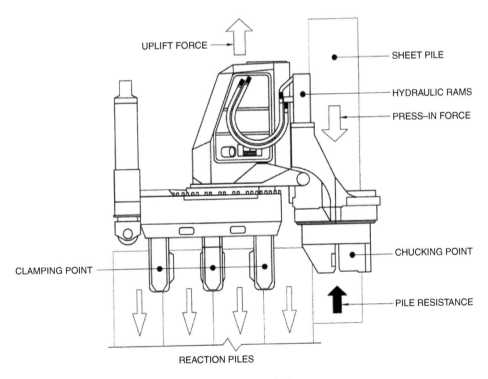

Figure 3.14 Giken pile driver.

pile failure loads can be obtained during installation. The system is illustrated in Figure 3.14.

3.4 Non-displacement piles

With the excavation of a borehole in the ground, a pile can be produced by casting concrete in the void. Some soils, such as stiff clays, are particularly amenable to the formation of piles in this way, since the borehole walls do not require support, except close to the ground surface. In unstable ground, such as gravels, the ground requires temporary support, from casing, a bentonite slurry or a polymer fluid. Alternatively, the casing may be permanent, but driven into a hole that is bored as the casing is advanced. A different technique, which is still essentially non-displacement, is to intrude a grout or fluid concrete from a continuous flight auger that is rotated into the ground, and hence produce a column of concrete which can be reinforced. The three non-displacement methods are therefore:

(i)   bored (i.e. rotary or percussion formed) cast-in-place piles
(ii)   partially pre-formed piles
(iii) continuous flight auger (CFA) piles.

Particularly in pile types (i) and (ii), little or even negative soil displacement occurs during boring, and indeed, granular deposits may be loosened by the boring action. This is in contrast to a displacement pile, which produces (sometimes beneficial) ground compaction in granular deposits. On the other hand, the rough and irregular interface between the pile and the soil which is formed during the boring action tends to improve load transfer, and the difference in skin friction between the two types is less than it might be. In type (iii) the auger is usually made to run full throughout the process, and relaxation of the walls of the bore may be inhibited.

3.4.1 Bored cast-in-place piles

The range of pile sizes and hence design loads is considerable for this form of pile. The excavation process generally (but not exclusively) produces a borehole of circular cross-section, and piles are referred to as small-diameter when less than 600 mm, and large-diameter when greater than this nominal size. In the past, tripod rigs employing percussion methods were usually used to create the smaller diameter piles but today, it is usual to employ continuous flight auger (CFA) rigs for the smaller diameter piles. The larger sizes are generally bored with rotary or sometimes percussive methods. Nominal diameters (mm) for bored piles are as follows:

(i)   300, 350, 400, 450, 500, 550, 600 (small-diameter)
(ii) 750, 900, 1050, 1200, 1350, 1500, 1800, 2100 (large-diameter).

3.4.2 Small-diameter percussion bored cast-in-place piles

Simple cable tool tripod rigs employing percussion methods were generally used in the past to form small diameter pile bores even when significant lengths of temporary casing were required. The method is however labour-intensive and in the larger diameters requires the manual handling of heavy screwed casing sections. Tripod rigs and boring tools for piling are essentially heavier versions of the rigs employed for site investigation. Using tripod rigs, minimum pile diameters start at around 300 mm, which is about the maximum for a site investigation borehole. In cohesive soils, a 'clay cutter' is used, often weighted to improve penetration. The winch for raising the cutter and spoil to the surface may be driven by a diesel or compressed air motor. The hole is advanced by repeatedly dropping the clay cutter in a cohesive soil and winching the cutter to the surface with its burden of spoil. The cutters may consist of an open cylinder, with a hardened cutting edge or sometimes a cruciform section. Clay is extruded past the cutting ring, and adheres to the cruciform blades and may be pared away at the surface, or may pass into the body of the cylindrical type of cutter, and be shaken out at the surface. To assist the cutter in stiff clays, a little water may be added to the borehole. It is unavoidable that some of this water will be drawn into the pore space in the clays as they swell on stress relief from the boring action. However, the reduction in soil strength does not appear to be significant from this cause and it is more important to bore and concrete the shaft rapidly.

Tripod rigs are light, easily transported and have been widely used for installing piles in congested sites or even from inside buildings. Modified 'short' rigs can operate in limited headroom with a slight loss in efficiency. Very little site preparation is required. Minimal soil disturbance occurs when boring through clays and piles can be placed close to buried services. Care is however needed in granular soils below the water table (described below) because there is a possibility of removing material from around the bore leading to undermining of adjacent footings. During boring for a pile the temporary casing may be advanced to considerable depths although the single line pull of the winch may then be inadequate to extract it. By a pulley system, the standard winch pull of around 20 kN may be increased to 80 kN or more. Even this may be insufficient to extract long lengths of casing driven into dense soils (especially if the casing has been in the ground for some time), and hydraulic jacks acting against a collar at the top of the casing sometimes need to be used to initiate extraction.

In granular soils, a 'shell' is used for boring. This comprises a heavy steel tube with an open top and a flap valve at its base. The removal of soil by a shell relies on water being present. Dropping the shell into a granular soil below the water table causes the flap at the base to open, the flap then closes as the shell is raised. By working the shell up and down, the granular soil is loosened by liquefaction as a low pressure is produced under the shell as it is rapidly raised, and this material then flows into the shell as it is dropped. Spoil from the shelling action is simply tipped out at the surface by up-ending the shell. There is a danger during shelling that the soil may be over-loosened and material drawn from the sides of the bore, and in extreme cases, the shell may become buried below the casing. To prevent such over-break, the temporary casing is advanced by driving it into the ground as the borehole advances.

With a long length of heavy casing in loose sandy gravels, it is possible for rapid shelling action to so loosen the ground that the casing falls under its own weight. Whilst rapid progress is achieved, there is always a risk of over-break by this means, and the ground is loosened around the bore to a significant degree. It is important that a positive head of water is kept in the pile bore when boring through granular deposits to minimize the loosening effect and to prevent piping up the pile bore. Water may have to be added in order to achieve this. Obstructions, either natural or man-made, are cleared by the use of a heavy 'I' section chisel, with a hardened flat point or blade.

3.4.3 Large-diameter percussion bored piles

Large-diameter percussion bored piling equipment is well suited to the penetration of hard strata, which may include weak sedimentary rock. Where casing is advanced as the drilling proceeds, unstable strata are supported and may be interspersed with harder strata without compromising the boring action.

An early method of forming large diameter piles in difficult ground using a form of percussion equipment was the 'Benoto' system. This employed a heavy rig to handle large diameter casing that was pressed into the ground using an oscillatory semi-rotary movement with an internal grab to remove the soil.

The casing was sectional, but because of the semi-rotary driving action, was not screwed together, but employed a flush quick-action circumferential joint within the twin-walled casing thickness. The boring tube had a hardened cutting edge and was hydraulically clamped in a collar that transmitted the semi-rotary action and vertical motion provided by hydraulic rams. In hard ground, rock jaws were fitted to the hammer grab. The piles were reinforced and concreted in the usual way, the casing being hydraulically jacked out of the ground during concreting. Typical working loads were 2000 to 5000 kN in suitable ground for pile diameters of between 670 and 1200 mm respectively. Pile lengths of up to 40 m were possible. Working on a similar principle, but with the oscillatory movement and the down-thrust carried out as separate processes, are the Bade and Hochstrasser-Weise systems which can give higher rates of progress. During the concreting stage of the latter system, compressed air is applied to the air-space in the upper part of the casing which is sealed off. This assists in lifting the casing that is oscillated as it is withdrawn but the air supply needs to be carefully controlled to prevent an unstable condition.

More recently, semi-rotary down-hole percussive hammer rigs powered by compressed air, of the type used for well boring, have become available in sizes which approach those of large-diameter piles. Large quantities of compressed air are required to power these hammers and space is therefore needed on a site to house the compressors although the set-up is adaptable in its arrangement. Generally speaking, a down-hole hammer will require a separate reaming and casing operation from an attendant auger rig. However, the plant will bore through weak rock with little difficulty and in certain circumstances may be suitable for forming deep rock sockets (see Figure 3.15). The Atlas Copco 'Symmetrix' system is a patented development of the down-hole percussive method of drilling in which casing follows the bit and broadens the scope of this type of hole boring. It is available for drilling piles into broken or hard strata that may require temporary or permanent casing in diameters up to 914 mm. Further developments include the use of hydraulically powered down-hole hammers that will work with drilling muds thereby combing the hole forming with a borehole wall support system.

Figure 3.15 Down-hole hammer bit. (Photograph courtesy of Ritchies.)

3.4.4 Rotary bored cast-in-place piles

The majority of large-diameter piles are bored using rotary methods, generally by augers in the United Kingdom, although in other parts of the world, such as the USA, large-diameter drilling plant may be used to construct 'caisson' piles.

Augering rigs are usually crawler mounted, although truck mounted rigs are more common in the United States where they are manufacture by such companies as Watson or Bay Shore Systems. The truck mounted rigs have maximum ease of transport between sites and pile positions but there are limitations with respect to auger clearance height. Crawler mounted rigs are often purpose-built by manufactures such as Bachy, Bauer, Casagrade, Soilmec or Wirth. Although the smaller rigs are generally dedicated machines, in the larger sizes, standard crane bases still tend to be used with an auxiliary power pack to drive a ring gear within a rotary drilling table fixed to the crane base (Figure 3.16).

In operation, the auger is driven from the ring gear at the drilling table via a sliding square or keyed circular-sectioned Kelly bar, usually telescopic to enable the deep piles to be bored. The auger and Kelly bar are suspended from the crane by a winch rope and following ground penetration by the screw action of the auger, the spoil-laden auger is raised and the spoil removed by 'spinning off' (see Figure 3.17). An attendant

Figure 3.16 Auger Rig with Kelly bar drive. (Photograph courtesy of Bachy-Soletanche.)

crane is generally necessary for handling temporary casing, boring equipment, etc. Some specially developed rigs, usually for the smaller diameter piles, can handle boring tools and casing from a second winch line running in a double cathead.

The usual crane-mounted auger will bore to a diameter of at least 3000 mm. Depths of boring vary from around 25 m from cased auger rigs to as much as 60 m for the larger crane-based auger rigs, with telescopic, triple extension Kelly bars. Typical design loads vary from 1000 to 20 000 kN, in suitable ground conditions. Although the auger rigs can bore raked piles at up to 1:3.5 (more usually 1 in 4), it is considered better practice to take lateral and inclined loads on a suitably reinforced vertical pile of sufficient cross-section.

The augers fitted to this type of rig are usually of the 'short' flight type for large diameters. Alternatively, in ground that breaks up excessively after cutting, a bottom-opening toothed bucket may be used. Both augers and buckets can be equipped with hardened teeth for breaking up lightly cemented deposits. A slight, if irregular, increase in bore diameter can be obtained using reaming cutters attached to the periphery of the auger. Various items of auger equipment are illustrated in Figure 3.18.

Figure 3.17 Auger rig 'spinning off' spoil. (Photograph courtesy of Bauchy-Soletanche.)

Figure 3.18 Boring tools used in the construction of rotary bored cast-in-place piles. Foreground – heavy duty augers, background-drilling buckets. (Photograph courtesy of Carillion Piling.)

Figure 3.19 Heavy chisel for breaking obstructions. (Photograph courtesy of Carillion Piling.)

A heavy chisel for breaking obstructions is shown in Figure 3.19.

Although not possible in flooded bores, a special feature of auger bored piles is the possibility of creating an underream, which enables the bearing capacity of suitable strata to be exploited by providing an enlarged base with a maximum diameter of approximately 6000 mm. The soil has to be capable of standing open unsupported to employ this technique and therefore stiff to hard clays, such as the London Clay, are ideal. An underreaming tool is shown in Figure 3.20. In its closed position it will fit inside the straight section of a pile shaft, and can then be expanded at the base of the pile to produce the underream.

The shaft and underream of such a pile would normally be inspected before concreting is carried out. In the past this was usually carried out by lowering operatives in a man-carrying cage down the pile bore provided it had a diameter of 750 mm or more. Safety precautions in this operation included the provision of an air line and a running casing sleeve within the pile bore. Today, the inspection is more likely to be carried out by remotely operated cameras, thereby obviating the need for personnel to

Figure 3.20 An underreaming tool in the open or working position. (Photograph courtesy of Stent Piling.)

enter the bore. It is however important that loose material does not remain at the pile toe and a suitable means of removing this is required if remote inspection is used. The sequence of construction for a large diameter rotary bored pile with an underream is shown in Figure 3.21.

Although the sliding Kelly bar and fixed rotary table is a common arrangement, an hydraulically driven motor which slides in vertical guide rails can be mounted on the top of the drilling stem. Because of the weight of such a motor, this system is more popular in the smaller-diameter continuous flight auger (CFA) rigs (see section 3.5).

At one time there was interest in a remotely controlled wire line-suspended rotary drill in which a submersible electric drive motor is mounted directly over a triple cutting head. There is no torque to be resisted by the rig with this arrangement, and in suitable ground deep holes could be drilled without any drill stem handling or extension problems. The Tone Boring Co. Ltd. of Japan manufactured the drill (Figure 3.22).

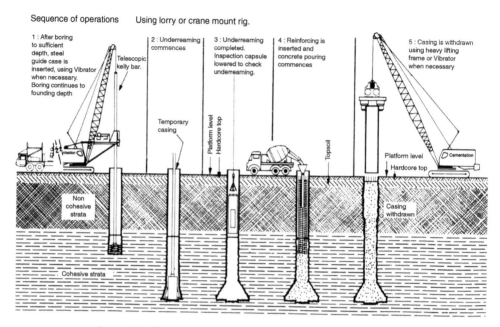

Figure 3.21 Sequence of large-diameter rotary pile construction.

Recent improvements in the power of heavy auger rigs has meant that large diameter pile bores can be achieved in weak sedimentary rocks, whereas at one time, rotary drilling methods with large core barrels or rock roller bits were employed. These drills, which were mounted on heavy base cranes have all but disappeared. A further development of drilling methods is the use of air-lift drilling in diameters up to about 4 m. For this procedure, top-drive rigs are employed where the rotary drive head is attached to the top of a casing which is pulled into the ground via hydraulic rams as a rotary drill bit forms the bore ahead of the casing such as produced by Wirth Maschinen.

3.4.4.1 Temporary ground support with casing

The upper portion of a pile bore is likely to be in loose or weak ground so that some form of temporary support becomes necessary. It is usual for a steel casing to be employed for this, even when the remainder of the bore is made under the stabilizing action of a bentonite suspension or polymer fluid, described in a later section. The temporary casing also seals off water below the water table in permeable ground.

Shallow, dry, granular deposits are loosened by the auger, and will generally stand open long enough for a short length of casing to be inserted. At the level of an impermeable soil, such as a stiff clay, the casing can be forced into the ground to form a seal against entry of water by 'crowding', i.e. a downward force applied to the casing. The 'crowd' force is applied via the Kelly bar by clamps gripping the casing and applying an axial thrust via retracting hydraulic rams against the reaction of the rig. To advance the bore further in granular soils, the auger can be used to loosen the soil, followed by

Figure 3.22 Tone RRC-U rope-suspended rotary drill for variable-diameter boreholes/pile shafts. (Photograph courtesy of Tone, Japan and Geo Publications Ltd., Brentwood.)

driving the casing with a drop hammer. Alternatively, the casing can be advanced with an hydraulic casing vibrator suspended from the attendant crane. The process can be made more efficient in some instances by adding bentonite powder to the granular soil, and 'mudding-in'. Water may also need to be added. The effect is to produce a column of loosened soil within a stable bore through which the casing can be installed to depths of about 10 m. A casing vibrator can increase the depth of insertion to around 25 m. Casagrande hydraulic casing oscillators provide an alternative means of installing casing (Figure 3.23) and can be of use in freeing-up long lengths of casing at the start of extraction.

3.4.5 Boring and concreting

Concreting in 'dry' bores is usually a straightforward procedure. On completion of the borehole, the reinforcement is placed with suitable spacers to locate it centrally in the bore. During concreting, the casing is withdrawn, always maintaining a head

Figure 3.23 Casagrande hydraulic casing oscillator with diesel powered pump.

of concrete within the casing. When pouring concrete in a dry vertical bore, a hopper and tube is used to ensure that the mix is not directed towards the reinforcement. In raking piles, a chute may be required or a first charge of grout placed.

Any sudden drop in concrete level as the casing is withdrawn would probably indicate a zone of over-break. Intermixing of soil, water and concrete can then take place to create a defect in the pile shaft (see Chapter 7). This occurrence is sometimes accompanied by a loss or distortion of pile reinforcement, the steel cage being carried downwards and outwards at the overbreak zone. To prevent this, boring ahead of the pile casing in unstable ground should be avoided. Once a zone of over-break has been created, casing it off then leads to the conditions described above as the casing is withdrawn during concreting. When such a defective shaft is examined, its occurrence is sometimes mistakenly ascribed to the cross-flow of groundwater.

With a self-compacting mix, it is not necessary to vibrate the concrete, and a vibrator should not be used inside the pile casing. It is sometimes considered beneficial to vibrate the top 1 to 2 m of a pile after the casing has been withdrawn, but this should not be overdone, or soil mixing can occur.

In weaker soils, such as alluvial clays, it is possible for the fluid pressure of the wet concrete in the pile bore to fail the soil, and a slump occurs with a bulge in the pile shaft. An undrained shear strength of around 15 kN/m^2 or less has been observed to be associated with this effect and where soils of such low strengths are present at depths typically of between 3 to 5 m, a full head of concrete in a pile bore can fail the ground. Extracting temporary casing at a steady rate rather than pulling in short, rapid tugs can lessen the possibility of soil failure from this cause. The head of concrete in the casing may therefore need to be limited, provided this can be done without risk of the concrete slumping below the bottom of the casing in the bore.

3.4.5.1 Boring and concreting in water-bearing ground

In a stratigraphical succession in which water-bearing gravel overlies stiff clay, the temporary casing is used to support the borehole through the gravel, and is then driven a short distance into the clay in order to seal off the water. Boring then continues under dry conditions. This may not always be possible, and the base of the pile may be in a granular deposit below the water table. The tremie method of concreting is then employed. This is a practical method of forming a pile shaft provided certain precautions are taken. It is essential that a total collapse slump mix is used (i.e. greater than 175 mm – see also CEN EN 15336 for European Standards), and that the tremie pipe is always well below the water/concrete interface during concreting. The tremie pipe and hopper connections should be watertight and in clean condition to permit free flow of the concrete. A minimum internal diameter for the tremie pipe of 150 mm is suggested for use with concrete having a maximum aggregate size of 20 mm, increasing for larger aggregates. A vermiculite or expanded polystyrene plug should be used in the tube between the water and first batch of concrete, and immediately before placing the tremie pipe in the pile bore, a check on sedimentation should be made, and any significant sediment should be air-lifted out. The procedure is illustrated in Figure 3.24. Where the pile cut-off level is above ground, concrete should overflow from the pile head on completion. The upper part of the pile is composed of weak concrete from an excess of water and may also be contaminated from boring detritus. Where there is a deep cut-off, the concrete level should be raised to allow for around 0.5 to 1.0 m for trimming off the weak concrete.

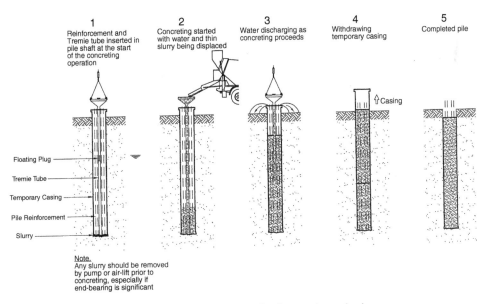

Figure 3.24 Placing concrete by the tremie method.

3.4.6 Concrete mixes for bored piles

The integrity of the pile shaft is of paramount importance, and the concreting mixes and methods that have evolved for bored piles are directed towards this, as opposed to the high-strength concrete necessary for pre-cast piles or structural work above ground.

This prerequisite has led to the adoption of highly workable mixes, and the 'total collapse' mix for tremied piles has been mentioned. In order to ensure that the concrete flows between the reinforcing bars with ease, and into the interstices of the soil, a high-slump, self-compacting mix is called for. A minimum cement content of $300 \, \text{kg/m}^3$ is generally employed, increasing to $400 \, \text{kg/m}^3$ at slumps greater than $150 \, \text{mm}$, with a corresponding increase in fine aggregate content to maintain the cohesion of the mix. Three mixes recommended by the Federation of Piling Specialists are given in Table 3.9. However, these are now superseded by CEN EN 1534 European Standards which quote minimum cement contents of $325 \, \text{kg/m}^3$ under dry placement conditions and $375 \, \text{kg/m}^3$ under submerged conditions, both at a water–cement ratio of 0.6. Other requirements are quoted with respect to slump according to the placing conditions, i.e. slumps between 130 to $180 \, \text{mm}$ and >$180 \, \text{mm}$ depending on whether the concrete is placed in a dry bore or tremied.

3.4.7 Reinforcement for bored piles

For piles loaded in compression alone, it is generally only necessary to reinforce the shaft to a depth of $2 \, \text{m}$ greater than the depth of temporary casing, to prevent any tendency for concrete lifting when pulling the casing. Piles subject to tension or lateral forces and eccentric loading (possibly by being out of position or out of plumb) do however require reinforcement suitable to cope with these forces. Typical *nominal* reinforcement for piles in compression is shown for guidance in Table 3.10. *The restrictions that apply to the use of this table have to be carefully considered in any particular application.*

It should be noted that the minimum amounts of reinforcement to comply with European Standards are prescribed by CEN EN 1536. The assembled cage should be sufficiently strong to sustain lifting and lowering into the pile bore without permanent distortion or displacement of bars and in addition bars should not be so densely packed that concrete aggregate cannot pass freely between them. Hoop reinforcement

Table 3.9 Recommended concrete slumps for cast-in-place piles (based on Federation of Piling Specialists' Specification, for cast-in-place piling)

Piling mix	Typical slump (mm)	Conditions of use
A	125	Poured into water-free unlined bore. Widely spaced reinforcement leaving ample room for free movement of the concrete between bars.
B	150	Where reinforcement is not spaced widely enough to give free movement of concrete between bars. Where cur-off level of concrete is within casing. Where pile diameter is < 600 mm
C	>175	Where concrete is to be placed by tremie under water or bentonite in slurry.

Table 3.10 Guidance on the minimum reinforcing steel for bored cast-in-place piles

Pile diameter (mm)	Main reinforcement		Secondary reinforcement	
	No	Bar	Bar	Pitch (mm)
400	4	H12	H6	200
450	5	H12	H6	225
500	6	H12	H6	250
600	6	H12	H8	300
750	6	H16	H8	300
900	8	H16	H10	300
1050	11	H16	H10	300
1200	12	H16	H10	300
1500	12	H20	H10	400
1800	12	H20	H10	400
2100	16	H20	H10	400
2400	14	H25	H10	500

'H' steel denotes yield stress (f_y) 500 N/mm^2.

Notes
(a) The above guidelines are for 'build-ability' only. They are not appropriate where:

   (i) BS5400 or associated Highways Agency codes of practice are specified (UK).
   (ii) piles are required to resist any applied tensile or bending forces – the reinforcement has to be designed for the specific loading conditions.
   (iii) piles are required to accommodate positional and verticality tolerances, or where they are constructed through very soft alluvial deposits (c_u < 10 kN/m^2). Specific reinforcement design is then necessary.

(b) Minimum depth of reinforcement is taken as 3 m below cut-off for simple bearing only. Any lateral loads or moments taken by the pile will require reinforcement to extend to some depth below the zone subjected to bending forces. This zone may be determined from a plot of the bending moments with depth. Furthermore, the reinforcement would normally extend at least 1 m below the depth of any temporary casing.
(c) Even with the appropriate reinforcement care will still be required to prevent damage to piles by construction activities especially during cutting-down or in the presence of site traffic.

(for shear) is not recommended at closer than 100-mm centres. Concrete cover to the reinforcement periphery is generally a minimum of 75 mm.

It is frequently necessary to lap bars in long piles, and it is here that trouble can arise, with aggregate blocking at the laps and preventing concrete flowing to the borehole walls. Such an occurrence results in a defective pile, but is sometimes not appreciated by designers. To over-design the steel reinforcement in a bored pile can lead to practical difficulties in formation of the pile shaft. Where large steel cross-sections are unavoidable, consideration can be given to staggering the laps, although this can be difficult to arrange in pre-fabricated cages. If bar connectors are used, staggering is not such a problem as the lengths involved are shorter, but these are designed for use either in tension or alternatively in compression, and it is not generally possible to pre-determine the orientation of the cage in the pile bore. Steel bars can be butt joined by welding. Provided high-quality welds are used and the welding process carried out with strict quality control, taking into account the type of steel employed, etc., this can be an effective solution. Finally, provided the cage can be oriented with certainty, maximum steel need only be placed over that part of the pile subjected to maximum stress (if this is accurately known), and a reduced density can be used in the plane of the neutral axis.

3.4.8 Excavation using a bentonite suspension

A method of borehole wall support developed from slurry trench techniques is the use of a bentonite suspension. Where pile bores are required to penetrate considerable depths of unstable soil, the installation of long lengths of temporary casing is a time-consuming process.

By employing a bentonite suspension of around 6% by weight of bentonite, the borehole walls may be effectively supported, provided there is a positive head of bentonite slurry above the groundwater table of at least 1.5 m. In permeable (granular) soils, there is an initial flow of slurry into the borehole wall, where a 'filter cake' rapidly builds up which then prevents further flow, and the excess hydrostatic pressure in the borehole supports the ground. An additional benefit is the holding in suspension of fine detritus from the boring operations. Before concreting, the bentonite in the pile bore is checked for contamination and if necessary is exchanged for fresh bentonite. Bentonite slurry is returned to the surface for settling and de-sanding. Cleaned slurry may be re-used, with adjustment of bentonite content by the addition of new bentonite or mains water as necessary. After some re-uses, the bentonite slurry may become too contaminated for effective cleaning, and it is then discarded.

Concreting is carried out by the tremie method, using concrete with a slump in excess of 175 mm. With clean bentonite slurry in the bore, the suspension is effectively scoured and displaced by the fluid concrete. No decrease in pile adhesion has been observed in properly constructed bentonite piles, whether the shafts are in granular soils or cohesive clays. A full description of the constructional method, and the results of many pile tests on bentonite piles, are given by Fleming and Sliwinski (1977). The process is illustrated in Figure 3.25. To comply with CEN EN 1536, longitudinal reinforcement for bores stabilized with bentonite should be of ribbed bar.

3.4.9 Excavation using a polymer fluid

The use of a polymer fluid to stabilize a pile bore may be considered as an alternative to the use of bentonite suspension although a polymer fluid is not as dense as a bentonite suspension and therefore it is not suitable for pile bores with a water table less than 2 m from the ground surface. Polymer support fluids are suitable in clays and granular deposits except perhaps those that are the most permeable being poorly graded and/or of coarse particle size but they will work in salt water. In salt water approximately twice the quantity of polymer will however be needed which could be a deciding factor as polymer fluids are more expensive than bentonite muds. The polymer consists of long chain molecules that keep small particles of soil detritus in suspension and is generally biodegradable. Polymer 'threads' can be seen in the spoil as it is spun off from an auger (Figure 3.26). The action of the polymer fluid is not the same as bentonite mud, as no filter cake is formed, but instead a high seepage gradient is maintained over the bore/soil surround.

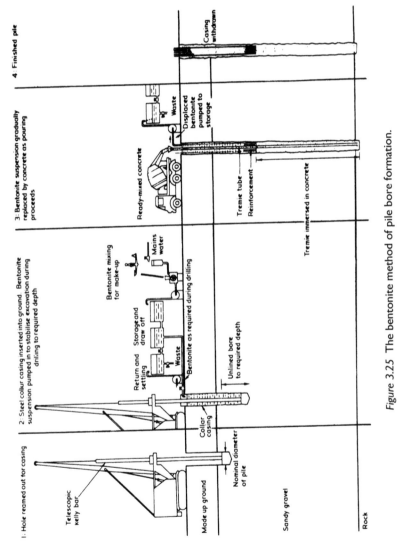

Figure 3.25 The bentonite method of pile bore formation.

Figure 3.26 Polymer strands attached to excavated spoil on an auger. (Photograph courtesy of Carillion Piling.)

The first stage of the polymer process is to bring the pH of the mixing water tank to an alkaline value of around 8 to 10, usually by the addition of sodium carbonate ('soda-ash') to the mixing tank. There is no benefit in raising the pH value further and the liberal use of caustic soda to achieve this is not recommended. The polymer, either as a granular solid or a ready mixed solution is then added to achieve a suitable viscosity. Checks on the properties of the polymer need to be frequently carried out and in a pile bore to say 20 m, the viscosity would be checked at each 5 m stage during boring with density, sand content and pH finally checked just prior to concreting. After use the polymer fluid can be treated with sodium hypochlorite ('bleach') to break it down to a harmless liquid that can be disposed of into a foul sewer or following favourable results of chemical analyses, into water courses. It is possible for the polymer to become contaminated with pollutants in the ground during the boring operation and this can make disposal more difficult. In general however, the use of polymer fluid leads to cleaner sites with considerably less space being taken up with its production compared with bentonite mud.

Whilst polymer fluids have been in use for well drilling for some time, they are relatively recent additions to the piling market and as yet, their long-term properties in this application have not been fully established although no special problems are envisaged. The skin friction of bored pile shafts formed with polymer fluids appears not to be reduced relative to bentonite mud supported bores although occasionally some reduction in end bearing has been observed, sometimes ascribed to a perceived inability of a polymer fluid to keep detritus in suspension. In this respect the cleaning of the toe of a pile needs to be thorough and may require the more aggressive action of a sand-pump rather than relying on air-lift equipment. It is possible that there may also be some slight reduction in the bond between the reinforcement and the concrete and further research into this aspect of polymer use is required although as with bentonite mud, ribbed longitudinal pile reinforcement is recommended. There is however evidence to suggest that with polymer fluid, the softening of clay does not occur as rapidly as with support with bentonite mud so that pile bores formed with polymers may be left over-night prior to concreting.

3.4.10 Base grouting of bored piles

Base grouting of bored piles via a grout pipe installed at the main concreting stage allows a form of shaft pre-stress to be applied, i.e. the direction shaft friction is reversed by the injection pressure and therefore when the pile is loaded, an enhanced load/settlement behaviour is achieved. It is of course necessary that sufficient shaft friction is available for the procedure to work, but by this means, poor end bearing conditions can be improved as well as improving the stiffness of a pile in the early part of the loading curve. It should be noted however, that the procedure is not a cure for excessive settlement of piles within a group that is the result of deep-seated consolidation of a compressible layer.

3.4.11 Diaphragm piers (barrettes)

A form of non-displacement piling suitable for exceptionally large working loads is the diaphragm pier or 'barrette' which is in effect a short section of diaphragm wall and is usually of rectangular cross-section. Excavation of barrettes is as for diaphragm walls, i.e. grabbing methods, with trench support by bentonite slurry or polymer fluid (Figure 3.27). However, excavation in rock formations will generally require the use of rotating hard-faced cutters or rock roller bits (rock mills) rotating in a vertical plane at the base of a drive column (Figure 3.28). Barrettes may be formed in rock with uniaxial compressive strengths of up to about $50 \, MN/m^2$ with rotating cutters, increasing to around $100 \, MN/m^2$ using rock mills. In practice, an excavation is made to say 1.5 to 2 m depth and of slightly larger plan size than the required cross-section of the barrette. Guide walls are then formed by back-shuttering methods, generally to a size to suit the grab or cutter to be used. Excavation to final depth continues as for a normal diaphragm wall, with concreting by identical techniques. The advantages of the large cross-section of the units is best exploited in end-bearing (especially in rock), but support from the soil may be derived from a mixture of shaft friction and end-bearing, as for other piles. This type of pile is capable of supporting large lateral loads and is therefore an attractive solution in land-slipped ground where 'spill through' soil displacement is planned to occur around the barrette. In a recent application of this type in Cwm, Mid Glamorgan, barrettes support a viaduct across a post-glacial

Figure 3.27 Heavy diaphragm wall grab in use for barrette construction. (Photograph courtesy of Carillion Piling.)

land-slip and are anchored back into the underlying rock within the hillside to increase their resistance to lateral forces.

3.5 Continuous flight auger piles

Whilst piles formed using continuous flight auger (CFA) rigs are essentially non-displacement bored piles, the method of hole formation and concreting is sufficiently different to justify separate discussion.

3.5.1 Basic method of pile formation

CFA piles are now widely available in sizes up to 1000 mm diameter, and exceptionally 1200 mm diameter. The commonly used diameters are 300, 450, 600, and 750 mm. These piles offer considerable environmental advantages during construction. Vibration is minimal, and noise outputs are low. In permeable soils with a high water table, their use removes the need for concreting by the tremie method, and temporary support for the borehole walls using casing or bentonite slurry is not necessary. The method of pile construction is suitable in sands, gravels and clays. The auger is full length

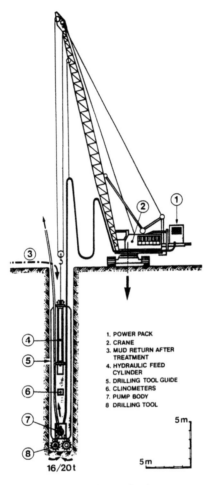

1. POWER PACK
2. CRANE
3. MUD RETURN AFTER
   TREATMENT
4. HYDRAULIC FEED
   CYLINDER
5. DRILLING TOOL GUIDE
6. CLINOMETERS
7. PUMP BODY
8 DRILLING TOOL

5 m

5 m

16/20 t

Figure 3.28 Milling rig for barrettes.

and has a hollow stem. With an hydraulic supply from the base machine which can be a crane or purpose-built crawler unit, the auger is top driven and is guided by long leaders suspended from the mast of the rig. A typical CFA rig is illustrated in Figure 3.29.

The piles are constructed by rotating the continuous-flight auger into the ground to the required depth, which is restricted to approximately 30 m. During this stage, the hollow stem of the auger is blanked off with a disposable plug. At the full depth, a highly workable concrete mix is pumped down the hollow auger stem to the tip of the auger thereby displacing the blanking plug. The auger is then rotated out of the ground as the concrete continues to be pumped down the central stem of the auger. Grout rather than concrete intrusion can be used for small diameter piles carrying light loads (e.g. the Cemcore pile). Short lengths (say up to 10 m)

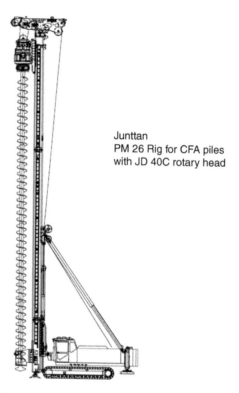

Junttan
PM 26 Rig for CFA piles
with JD 40C rotary head

Figure 3.29 Continuous-flight auger rig (CFA). (Illustration courtesy of Junttan Oy.)

of reinforcing cage with suitable centralizing spacers attached can be installed by the manual means of pushing the cage into the wet concrete, but longer cages will require the use of a vibrator in which case it is essential that the reinforcement cages are welded. As high tensile steel is employed for the reinforcement, special measures are required to prevent embrittlement being caused by the welding procedure.

The construction process for the Cementation Foundations' grouted 'Cemcore' pile, is shown in Figure 3.30. The more widespread use of a highly workable concrete mix rather than a cement grout does not alter the procedure.

In early attempts at using this method of construction, there was difficulty in controlling the relative rates of auger extraction and concrete (grout) intrusion. There could also be difficulties in unfavourable sequences of hard and soft (or loose) strata. Therefore, an important feature in the formation of CFA piles is the use of comprehensive instrumentation to monitor the performance of the rig at the time of boring including auger rotation per unit depth, boring rate and the pressure in the rig hydraulic system. Once the auger is at a suitable depth, the concreting of the pile is similarly monitored as the auger is withdrawn. The monitored parameters include the pressure in the concrete pumping main, the volume of concrete pumped per unit pile length and the extraction rate of the auger. Data from the sensors monitoring these parameters

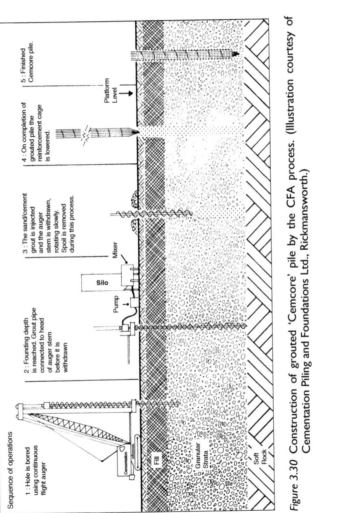

Figure 3.30 Construction of grouted 'Cemcore' pile by the CFA process. (Illustration courtesy of Cementation Piling and Foundations Ltd., Rickmansworth.)

Figure 3.31 Heavy-duty CFA rig with casing. (Photograph courtesy of Cementation Foundations.)

are fed to an on-line computer with a display in the cab such that the rig operator is in full knowledge of the augering and shaft-forming process. In concreting such a pile, a small excess over the theoretical column volume is applied to ensure full shaft formation. All the main UK piling contractors offering a CFA pile offer such a service, which is an extension of the early basic instrumentation provided by Kvearner Cementation Foundations when the system was first introduced in 1981. Such rig-based instrumentation can be relayed directly to a Company office as well as the site office and can include a link to the client as in the Stent Piling system (SIRIS). For a full discussion of CFA rig instrumentation, reference may be made to Unsworth and Fleming (1990).

For secant piling with CFA rigs, heavy duty augers and high-torque motors are necessary. Where the infill (male) piles partly intercept 'hard' concrete adjacent piles, a casing may be needed – see Figure 3.31.

3.5.2 Typical CFA pile loads

Concrete with a slump of about 150 mm is generally used with a crushing strength of at least 30 N/mm^2 at 28 days. If grout is employed, a typical grout mix is sand:cement in a ratio of 1 to $1\frac{1}{2}$ or 1 to 2 (depending on the quality of the sand) by weight with possibly an additive to improve workability. Crushing strength may be between 21 and 30 N/mm^2 at 28 days.

Typical pile diameters and pile loads are given below.

Pile diameter	Design load (approx.)
mm	kN
300	350
400	500
450	500 to 750
500	750 to 1000
750	1000 to 2500

3.6 Micro-piles

The term 'micro-pile' is commonly used to denote piles of 250 mm in diameter or less although the alternative term 'mini-pile' is sometimes used. In BS EN14199 (BSI, 2005), micro-piles are defined as having a maximum shaft diameter of 300 mm for drilled piles and a maximum diameter of 150 mm for driven piles. Uses of micro-piles include the foundations of domestic properties or other lightly loaded structures where settlement would be excessive using traditional shallow foundations but they also have a number of specific applications as described in the following sections.

3.6.1 Uses of micro-piles

Three frequent uses of micro-piles can be identified, i.e. as bearing piles, for soil reinforcement and to provide temporary lateral ground support in excavations of limited depth. Their use in these applications is outlined below. Whilst micro-piles can also be used in tension (with full length reinforcement), it should be noted that in this application their performance/economies may be unfavourable compared with anchors that are specifically intended for tensile operation.

3.6.1.1 Bearing piles

Typical uses as bearing piles include:

(i)   Light industrial or domestic dwellings in (a) weak soils, (b) swelling soils, (c) shrinkable soils (see Figure 3.32) or (d) to reduce differential settlement between old and new construction.
(ii) For underpinning combined with strengthening of the structure as necessary (see Figure 3.33).
(iii) For support of isolated machine bases (see Figure 3.34).

(iv) For support of floor slabs in industrial buildings combined with conventional piling for the structure. This is often carried out as a remedial measure (see Figure 3.35).

(v) As 'settlement reducers'.

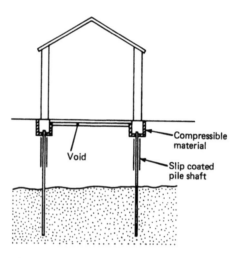

Figure 3.32 Construction in shrinkable clay using micro-piled ring beam and suspended floor slab. The beam is isolated from the soil by compressible material. (Illustration courtesy of *Ground Engineering*.)

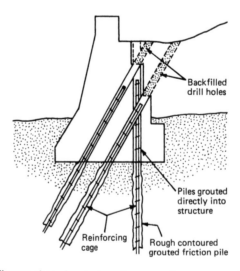

Figure 3.33 Retaining wall strengthened against settlement and overturning. Load is transferred to the micro-pile by grouting into the structure. The structure is reinforced and no pile cap is required. (Illustration courtesy of *Ground Engineering*.)

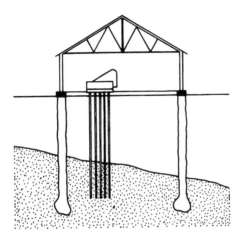

Figure 3.34 Support of isolated machine base using micro-piles. (Illustration courtesy of *Ground Engineering*.)

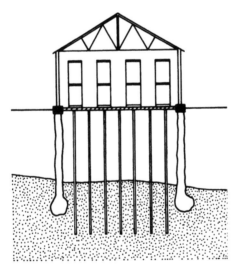

Figure 3.35 Support of floor slab with uniform load from storage racks, with micro-piles. (Illustration courtesy of *Ground Engineering*.)

3.6.1.2 Soil retaining and reinforcement

(i)   To form a reinforced block of soil (see Figure 3.36).
(ii) Stabilization of slopes (see Figure 3.37).
(iii) Bored micro-pile wall (see Figures 3.38 and 3.39).
(iv) Strengthening existing retaining structures (see Figure 3.33).

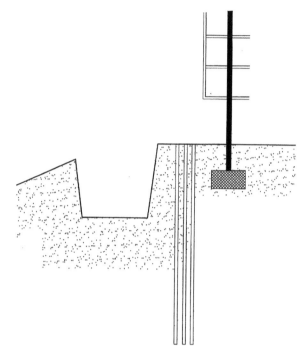

Figure 3.36 Micro-pile wall to provide resisitance against soil movement adjacent to an excavation.

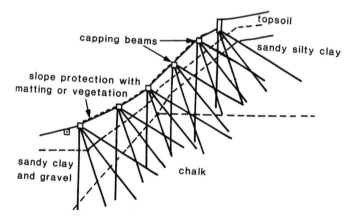

Figure 3.37 Fondedile 'Pali radice' reticulated structure to provide resistance in unstable ground.
(Illustration courtesy of *New Civil Engineer.*)

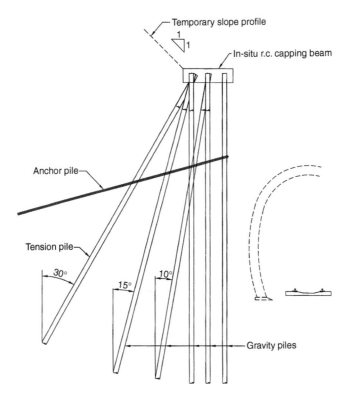

Figure 3.38 Cross section of Horsfall tunnel showing ground support with micro-piles. (Illustration courtesy of White Young Green Environmental.)

3.6.2 Installation methods

As for larger piles, there are two main methods of installation: non-displacement (i.e. drilling/boring cast-in-place methods) or displacement (e.g. driven) methods.

3.6.2.1 Non-displacement micro-pile installation

(i) *Drilled or bored cast-in place methods.* Such methods vary according to the type of soils encountered. It is usual to employ a small crawler mounted rig which has a low mast and hydraulic rotary head although the motive power for the drilling or boring (using an auger) may also be electric or pneumatic. Micro-pile rigs need to be versatile because of the variety of uses to which the piles may serve. Therefore rigs need to be capable of forming boreholes in diameters ranging from around 100 to 250 mm (or even 300 mm) though a variety of natural materials in areas of confined space or limited headroom. Also however, materials such as concrete, steel, timber or natural obstructions, may be encountered, and if the rig itself cannot form a hole through such obstructions, especially reinforced concrete forming an existing foundation, a diamond coring facility or a down-hole percussive rig will be needed as an adjunct. In stiff cohesive soils, continuous flight

Figure 3.39 Photograph of Horsfall tunnel showing ground support with micro-piles. (Photograph courtesy of Skanska UK Plc.)

augers can drill unsupported holes whilst in cohesionless soils a steel temporary casing fitted with a cutting shoe at its leading edge is drilled-in using a drill fluid. The fluid is usually water, but can consist of a drilling mud. In a typical system, such as that operated by Fondedile Foundations Limited, the fluid is transmitted down the inside of the drill casing to return cuttings to the surface via an annulus that is formed between the casing and the soil. Casing is added as the drilling advances, often using short lengths as headroom is frequently limited in the applications where micro-piles are employed.

At the required depth, the borehole is flushed clean and grouted with a sand cement mix and the tremie method is used to introduce the grout at the base of the borehole. Reinforcement is added and the casing withdrawn with further grout being tremied into the bore to make up for the casing volume. In self-supporting cohesive soils, it is not necessary to employ a temporary casing and drilling can be carried out using an auger or drag bit. Micro-piles can also be installed to bear into weak rock by using rock roller bits or down-hole hammers to advance the bore. The steps involved in this method of micro-pile formation are illustrated in Figure 3.40. When installing micro-piles through existing foundations, it may be necessary to step-down the temporary casing size as drilling

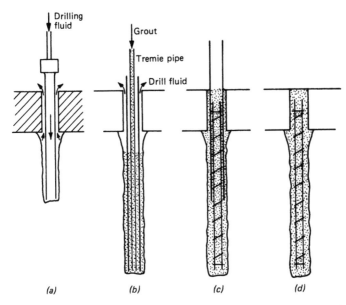

Figure 3.40 Method of installing typical drilled cast-in-place micro-piles. (a) Drilling with external flushing; (b) casing drilled to full depth, grout being introduced via a tremie pipe and displacing drilling fluid; (c) reinforcement placed, and casing being withdrawn; and (d) completed pile. (Illustration courtesy of *Ground Engineering*. Thomas Telford Ltd., London.)

proceeds whilst proving protection from the flushing fluid to the soil immediately beneath.

(ii) *Jet grouted methods*. Using the jet grouting process, micro-piles can be formed by combining the *in-situ* soil with an injected cement grout. In practice a jet grout pile, similar to a hollow drill rod but with exit ports over its lower section, is jetted down to the required depth of the pile using water under high pressure. At this depth the water is exchanged for a fluid cement grout also under high pressure and the jet pipe is retracted to the surface at a controlled rate combined with some rotation. This forces the grout into the soil if granular or causes some displacement in more cohesive deposits, the net effect being to produce a grouted column of a diameter considerably greater than that of the jet pipe. There is scope for considerable variation in hydraulic pressure, rates of installation and withdrawal, grout mixes nozzle configuration etc. and it is prudent to carry out trials in any soil to achieve a suitable and reliable process. Column strengths in excess of 5 MN/m^2 can generally be achieved without difficulty. Because the jet-grouted micro-piles can overlap, a frequent use of these piles is to produce a cut-off wall where there is confined access and driven sheet piles cannot be employed.

3.6.2.2 Installation of displacement micro-piles

(i) *Jacking methods*. Micro-piles lend themselves to installation using hydraulic jacking. Using this method, a small diameter tube or pre-formed pile section can be pressed into the ground in an almost vibration-free and quiet manner sometimes using the weight of the structure to be supported as reaction. If appropriate instrumentation is provided, the micro-piles are effectively load-tested as they are installed.

This type of system is available from companies such as Roger Bullivant in the United Kingdom. In a displacement system evolved by Ménard Techniques Ltd., a tapered steel mandrel is driven into the ground by vibratory methods. The pre-formed hole is then filled with a dry concrete mix and the mandrel re-driven. Some expansion of the pile occurs against the soil during this operation and a pile 'skin' develops. Filling with concrete, and finally inserting the reinforcement cage completes pile installation.

(ii) *Down-hole hammer methods.* It is common practice to form micro-piles by driving thin-walled steel tubes into the ground using a down-hole percussive hammer. Typical pile diameters for this form of pile installation range between 100 and 250 mm, with wall thicknesses of 3 mm. A 2 or 3 m long tubular lead section with a crimped cruciform point is first set vertically at the pile location and plumbed into position with the down-hole hammer in place. No special guidance equipment is necessary. The hammer is usually air operated from a small compressor and bears onto a plug of dry-mix concrete in the base of the pile. The hammer is set into action to advance the tube until a further open-ended length is added. Each tube is slightly expanded over a length about a diameter at the upper open end so that the next tube can be lightly tapped inside and welded in place. The tubes may be left open or cement-grouted, in which case a structural steel section can be incorporated in order to improve the bending resistance of the piles (see Figure 3.41).

Maximum lengths of driving are about 10 m and the pile can equally be employed at a raking angle. Unless the pile is simply driven to a prescribed depth, empirical means of determining a suitable 'set' are employed, which may be in the form of a maximum penetration per timed interval, e.g. no more than 6 mm in 10 seconds. The down-hole system with permanent steel casing is referred to as the 'Grundomat' pile, and in the United Kingdom is operated by companies such as Systems Geotechnique Limited and has proved an adaptable system that can be used in temporary works to provide support for access platforms for heavier rigs or as lateral support to excavations in clays. In one application by the author, because of the simple means by which Grundomat piles can be installed, thin-walled 100 mm diameter grouted micro-piles were used as a means of installing inclinometers into a landslip where there was very difficult rig access. The pile casing was not found to impede the mass movement of the soil and whilst it was accepted that a minor vertical displacement of the true shear zone could be recorded, a reliable record of the slip movement was achieved.

It should be noted that the installation of displacement micro-piles is not practicable in ground with significant obstructions. Where they are used in underpinning and intercept existing foundations, a hole in the footing has to be drilled as a separate process.

3.6.3 Sizes of micro-piles and load capacities

Many types of proprietary micro- and 'mini-' piles are available, in a range of sizes, from approximately 75 mm to (say) 300 mm in diameter. Design load capacities are

Figure 3.41 Small diameter (150 mm) Grundomat pile with hollow structural steel section centrally grouted in place.

clearly difficult to state with any certainty, but generally fall in the range of 20 to 500 kN. The systems are frequently specialized with respect to a particular contractor, and each system requires careful consideration in respect of capacity, installation methods and long-term performance before use. Some micro-pile installations are tailored to particular applications, such as underpinning, and may not be suitable in other situations. In general, individual micro-piles are not able to provide much lateral resistance, and large groups of piles or raking piles are usually necessary to deal with horizontal forces.

3.6.4 Special considerations relating to micro-piles

Because micro-piles are frequently employed in combination with existing structures or foundations, local effects of pile installation associated with the process are relatively more important than with conventional piling. Drilled micro-piles are generally most suited to installation where vibration or soil displacement is unacceptable although unsuitable drilling methods could lead to loss of support to adjacent structures if over-extraction of soil were to occur. Spoil is, however, produced and if contaminated, can lead to expensive disposal. Displacement methods involving jacked-in piles or pile sections create little vibration but produce lateral soil displacement and potentially, some heave. Therefore they should not be employed close to the foundations of an adjacent structure sensitive to any lateral loads. The noise and vibration associated with driven micro-piles may be unacceptable in some locations and such piles have the common characteristic of causing lateral displacement.

Micro-piles do not have large end-bearing areas although they may be installed to depths comparable with piles of larger diameters. As a consequence, they work primarily by shaft friction. However, should the shaft friction be fully developed, there will be relatively little base resistance to mobilize and unless a composite action between a reinforced slab foundation and the piles is deliberately planned, settlements can be excessive. In non-cohesive soils or weak rock, higher than normal unit shaft resistance has been noted in the case of drilled grouted-in-place micro-piles, whilst displacement types may not develop resistances comparable with conventional piles as the volume displacement is less. It is, however, usual practice to adopt empirical methods of design for granular soils, modified in the light of experience by the specialist contractor. Settlements are generally found to be similar to, or slightly less than, those associated with traditional friction piles where the piles are used in cohesive deposits. Cooke and Price (1978) report the magnitudes of strains and displacements associated with small-diameter piles in London Clay. For end-bearing micro-piles, larger than usual settlements can occur, as working stresses within the shafts of such piles tend to be high, leading to relatively large elastic compression.

Unless reinforced with a steel section, micro-piles are weak in shear. However, the ability with which most micro-piles can be raked means that inclined loads can be taken axially, to obviate the need for shear resistance in many applications.

The costs of load testing to check design assumptions can be prohibitive for a small number of micro-piles and some systems of integrity testing may be impractical because of their small diameter. It is therefore particularly important that strict quality control is maintained and, according to the sensitivity of the project, generous factors of safety applied. Those systems which are to some extent self-checking (i.e. instrumented jacked piles) and those employing a permanent former such as a tube, which can be inspected after installation may have an advantage over a drilled pile where concreting or grouting may take place under unfavourable conditions. However there is no record of micro-piles being particularly prone to poor performance and in general the workmanship of the established operators in this specialized field is good.

A further important factor to consider is that of corrosion, which can lead to serious problems more rapidly than in conventional piles in view of the small sizes of micro-piles. Because of this, small corrosion pits in a permanent steel casing could be of significance relative to the wall thickness of the pile. In pre-cast or pre-formed piles, the shaft can be inspected prior to use and coatings can be applied to combat corrosion (or to reduce down-drag or heave effects). It is however common practice to employ a central reinforcing tube or bar grouted in place so that the steel outer casing becomes sacrificial.

3.7 Corrosion

3.7.1 Corrosion of steel piles

The corrosion of steel piles is most likely to be a serious problem in two situations:

(a) piles driven into disturbed ground (i.e. fill)
(b) piles in a marine environment.

Steel piles and the reinforcement in reinforced concrete piles installed in undisturbed ground are not generally subject to significant corrosion below the water table. Above the water table, or in a zone of alternate wetting and drying from tidal action, corrosion rates can be increased, depending on temperature, pH and chemistry of the aqueous environment.

Corrosion rates for steel piling are given in BS 8002:1994 (BSI, 1994) as shown in Table 3.11.

A recently reported occurrence is that of Accelerated Low Water Corrosion (ALWC), in which rapid corrosion of steel piles has taken place at the low-water line in marine environments. It is identified by a bright orange/brown discoloration on the pile that conceals a thin layer of black 'sludge'. Beneath the sludge is clean bright steel but with corrosion pits such that complete penetration of a pile section can be present locally. Typically, different types of pile have a characteristic form of attack. It has been observed that 'Z' profile Frodingham piles are corroded through in a circular patch in the centre of the seaward pans whilst 'U' section Larssen piles are typically holed in elongated slits at the corners of the seaward pans. The cause of ALWC is not fully understood but it is clear that its presence needs to be identified if failure by holing at the low water line is to be avoided in marine structures. The occurrence of ALWC is not of course limited to sheet piling but can occur in any tubular steel or 'H' section steel piles, etc.

Table 3.11 Corrosion rates for steel piling in natural environments (after BS 8004: 1994)

Pile environment	Corrosion rate mm/side/year
Embedded in undisturbed soil	0.015 (maximum)
Exposed to the atmosphere	0.035 (average)
Immersed in fresh water	< 0.035
Exposed to a marine environment	
– below bed level	0.015 (maximum)
– seawater immersion zone	0.035 (average)
– tidal zone	0.035 (average)
– low water zone	0.075 (average)
– splash and atmospheric zones	0.075 (average)

Note
Corrosion rates for steel piles are prescribed in Eurocode EC3.

There are a number of measures which may be taken to combat corrosion in steel piles as follows (Table 3.11).

1   The use of copper-bearing steel is effective against atmospheric corrosion only.
2   The use of high-yield steel (Grade Fe 510A steel to BS EN 10025:1990 in a structure designed for mild steel at BS EN 10025:1990 Grade Fe 430A stresses) (BSI, 2004) permits a greater loss of section before stresses become critical.
3   Increasing the steel thickness by using the next heavier section in the range ('H' section and tubular piles).
4   The use of a protective coating, which may be restricted to the exposed section of the pile. The coating may be of metallic type (e.g. metal spraying or galvanizing) or a paint system. A better coating can be achieved under workshop conditions but such coatings have to survive the pile installation process.
5   Cathodic protection of piles in soils below the water table or for marine piles. This type of protection is complex and tends to be costly.
6   Concrete encasement of steel piles above the waterline. The inherent impact absorbing properties of the pile are lost by this method.

Experience in the use of low-alloy steels, containing 0.2 to 0.35% copper, is not well established, but the method could be effective in reducing splash zone rates of thickness loss. The danger in this method is that corrosion does not take place evenly, particularly after long periods of time, and deep corrosion pits can occur, locally reducing the strength of the pile.

If it is considered necessary to use coatings to provide protection below ground level these should be able to resist abrasion during driving. Coal tar epoxies are considered to be durable in marine conditions, and there is a view that isocyanate-cured pitch epoxy paints are especially effective. When such systems eventually fail, however, they cannot be replaced. The shaft friction of piles coated in this way may well be different to that of untreated steel, and comparative pile tests should be carried out to investigate this possibility. Strict control of the coating procedure will be necessary, and it may be better that this is not performed on site.

If fill is to be placed around or against steel piles, mixing-in lime or using limestone if available can beneficially increase the pH value of the ground. In permeable, low-pH fill, corrosion can be rapid, and some form of protection would be necessary.

3.7.2 Attack on concrete piles

Attack on both pre-cast and cast-in-place concrete piles can occur, in soils and groundwater having high sulphate and chloride concentrations and low pH. The degree of concrete attack is however a function of the availability of harmful chemicals in the groundwater. The objective should be to provide a dense concrete for all piles in this environment.

3.7.2.1 Sulphate attack

Attack on concrete by sulphates is well documented. Reaction between sulphate ions in groundwater and Portland cement results in a chemical product of greater volume, and hence cracking occurs, leading to spalling and an accelerated rate of attack. In order to cause corrosion, sulphates must be in solution, and therefore the solubility of the particular sulphate in question has a considerable bearing on the severity of attack. Groundwater mobility (and therefore soil permeability) is a relevant factor. The sulphates commonly encountered are those of calcium, magnesium and sodium.

Various standards are available for assessing the aggressiveness of the ground-water, based on its sulphate concentration. Commonly used guidance in the United Kingdom is that contained in BS 8500:2002. This document gives general guidance regarding the sulphate resistance of buried concrete. The document has been augmented by BRE Special Digest 1 *Concrete in aggressive ground* (SD1: 2005). The BRE Digest takes into account the type of site (i.e. 'Greenfield' or 'Brownfield'; Nixon *et al.*, 2003) and whether the groundwater is mobile or static. Analysis for magnesium, chloride, nitrate and total sulphur together with sulphate and pH will form part of the assessment. Deterioration of concrete by thaumasite is a form of particularly damaging sulphate attack. It is discussed by Crammond and Nixon (1993). Special considerations that may apply to buried concrete in the Middle East are discussed by Fookes and Collis (1975).

3.7.2.2 Chloride attack

Chlorides are encountered in particular in Middle Eastern soil conditions but could well also be present in industrial waste. Concrete itself is not usually subject to attack by chlorides in solution, but at even very low concentrations, chlorides can cause accelerated corrosion of steel reinforcement with consequential bursting of the concrete as the products of corrosion are formed. The corrosion can then progress at an enhanced rate. Protection of the reinforcement is therefore important, and increasing the cover of dense concrete is one way of improving matters. Stainless steel, galvanized or epoxy-coated reinforcement provide alternatives. A further measure is the use of the protective membranes around the pile shaft. Bartholomew (1979) considers permanent sleeving of pile shafts is necessary under tropical conditions for chloride concentrations exceeding that of sea water (3% salt by mass). Further detailed discussion of the corrosion of reinforcement steel in concrete (not specifically in piling applications) is available in the a publication by the ACI Committee 222, 1996, 'Corrosion of Metals in Concrete'.

3.7.2.3 Acidic groundwater

This condition, represented by a low-pH value, is most frequently met in peaty soils, but can also be present because of industrial contamination. In the former situation, the condition may not be severe. Reinforced concrete piles extracted from peaty Thames Alluvium at Littlebrook Power Station after approximately 40 years in the ground did not exhibit any obvious deterioration (author's observations). In the latter situation of industrial contamination, aggressive conditions can be extreme and localized, perhaps affecting only a small number of piles. In this case the maximum likely concentrations should be determined, and an assessment made of the distribution and extent of the contamination which can then be assessed using the criteria in the BRE Special Digest 1 (*loc. cit.*). External protective measures to a pile shaft may be necessary and the selection of the pile type may need to be carefully considered.

Table 3.12 Miscellaneous aggressive agencies that may cause attack of concrete

Agent	Attack
Inorganic acids (e.g. sulphuric, nitric, hydrochloric acid)	Can dissolve components of concrete; severity depends on concentration.
Organic acids (e.g. acetic, lactic, formic, humic acid)	Can slowly dissolve concrete. The attack is generally slow.
Alkalis (e.g. sodium and potassium hydroxide)	Alkalis such as these will dissolve concrete when highly concentrated.
Plant and animal fats (e.g. olive oil, fish oil, linseed oil)	Bond strength may be reduced.
Mineral oils and coal tar distillates (e.g. light and heavy oils, paraffin)	If of low viscosity, these agents can degrade concrete. Phenols and creosols corrode.
Organic matter (e.g. landfill refuse)	Organic matter can degrade concrete if hydrolysis results in lime removal.

3.7.2.4 Miscellaneous agencies aggressive to concrete

Nixon *et al.*, (2003) lists various miscellaneous aggressive agencies that may cause attack of concrete (Table 3.12).

3.7.3 Protection of concrete piles

Although the provision of dense and impermeable concrete reduces the attack of concrete by aggressive agencies, other means that can be considered include the use of coatings on pre-cast piles but ultimately, for cast-in-place piles in highly aggressive ground, a permanent casing may provide the best alternative. The permanent casing or sleeve may be flexible or rigid (usually uPVC). The sleeve may be placed inside the temporary casing, which has been advanced into the ground and may be fixed to the reinforcing cage. Alternatively, a flexible sleeve may be attached to the outside of the temporary casing, and remains behind when the casing is withdrawn. In granular soils, it is possible for the sleeve to tear, and a test drive of the system should be carried out. A more certain measure, despite the cost, is to employ permanent stainless steel casing for the upper part of a pile.

Chapter 4

Design of single piles

Although entitled 'Design of single piles', this chapter deals mainly with the analytical methods used in design, rather than with the actual design process. Practical and economic factors play an essential role in choosing the type and relative size of pile for a given application. For each pile considered, analysis is then required in order to assess whether that pile will satisfactorily fulfill its function as a foundation element. This analysis will generally involve calculation of the ultimate capacity of the pile under working load conditions, and an assessment of any other potential problems, for example, whether the pile can be installed with a given size of pile driving rig.

Analytical methods in pile design have advanced very rapidly over the last 25 years. This advance has enabled many of the traditional empirical approaches to be substantiated, or in some cases replaced by more soundly based theory, particularly as regards estimating the likely deformations – vertical and horizontal displacements, and rotation – that a pile may undergo. Economic pressures have led to more efficient design of pile groups, with a greater emphasis on limiting deformations (particularly settlement) to an acceptable level. This has led to an increasing trend towards piled raft foundations, where the design considers load transferred directly from the pile cap (or raft) in parallel with load transferred to the piles. The aim of the authors has been to present some of these newer methods of analysis, in a practical manner, with less emphasis on more traditional methods that are adequately covered elsewhere.

The main sections of the chapter deal sequentially with methods of estimating the axial capacity and deformation of single piles, followed by lateral capacity and deformation. In the later part of the chapter, specific problems that involve both axial and lateral response of the pile are considered. These include the use of raking piles, problems associated with piles in soil which is consolidating, and consideration of the buckling potential of piles, both during driving and under static loading.

4.1 Axial capacity of piles

A pile subjected to load parallel to its axis will carry the load partly by shear generated along the shaft, and partly by normal stresses generated at the base of the pile (see Figure 4.1). The ultimate capacity, Q, of the pile under axial load is equal to the sum of the base capacity Q_b, and the shaft capacity Q_s. Thus

$$Q = Q_b + Q_s = A_b q_b + A_s \bar{\tau}_s \qquad (4.1)$$

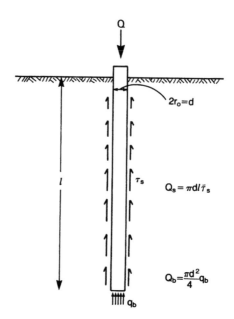

Figure 4.1 Axially loaded pile.

where A_b is area of the pile base.
        q_b is end-bearing pressure.
        A_s is area of the pile shaft.
        $\bar{\tau}_s$ is average limiting shear stress down the pile shaft.

The relative magnitude of the shaft and base capacities will depend on the geometry of the pile and the soil profile. Piles which penetrate a relatively soft layer of soil to found on a firmer stratum are referred to as 'end-bearing' piles and will derive most of their capacity from the base capacity, Q_b. Where no particularly firm stratum is available to found the piles on, the piles are known as 'friction' or 'floating' piles. In cohesive soil, the shaft capacity is generally paramount, while in non-cohesive soil or where the pile base is underreamed, the overall capacity will be more evenly divided between shaft and base. In cohesive soil, it is generally assumed that the shaft capacity in uplift is similar to that under downward loading, but it is now acknowledged that the shaft capacity in uplift for piles in non-cohesive soil is significantly less than that in compression (see section 4.1.1).

The shaft capacity of a pile is mobilized at much smaller displacements of the pile (typically 0.5 to 2% of the pile diameter) than is the base capacity. The latter may require displacements as large as 5 to 10% of the pile base diameter (even larger for low-displacement piles in granular soil) in order to be fully mobilized. This difference between the load deformation characteristics of the pile shaft and that of the pile base is important in determining the settlement response of a pile, and the sharing of load between shaft and base, under working conditions. To illustrate this, Figure 4.2 shows results from a load test on a bored pile, 0.8 m diameter by 20 m long. The measured response has been simulated numerically using a load transfer approach

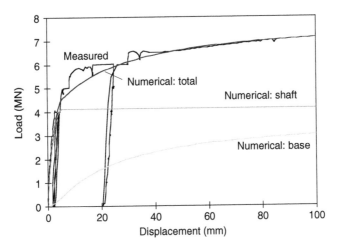

Figure 4.2 Idealized load settlement response.

(Randolph, 2003), and the separate response of the pile shaft and pile base is indicated in addition to the total response. The shaft capacity of 4.1 MN is mobilized at a displacement of less than 6 mm (0.75% of the pile diameter), while the base response is still rising for displacements in excess of 80 mm (10% of the pile diameter).

The nominal ultimate capacity of the pile may be taken as 7 MN, of which 60% is provided by the shaft. However, at a working load of say 3.5 MN, Figure 4.2 shows that 95% of the load is carried by the pile shaft, with only 5% reaching the base. Choice of factor of safety for such a pile must be made in the light of the different response of pile shaft and base. Once the pile shaft capacity has been fully mobilized (at a load of 4.1 MN), the 'stiffness' of the pile-soil system reduces to that of the pile base, and the displacements start to increase rapidly. This type of consideration is of particular importance in the design of underreamed or partially end-bearing piles, where the base capacity is large. Settlements under working load conditions will be larger than for comparable straight-shafted friction piles.

The above example illustrates an important approach to the design of piles. In the first instance, unit values of 'shaft friction' (the limiting value of shear stress between pile and soil) and of end-bearing pressure must be estimated in order to calculate the shaft and base capacities. The same overall capacity may be achieved with a variety of combinations of pile diameter and pile length. However, before the final geometry of the pile is determined, some consideration must be given to the load settlement characteristics of the pile – and also, of course, to other factors such as the economics of installing piles of one size or another.

In general, long slender piles may be shown to be more 'efficient' than short stubby piles, both as regards their capacity per unit volume installed, and also in terms of the stiffness of the pile (Randolph, 1983). Limitations on the slenderness of a pile come from economic reasons (since, once a large boring or pile driving rig is mobilized, the rate of pile installation becomes largely independent of the pile diameter), and also from considerations of its stability during driving (Burgess, 1976) and of the axial compressibility of the pile. In the extreme case, piles may be so long that no load at all reaches the base under working conditions. Working settlements will probably be sufficient to mobilize full shaft friction in the upper part of the pile shaft (Randolph, 1983), affecting the load settlement characteristics and bringing the possibility of degradation of the shear transfer between pile and soil.

The following sections discuss methods of estimating the ultimate capacity of piles in different soil types. In all cases, the capacity may be assumed to be the sum of shaft and base capacities (see equation (4.1)). Methods of estimating values of the end-bearing pressure, q_b, and the shaft friction, τ_s, will be broadly divided into two categories, those based on fundamental soil properties such as shear strength or angle of friction, and those based directly on in-situ measurements such as standard penetration tests (SPTs) or cone penetration tests (CPTs).

It should be emphasized that estimation of pile capacity is still largely based on empirical methods, derived from correlations of measured pile capacity with soil data of variable quality. There is generally a wide scatter in the correlations, and different approaches suit different soil types better than others, albeit still with a significant margin of error. Many soils do not fit neatly into the categories discussed in the following sections of non-cohesive (implying free-draining, granular), or cohesive (implying less permeable, fine-grained) soil, and of rock. Pile design must thus be seen as a compromise, adopting the most appropriate method available for the particular soil types and site conditions prevailing.

Where a pile penetrates layers of differing soil types, the shaft capacity in each layer is cumulative, with minimal 'interaction' effect between adjacent layers. However, experimental evidence from instrumented piles (Lehane et al., 1993; Jardine and Chow, 1996) has demonstrated significant degradation in local values of shaft friction during the installation of jacked or driven piles. Hence the relative position along the pile shaft can have an important bearing on the design value of shaft friction.

In a given soil, design values of shaft friction and end-bearing pressure will vary for different types of pile. While it is possible to identify trends of these parameters with broad categories of construction technique (driven precast piles, bored piles, continuous flight auger (CFA) piles), the actual values will also be affected by the construction parameters for any given pile. Thus, excessive hard driving will lead to greater degradation of shaft friction, as will excessive auger-turns for a CFA pile (so-called 'overflighting' resulting in overexcavation of soil). For bored piles constructed under bentonite, the shaft friction can be severely reduced if the quality of the bentonite is allowed to deteriorate, or if an excessive head of bentonite leads to caking of mud on the borehole wall. It is important, therefore, to ensure good quality control during construction of piles.

4.1.1 Capacity in non-cohesive soils

4.1.1.1 End-bearing pressure

It might be expected that the end-bearing pressure beneath a pile in a uniform deposit of non-cohesive soil would be directly proportional to the local vertical effective stress, and thus would increase approximately proportionally with depth. However, extensive research by Vesic (1977) showed that the end-bearing pressure appeared to reach a limit, beyond which no further increase was observed with further penetration of the pile. This led to the introduction of limiting values of end-bearing pressure, with the limit depending on the relative density and type of soil. Even for dense sand, limiting values as low as 11 to 12 MPa have been proposed (Tomlinson, 1977; American Petroleum Institute, 1997), although Coyle and Castello (1981) quote measured end-bearing pressures of 15 MPa for piles in clean dense sand.

The design end-bearing pressure is normally taken as that mobilized within a limited displacement, typically 10% of the pile diameter, rather than a true ultimate value (which might need displacements in excess of one diameter to mobilize). Even for a solid (or closed-ended) driven pile, the design value of end-bearing will be a fraction of the cone resistance obtained during continuous penetration.

Modern approaches to pile design have generally moved away from limiting values of end-bearing pressure, but accept that in a uniform sand deposit there will be a gradually decreasing gradient of design end-bearing pressure with depth. Two effects contribute to this: firstly the friction angle, ϕ', decreases with increasing confining pressure (Bolton, 1986); secondly, the rigidity index (ratio of shear stiffness to strength) decreases with increasing stress level. This latter aspect has been discussed by Kulhawy (1984), and Randolph et al. (1994).

Here, attention will be focused on the variation of friction angle, which has a dominant effect except in very loose soils. Bolton (1986) discusses the strength and dilatancy characteristics of sand and shows that, when dense sand is sheared at high confining stresses, the peak friction angle will be that associated with shearing at constant volume. As will be shown in this section, the implication of this for bearing capacity is that, as the embedded length of pile increases, values of end-bearing pressure increase at a gradually decreasing rate, and, for practical pile lengths, appear to asymptote towards values in the range 10 to 20 MPa, depending on the relative density of the soil, and also the constant volume (or critical state), friction angle.

The end-bearing pressure, q_b, may be expressed in terms of the effective vertical stress, σ_v', and bearing capacity factor, N_q, as

$$q_b = N_q \sigma_v' \tag{4.2}$$

Values of N_q quoted in the literature vary considerably, but those derived by Berezantzev et al. (1961) are used most widely for the design of deep circular foundations. Figure 4.3 shows the variation of N_q with friction angle ϕ'. It is then necessary to choose an appropriate value of ϕ' consistent with the type of non-cohesive material, its relative density, and the average stress level at failure. Following Bolton (1986), ϕ' may be related to the relative density of the sand, corrected for the mean stress level, p', and a critical state angle of friction, ϕ_{cv}, which relates to conditions where the soil shears with zero dilation (i.e. at constant volume). The corrected relative density, I_R, is given by

$$I_R = I_D \left[5.4 - \ln(p'/p_a) \right] - 1 \tag{4.3}$$

where I_D is the uncorrected relative density, and p_a is atmospheric pressure (100 kPa). A lower limit of zero may be taken for I_R at very high stress levels, while values of I_R greater than 4 should be treated with caution. In a subsequent discussion,

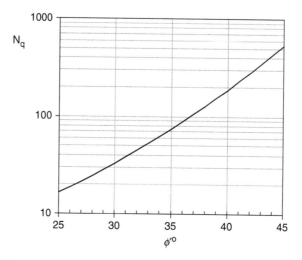

Figure 4.3 Variation of N_q with ϕ' (Berezantzev et al., 1961).

Bolton (1986) suggests restricting equation (4.3) to mean effective stress levels in excess of 150 kPa, below which the corrected relative density is taken as $I_R = 5I_D - 1$.

The appropriate value of ϕ' may then be calculated from

$$\phi' = \phi_{cv} + 3I_R \text{ degrees} \tag{4.4}$$

The average mean effective stress at failure may be taken approximately as the geometric mean of the end-bearing pressure and the ambient vertical effective stress, i.e.

$$p' \sim \sqrt{N_q \sigma_v'} \tag{4.5}$$

The end-bearing pressure, q_b, may now be calculated, for given values of ϕ_{cv}, I_D, and σ_v', by iterating between equations (4.3) to (4.5) and the chart for N_q shown in Figure 4.3. As an example, consider the case of $\phi_{cv} = 30°$, $I_D = 0.75$ and $\sigma_v' = 100$ kPa. Assuming an initial value of N_q of 50, equation (4.5) gives $p' = 707$ kPa from which ϕ' may be calculated as 35°. For this value of ϕ', Figure 4.3 yields a new value of N_q of 75. Further iteration yields a final value of $N_q = 66$, and an end-bearing pressure of $q_b = 6.6$ MPa.

Figure 4.4 presents charts of end-bearing pressure against ambient vertical effective stress, for different values of ϕ_{cv} and I_D The charts are plotted on logarithmic axes, which obscures the natural variation of q_b with depth. This variation is non-linear, showing a gradually reducing rate of increase with depth, similar to that found from pile load tests. Only at great depths and for dense deposits of sand do end-bearing pressures exceed 20 MPa. Bolton quotes values ϕ_{cv} for different types of sand ranging from 25° for mica up to 40° for feldspar. In practice, presence of silt particles will mean that ϕ_{cv} for most deposits will rarely lie much above 30°.

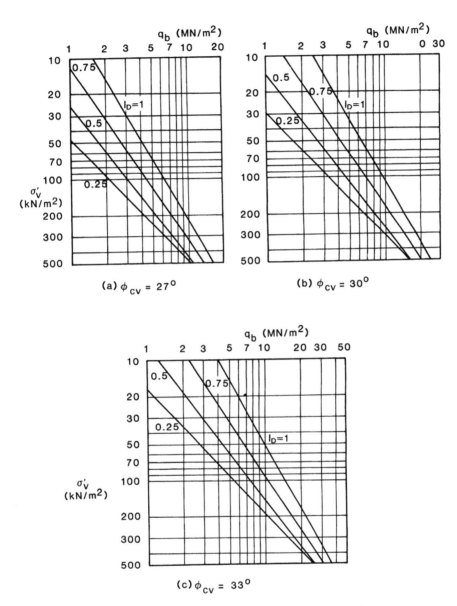

Figure 4.4 End-bearing pressures in non-cohesive soil.

Strong experimental support for the approach outlined above has been reported by Neely (1988), from measurements on expanded base piles. In a database of 47 pile test results, the mean ratio of predicted to measured capacity was found to be 0.92, with a coefficient of variation of 0.32.

The end-bearing pressures shown in Figure 4.4 are considered to be reasonable for closed-ended driven piles, in the absence of more specific information, such as field tests. Lower values are appropriate for conventional bored piles (including continuous flight auger piles), where design values of end-bearing will be 30 to 50% of those shown, and open-ended pipe piles where a reduction of 20% is recommended.

Lower values of end-bearing pressure should also be adopted in compressible soils, such as calcareous sands and very silty sands. Equally, certain dense sands may show end-bearing values much higher than those implied above, particularly if they are highly overconsolidated or cemented in any way. In view of the potential for such variations, it is strongly recommended that end-bearing pressures are assessed in relation to field penetration tests, as discussed below.

4.1.1.2 Correlations with penetration tests

Cone penetration testing allows design values of end-bearing pressures to be determined with greater confidence, because of the obvious similarities between the cone and an axially loaded pile. However, adjustments to the cone resistance must be made, due to (a) scale effects, and (b) the limited displacements appropriate for pile design. Scale effects are allowed for by appropriate averaging of the cone resistance, ignoring high 'spikes' in the cone resistance that may be due to small-scale effects (particles of gravel size or larger, or local variations of density of cementing).

Nordlund (1963) recommends taking an average value of q_c over a depth range of three pile diameters above the pile base down to two diameters below the pile base (see Figure 4.5). Fleming and Thorburn (1983) recommend more detailed schemes for averaging the cone readings, in order to put more weight on the minimum values.

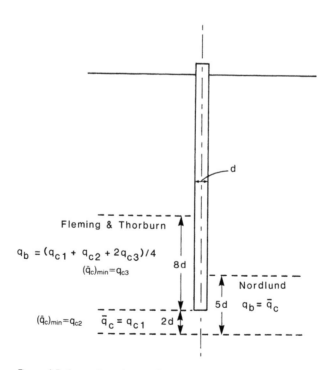

Figure 4.5 Averaging of cone data to give end-bearing pressure.

The range over which the average is taken is extended up to eight pile diameters above the level of the pile base. Thus, in homogeneous sand, the relevant cone resistance, $q_{cdesign}$ is estimated as (see Figure 4.5)

$$q_{cdesign} = (q_{c1} + q_{c2} + 2q_{c3})/4 \qquad (4.6)$$

where q_{c1} = average cone resistance over two diameters below pile base.
q_{c2} = minimum cone resistance over two diameters below pile base.
q_{c3} = average of minimum values lower than q_{c1} over eight diameters above pile base.

Design values of cone resistance must be reduced to allow for the limited displacements relevant for pile design. Adopting what is essentially a serviceability displacement limit of 10 % of the pile diameter, typical ratios of q_b/q_c recommended for closed-ended driven piles range from 0.2 (Jardine and Chow, 1996) to 0.5 (Kraft, 1990) or higher. De Nicola and Randolph (1997) noted a decreasing trend of the ratio q_b/q_c with depth, and suggested that q_b/q_c ratios may be taken as 0.5 at shallow depths, decreasing to 0.3 for effective overburden stresses greater than 200 kPa. Randolph (2003) has pointed out that the magnitude of end-bearing pressure mobilized within displacements of 10% of the pile diameter will depend on the magnitude of residual stress locked into the soil below the pile base during installation, but proposed a design value of 0.4 for q_b/q_c unless high residual stresses could be demonstrated (see Figure 4.6).

Further reduction values should be applied for other types of pile. For open-ended driven piles, Foray *et al.* (1993) and Bruno (1999) suggest a 25% reduction relative to a closed-ended pile. Lehane and Randolph (2002) have developed relationships for the *minimum* end-bearing resistance that can be relied upon as a function of

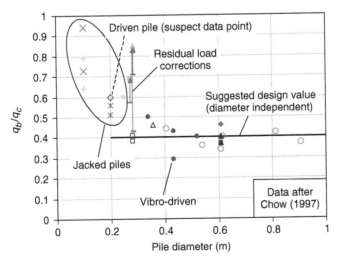

Figure 4.6 Normalized end-bearing resistance for driven closed-ended piles. (Annotations by Randolph, 2003.)

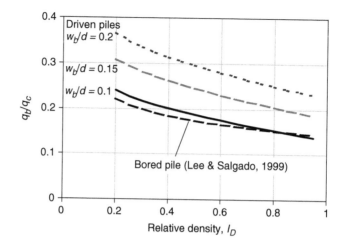

Figure 4.7 Normalized end-bearing pressures for driven open-ended piles and bored piles.

base displacement and relative density, as shown in Figure 4.7. Allowing for some densification of the soil below the pile base and within the soil plug, a reasonable design limit of 0.2 for q_b/q_c is probably realistic. Figure 4.7 also shows a design curve for bored piles (Lee and Salgado, 1999), which lies close to the curve for open-ended driven piles at a similar normalized displacement.

Lehane *et al.* (2005) reviewed the capacity of driven piles with particular reference to offshore design, and noted that more careful averaging of the cone resistance, allowing for limited penetration into strong founding layers, led to an improved correlation, with an end-bearing ratio q_b/q_c of 0.6 for closed-ended piles, rather than the value of 0.4 suggested by Figure 4.6. For open-ended piles, the end-bearing ratio reduced linearly with effective area ratio, A_r (ratio of net to gross area of pile, but allowing for partial plugging where this can be estimated) according to

$$q_b = (0.15 + 0.45A_r)q_c \qquad (4.7)$$

This gives an end-bearing ratio of 0.6 for closed-ended piles, reducing to 0.2 for typical open-ended steel piles with an area ratio of around 0.1.

Correlations have also been developed for estimating q_b directly from the results of standard penetration testing. The ratio q_b/N (or q_c/N) where N is the number of blows per 0.3 m penetration, varies with the composition of non-cohesive soil. As has been discussed in Chapter 2 (see Figure 2.7), there is an approximately linear relationship between q_c/N and average particle size of the soil, with the ratio varying from as high as 0.8 MPa for medium-size gravel, down to 0.25 MPa for silt.

Values of end-bearing pressure estimated from results of standard penetration tests should be assessed by means of equation (4.2), adopting estimated values of the various soil parameters where necessary, before being used in design. Correlations of results from standard penetration tests, with the relative density of the sand and effective overburden stress, may be used for this purpose, as indicated in Chapter 2 (Figure 2.9).

4.1.1.3 Shaft friction

The starting point for calculating values of shaft friction τ_s for piles in non-cohesive soil is the expression

$$\tau_s = \sigma'_n \tan \delta = K\sigma'_v \tan \delta \tag{4.8}$$

where σ'_n is the normal effective stress acting round the pile shaft after installation, and δ is the angle of friction between pile and soil. The latter quantity may be measured in interface shear tests for the particular pile material. Kishida and Uesugi (1987) have reported a detailed study of the effects of surface roughness, and shown how the interface friction angle may be related to the friction angle of the soil in terms of a normalized roughness coefficient, defined as the maximum roughness of the pile surface (over a gauge length of D_{50} for the soil) normalized by the value of D_{50}. For typical pile surfaces (oxidized mild steel or concrete), the normalized roughness coefficient will exceed 0.05, and the coefficient of friction at the interface will lie in the range 0.75 to 1 times that for the soil itself. An alternative assumption, where interface shear data are not available, is to assume that the interface friction angle, δ, may be approximated as ϕ_{cv}, the critical state angle of friction. This may be justified on the basis that no dilation is to be expected between the sand and the wall of the pile.

The normal effective stress may be taken as some ratio K of the vertical effective stress σ'_v, resulting in the second form of the expression in equation (4.8). The appropriate value of K will depend on (a) the in-situ earth pressure coefficient, K_0, (b) the method of installation of the pile and (c) the initial density of the sand.

For conventional bored piles it is common practice to adopt values of K of 0.7, while for piles bored with a continuous-flight auger, values of 0.9 (for sandy soils) down to 0.6 (for silts and silty sands) are taken. In general, the roughness of the hole will ensure that the full soil friction is mobilized at the pile edge. However, potential loosening of the soil during the installation process maybe allowed for by taking the friction angle δ between ϕ' and ϕ_{cv}.

For driven cast-in-situ piles, while the value of K may be 1.2 or higher outside the casing, some reduction in normal stress may occur during extraction of the casing. If wet concrete is placed, a value of K of 1.0 may be taken, while values up to 1.2 are appropriate where dry concrete is rammed into the pile shaft.

For continuous-flight auger piles, experience indicates that a value of $K = 0.9$ is appropriate in sands and gravels. Rather lower values, as low as 0.5 to 0.6, have been found with silts and silty sands, possibly due to entraining material from the sides of the pile into the spoil during augering, thus reducing the horizontal stresses in the ground. Close control is needed during construction in order to monitor the amount of material removed and the rates of concrete delivery and auger extraction as the pile is formed.

It should be noted that, particularly for relatively short piles, the shaft capacity of bored piles may be less than half the long-term capacity of driven piles of the same nominal size, once excess pore pressures generated during installation have dissipated. This factor, together with potential difficulties in construction, renders bored piles unattractive in soils with low shear strength. For driven piles in sand and other soils of high permeability, it has long been realized that the magnitude of shaft friction at a given depth can reduce as the pile is driven further, with the net effect that the average friction along the pile shaft can reach a limit and even reduce as the pile embedment increases (Vesic, 1969, 1977). However, this effect has only recently been quantified through carefully instrumented pile tests undertaken by the research group at Imperial College (Lehane *et al.*, 1993). The phenomenon of 'friction degradation' is illustrated in Figure 4.8 with profiles of shaft friction measured in the three instrument clusters at different distances, h, from the tip of a 6 m long, 0.1 m diameter, pile as it is jacked into the ground. For comparison, the cone profile is plotted on the same scale, but with q_c factored down by 100. The shaft friction measured at $h/d = 4$, in particular, follows the shape of the q_c profile closely, allowing for differences in cone and pile diameter. Comparison of the profiles from the instrument clusters at $h/d = 4$ and $h/d = 25$ shows that the friction measured at the latter position is generally less than 50% of that measured close to the pile tip.

The physical basis for friction degradation is the gradual densification of soil adjacent to the pile shaft under the cyclic shearing action of installation. This process is enhanced by the presence of crushed particles from the passage of the pile tip, which gradually migrate through the matrix of uncrushed material (White and Bolton, 2002). The far-field soil acts as a spring, with stiffness proportional to G/d (where G is the soil shear modulus), so that any densification close to the pile results in reduced radial effective stress. The operative value of G will be high, since the soil is heavily overconsolidated having moved through the zone of high stress close to the pile tip during installation and is being unloaded.

Different approaches to quantify the variation of the effective stress ratio, $K = \sigma'_n/\sigma'_v$, have been proposed by Randolph *et al.* (1994) and by Jardine and Chow (1996), although Randolph (2003) has shown that both approaches yield similar profiles of shaft friction. The former approach expresses K as an exponential function of h/d,

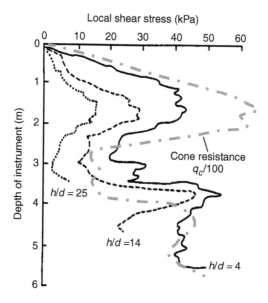

Figure 4.8 Measured profiles of shaft friction (Lehane *et al.*, 1993).

ranging between a maximum value, K_{max}, at the pile tip, and a minimum value, K_{min}, according to

$$K = K_{max} - (K_{max} - K_{min})e^{-\mu h/d} \qquad (4.9)$$

where the rate of degradation is controlled by the parameter, μ, which may be taken as ~ 0.05. K_{max} may be estimated as a proportion of the normalized cone resistance, typically 2% of q_c/σ_v' for closed-ended piles reducing to 1% for open-ended piles, and K_{min} lies in the range 0.2 to 0.4, giving a minimum friction ratio, τ_s/σ_v', of 0.1 to 0.25 (Toolan et al., 1990).

Where data from a cone friction sleeve are available, the friction ratio may be used to refine estimates of K_{max} for closed-ended piles, provided an appropriate value for the interface friction angle, δ between cone and soil is adopted. However, it should also be noted that data from friction sleeves are very sensitive to wear of the sleeve, and may therefore underestimate K_{max}.

Piles driven into loose to medium deposits of sand provide considerable compactive effect. As will be discussed in Chapter 5, the capacity of individual piles in sand tends to increase as neighbouring piles are driven in a group, due to this compactive effect. For dense deposits, the greater tendency of the sand to dilate will limit the amount of densification close to the pile shaft. Again, though, driving further piles close by will have a beneficial effect on the shaft capacity of a pile, due to vibratory compaction.

On the basis of instrumented pile tests, some authors have recommended that shaft friction for piles loaded in tension should be taken as half that for piles loaded in compression. However, many such pile tests took insufficient account of the residual stresses which exist after pile installation, and considerably underestimated the end-bearing capacity of the pile, thus overestimating the shaft friction in compression. Differences between shaft friction for piles loaded in tension and compression have been assessed by De Nicola and Randolph (1993), who noted two main effects. The first of these is due to contraction or expansion of the pile shaft due to Poisson's ratio effects, while the second (which dominates for short piles) is due to differences in effective stress changes in the soil as the pile is loaded in either direction. They proposed the ratio of shaft capacity in tension to that in compression may be estimated from

$$\frac{(Q_s)_{tens}}{(Q_s)_{comp}} \approx \left(1 - 0.2 \log_{10}\left(\frac{100}{L/d}\right)\right)\left(1 - 8\eta + 25\eta^2\right) \qquad (4.10)$$

where Q_s is the shaft capacity and $\eta = v_p(L/d)(G_{ave}/E_p)\tan\delta$, with G_{ave}, E_p and v_p being respectively the average soil shear modulus, Young's modulus of an equivalent solid pile and Poisson's ratio for the pile. Although other effects, such as local stress changes due to dilation, will influence the shaft capacity ratio, the expression in equation (4.10) provides a reasonable design basis for assessing the reduced shaft capacity for loading in tension, compared with that for loading in compression, with typical ratios in the range 0.7 to 0.85.

A simplified design approach linking the shaft friction of driven piles directly to the cone resistance, q_c, was suggested by Lehane et al. (2005) in their review of the capacity of driven offshore piles in siliceous sands. Their study led to an expression for the shaft friction of

$$\tau_s = 0.03 q_c A_r^{0.3} \left[\max \left(\frac{h}{d}, 2 \right)^{-0.5} \right] \tan \delta \tag{4.11}$$

for compression loading, with the shaft friction reduced by 25% for loading in tension.

The expressions given in this section for calculating the shaft capacity of driven piles should be used with caution in certain types of cemented sands where the in-situ voids ratio is high, or in sands containing crushable particles. Breaking down of the cementation and crushing of the soil particles during pile installation can lead to very low effective normal stresses acting on the pile shaft, and values of shaft friction that are less than 20% of those associated with silica sands possessing similar values of ϕ'. McClelland (1974) has recommended a maximum shaft friction of 20 kPa in such sands. Detailed discussions of pile capacity in calcareous soil have been presented by Murff (1987) and Randolph (1988), who emphasize not only the low values of shaft friction, but also the susceptibility of those values to cyclic loading. In general, bored (drilled and grouted) piles or post-grouted driven piles are to be preferred to driven piles in calcareous soils.

4.1.2 Capacity in cohesive soils

4.1.2.1 End-bearing pressure

The long-term, drained, end-bearing capacity of a pile in clay will be considerably larger than the undrained capacity. However, the settlements required to mobilize the drained capacity would be far too large to be tolerated by most structures. In addition, the pile must have sufficient immediate load-carrying capacity to prevent a short-term failure. For these reasons, it is customary to calculate the base capacity of piles in clay in terms of the undrained shear strength of the clay, c_u, and a bearing capacity factor, N_c. Thus the end-bearing pressure is

$$q_b = N_c c_u \tag{4.12}$$

For depths relevant for piles, the appropriate value of N_c is 9 (Skempton, 1951), although due allowance should be made where the pile tip penetrates a stiff layer (underlying soft soil) by only a small amount. A linear interpolation may be made between a value of $N_c = 6$ for the case of the pile tip just reaching the bearing stratum, up to $N_c = 9$ where the pile tip penetrates the bearing stratum by three diameters or more.

4.1.2.2 Shaft friction

Most piles in clay develop a high proportion of their overall capacity along the shaft. Whereas the ratio of end-bearing pressure to shaft friction may be typically 50 to 100 for piles in sand, the corresponding range is 10 to 20 for piles in clay.

As such, considerably more effort has been devoted to developing reliable methods for estimating values of shaft friction for piles in clay, than has been the case for sand. Historically, the shaft friction around a pile shaft has been estimated in terms of the undrained shear strength of the soil, by means of an empirical factor, α (Tomlinson, 1957) giving

$$\tau_s = \alpha c_u \tag{4.13}$$

The value of α deduced from pile load tests appears to reduce from unity or more for piles in clay of low strength, down to 0.5 or less for clay of strength greater than about 100 kPa. Figure 4.9 shows measured values of shaft friction from driven piles, plotted against the shear strength of the soil, illustrating this. This large amount of scatter apparent from the figure is a major cause for concern, and has prompted more sophisticated approaches. Two approaches are described here: the first is based on equation (4.11), but taking the parameter, α, as a function of the strength ratio, c_u/σ'_v; the second is a more fundamental approach based on estimating the effective stress state around the pile shaft after installation.

4.1.2.3 Driven piles: total stress approach

Attempts to follow, theoretically, the full stress history of soil elements in the vicinity of a driven pile have shown that the appropriate value of shaft friction will depend not only on the shear strength of the soil, but also on its past stress history and over-consolidation ratio (Randolph et al., 1979; Randolph and Wroth, 1981; Kraft, 1982). This is best reflected by the strength ratio, c_u/σ'_v, of the soil (Randolph and Wroth, 1982). Thus either of the parameters, $\alpha = \tau_s/c_u$ or $\beta = \tau_s/\sigma'_v$, should show a consistent

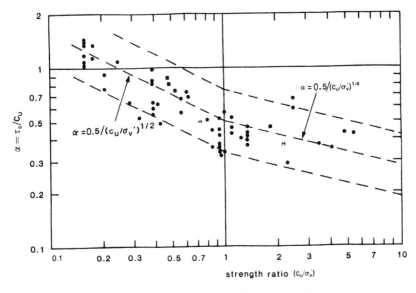

Figure 4.9 Variation of α with strength ratio.

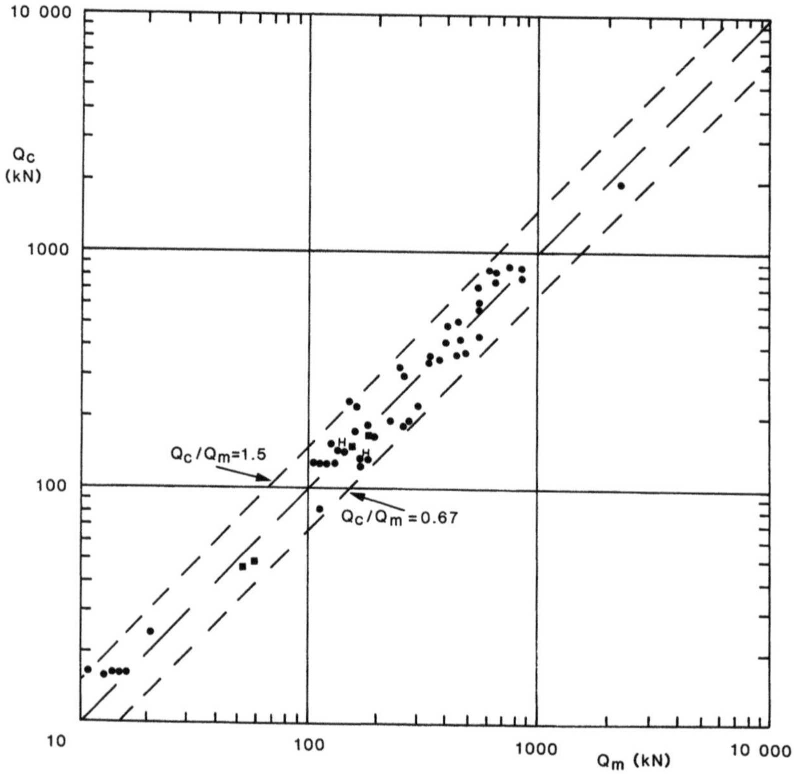

Figure 4.10 Comparison of calculated and measured pile capacities.

variation with the strength ratio for different soils. Figure 4.10 shows α, deduced from load tests on driven piles (Randolph and Murphy, 1985), plotted against the average in-situ strength ratio. The data fall in a band with a gradient of -0.5 at low values of c_u/σ_v', the gradient reducing to about -0.25 for c_u/σ_v' greater than unity.

If it is assumed that the value of α should be unity for normally consolidated clay, then the value of α may be given for the general case by the two expressions

$$\alpha = \left(c_u/\sigma_v'\right)_{nc}^{0.5} \left(c_u/\sigma_v'\right)^{-0.5} \tag{4.14}$$

for c_u/σ_v' less than or equal to unity, and

$$\alpha = \left(c_u/\sigma_v'\right)_{nc}^{0.5} \left(c_u/\sigma_v'\right)^{-0.25} \tag{4.15}$$

for c_u/σ_v' greater than unity. The subscript nc in these expressions refers to the normally consolidated state of the soil. Values of shaft friction may be deduced directly from similar expressions:

$$\tau_s = \left(c_u/\sigma_v'\right)_{nc}^{0.5} \left(c_u\sigma_v'\right)^{0.5} \tag{4.16}$$

for c_u/σ_v' less than or equal to unity, and

$$\tau_s = \left(c_u/\sigma_v'\right)_{nc}^{0.5} c_u^{0.75} \sigma_v'^{0.25} \tag{4.17}$$

for c_u/σ_v' greater than unity.

The above expressions contain no artificial parameters other than the assumption that τ_s equals c_u for normally consolidated clay. Since pile installation will remould the soil, the strength ratio for normally consolidated clay should be that for remoulded material. This may be estimated with reasonable accuracy from a knowledge of the friction angle ϕ' for the soil, using concepts of critical state soil mechanics (Schofield and Wroth, 1968). To within a few per cent, the strength ratio for normally consolidated, remoulded clay with friction angle ϕ', may be estimated as $\phi'/100$ (where ϕ' is given in degrees). The square root sign means that the variation will be small; for most soils $(c_u/\sigma_v')^{0.5}$ falls within the range 0.45 to 0.55. The value of undrained shear strength, c_u, occurs to the power of 0.5 or 0.75 in equations (4.16) and (4.17), thus reducing the sensitivity of the approach to accurate determination of this property.

In applying the above expressions, design profiles of c_u and σ_v' with depth should be determined, and then a profile of shaft friction with depth deduced. For most soils, apart from at shallow depths, the value of c_u will be less than σ_v' and equation (4.16) will apply, giving values of shaft friction that are proportional to the geometric mean of the shear strength and the effective overburden stress.

As an example of how the approach may be applied in practice, the capacity of a driven pile in stiff overconsolidated clay will be determined. Soil conditions and pile properties are taken from a group of six papers published in Ground Engineering in November, 1979, and January and March, 1980, in particular Parry (1980), describing the design of foundations for an offshore platform situated in the Heather Field in the North Sea. The piles were 1.524 m in diameter, 63.5 mm wall thickness, driven with a driving shoe of internal diameter 1.346 m to a final penetration of 45 m. The piles were to carry maximum working loads of 29.5 MN in compression, with a cyclic component of ±21.2 MN. For a factor of safety of 1.5 (a typical figure for piles where the amount of settlement is not critical), the required pile capacity should be 44.3 MN.

Profiles of σ_v', c_u and c_u/σ_v' are shown in Figure 4.11. The site showed strong sandy clay deposits, with shear strengths ranging up to 750 kPa in the upper 10 m, which was very heavily overconsolidated. Parry (1980) has discussed the magnitude of these strengths, and a probable profile of the overconsolidation ratio. He quotes values of ϕ' for the soil of about 32° for depths between 5 and 50 m. In deducing a profile of shaft friction, the upper 5 m of soil (3.3 pile diameters) has been ignored due to the likelihood of loss of shaft friction due to lateral loading (and resulting post-holing) of the pile. Below this depth, the relationships given above have been used to obtain the profiles of α, β, and thus of the shaft friction, τ_s, which are shown in Figure 4.11.

It is interesting to compare the deduced cumulative shaft capacity of the pile with that estimated by Parry (1980), using a different effective stress approach (see Table 4.1). At the design penetration of 45 m, the two approaches agree to within 5%. In addition

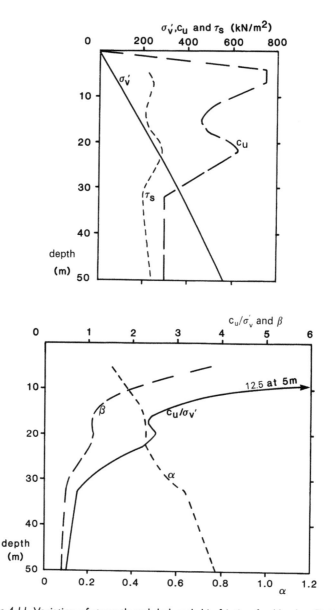

Figure 4.11 Variation of strength and deduced skin friction for Heather Field.

to the shaft capacity, the base capacity of the pile may be estimated in terms of an end-bearing pressure of 9 times the shear strength of 300 kPa, giving a total of 4.9 MN (assuming the pile is fully plugged under static loading conditions). Thus the overall predicted capacity is at least 48.1 MN.

Other approaches based on a correlation of shaft friction with undrained shear strength and the strength ratio have also included an allowance of the slenderness ratio of the pile. Semple and Rigden (1984) proposed the variation of α, and

Table 4.1 Shaft capacity with depth

| Depth (m) | Cumulative shaft capacity, MN | |
	Present approach	Parry (1980)
5	0	1.3
10	5.6	4.9
15	10.8	10.7
20	16.3	17.6
25	22.9	24.0
30	28.4	29.7
35	33.1	35.0
40	38.0	40.2
45	43.2	45.6
50	48.6	51.2

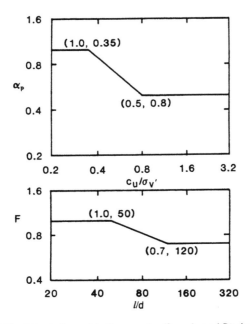

Figure 4.12 Criteria for axial pile capacity (Semple and Rigden, 1984).

a multiplicative correction factor, F, shown in Figure 4.12. An alternative correlation has been suggested by Kolk and van der Velde (1996), with the shaft friction given by

$$\tau_s = 0.55 c_u^{0.7} \sigma_v'^{0.3} \left(\frac{40}{L/d} \right)^{0.2} \tag{4.18}$$

4.1.2.4 Driven piles: effective stress approach

Chandler (1968) described a more fundamental approach, considering the bond between pile and soil as purely frictional in nature, with the resulting shaft friction a function of the normal effective stress, σ'_n and an interface friction angle, δ, in much the same way as for piles in free-draining soils. The normal stress was related to the effective overburden stress, σ'_v, by a factor, K, to give

$$\tau_s = \sigma'_n \tan \delta = K\sigma'_v \tan \delta = \beta\sigma_v \tag{4.19}$$

where $\beta = K \tan \delta$. The value of K will vary depending on the type of pile (driven or bored) and the past stress history of the soil. For piles in soft, normally or lightly overconsolidated clay, Burland (1973) and Parry and Swain (1977) suggested values of K lying between $(1 - \sin \phi')$ and $\cos^2 \phi'/(1 + \sin^2 \phi')$. Neither of these approaches takes due account of the stress changes that occur during and after pile installation. For more heavily overconsolidated soils, Meyerhof (1976) showed that the value of K consistent with measured shaft capacities varied between 1 and 2 times the in-situ horizontal stress, with an average ratio of 1.5 – a value that was supported by data from instrumented model pile tests (Francescon, 1983).

During the 1990s, detailed field measurements of pore pressure and normal stress variations around full-displacement piles were obtained by a research group at Imperial College, and these have led to a design approach that considers separately the three stages of: pile installation, pore pressure equilibration, and pile loading (Lehane *et al.*, 1994; Jardine and Chow, 1996). Chow (1997) assembled a database of results from pile load tests, and the data for radial total stresses (less the in-situ pore pressure, u_0, and normalized by σ'_v) immediately following pile installation, and the radial effective stress ratio, σ'_{rc}/σ'_v, after full equilibration of excess pore pressures are plotted against the overconsolidation (or more correctly yield stress) ratio, R, in Figure 4.13 for closed-ended piles.

Immediately after installation, the total radial stress ratio, $(\sigma_{ri} - u_0)/\sigma'_v$, increases linearly with the yield stress ratio from a value of just under 2 for normally consolidated soil (yield stress ratio of unity), with a gradient of about 0.4. As Lehane (1992) observed, the gradient is approximately parallel to the correlation of K_0 with

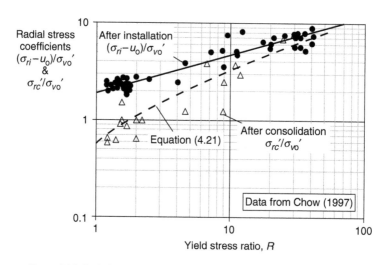

Figure 4.13 Radial stress measurements around full displacement piles.

overconsolidation ratio proposed by Mayne and Kulhawy (1982), with a radial total stress ratio of 3 to 3.5 times K_0.

The radial total stress (less the in-situ pore pressure) may be expressed as

$$\sigma_{ri} - u_0 = \sigma'_{ri} + \Delta u_{max} = (\sigma'_{ri} - p'_i) + p'_0 + \Delta p \qquad (4.20)$$

where Δu_{max} is the maximum excess pore pressure and p'_0 and p'_i are respectively the original in-situ mean effective stress and the value just after pile installation (adjacent to the pile shaft). The bracketed term has a relatively narrow range (negative, due to the slight unloading strains next to the pile according to the strain path method (Baligh, 1986; Whittle, 1992), but limited in magnitude to the current undrained shear strength, allowing for any remoulding that may have occurred as the pile is installed). The in-situ mean effective stress, p'_0, may be estimated through K_0, but it has proved difficult to estimate the increase in mean total stress, Δp, required to accommodate the pile, although approaches based on cylindrical cavity expansion and the more sophisticated strain path method have been attempted.

During equilibration, the excess pore pressure reduces to zero and the radial effective stress increases to a final value denoted by σ'_{rc}. The data for the final radial effective stress ratio, σ'_{rc}/σ'_v, in Figure 4.13 may be correlated with lines that lie nearly parallel to the trend of the installation stresses, but are offset by varying amounts, depending on the sensitivity of the clay (Lehane, 1992; Jardine and Chow, 1996), or fitted as shown by the expression (Randolph, 2003)

$$\frac{\sigma'_{rc}}{\sigma'_v} \approx \frac{\sigma'_{ri}}{\sigma'_v} + \frac{R}{5} \ln\left(1 + \frac{5}{R}\frac{\Delta u_{max}}{\sigma'_v}\right) \qquad (4.21)$$

The value of σ'_{rc} is affected to some degree by what assumption is made regarding the radial effective stress immediately after installation, σ'_{ri}. The curve in Figure 4.13 was obtained assuming this quantity to increase from around $0.1\sigma'_v$ for normally consolidated soil, to σ'_v at a yield stress ratio of 10. For a given total radial stress (for example, based on the upper curve in Figure 4.13), reducing σ'_{ri} leads to an increase in Δu_{max} and vice versa, and these adjustments partly compensate each other in the estimation of σ'_{rc} from equation (4.21).

The final stage to consider is what happens during loading of the pile. Data from the Imperial College field experiments have suggested that the radial effective stress adjacent to the pile shaft decreases by about 20% as the pile is loaded, giving $\sigma'_n \sim 0.8\sigma'_{rc}$ in equation (4.19). This value may be used, together with an interface friction angle that is either measured directly, or estimated from correlations such as those presented by Jardine and Chow (1996), in order to estimate the shaft friction.

Open-ended piles may be expected to generate lower radial stresses in the soil during installation, assuming they are driven in an 'unplugged' manner (with the soil plug moving progressively up the pile as it embeds). From cylindrical cavity expansion theory, the reduction in total radial stress (or excess pore pressure), relative to a closed-ended pile in the same soil, would be

$$\Delta\sigma_{ri} \approx c_u \ln(\rho) \qquad (4.22)$$

where ρ is the area ratio (ratio of annular area to total cross-sectional area). For steel piles, the ratio of diameter, d, to wall thickness, t, is typically around 40, corresponding to an area ratio of 0.1. This would therefore lead to a reduction in σ_{ri} of $2.3c_u$, with a corresponding reduction in the overall extent of the excess pore pressure field. However, while this approach gives reasonable adjustments for lightly overconsolidated clays, leading ultimately to a reduction in shaft friction of 15 to 20%, it appears to give unreasonably low radial stresses and shaft friction for more heavily overconsolidated clays.

4.1.2.5 Bored piles

Turning to bored piles, assuming that no chemical bond develops between the pile and soil, an effective stress approach is again more attractive than the empirical approach based purely on the undrained shear strength of the soil. Provided the pile is formed promptly after excavation of the shaft, little change in the in-situ effective stress state in the soil should occur, and equation (4.19) may be applied with $K = K_0$. In heavily overconsolidated clay, where the value of K_0 is large, some allowance for stress relaxation should be made, reducing the value of K by 20%. Alternatively, the mean stress between the in-situ horizontal stress and that due to the concrete poured into the pile shaft may be taken, replacing K by $(1 + K_0)/2$. It must be emphasized, however, that delays in pouring the concrete may reduce the in-situ horizontal effective stress considerably. Back analysis of tests of bored piles in heavily overconsolidated London Clay show deduced values of K below unity, where delays have occurred between augering and forming the piles (Fleming and Thorburn, 1983).

The question of what friction angle should be used in equation (4.19) is still subject to debate. The surface of most bored piles will be sufficiently rough to ensure that failure takes place in the soil immediately around the pile, rather than actually on the interface. However, there have been suggestions (notably Burland and Twine, 1988) that a residual angle of friction is appropriate, at least for bored piles in heavily overconsolidated clay. The friction angle mobilized on the (vertical) failure surface at peak shear stress will depend on the complete stress state. Randolph and Wroth (1981) have argued that, where the horizontal effective stress exceeds the vertical effective stress, the friction angle will be significantly lower than that obtained from a triaxial compression test, and will be similar to that mobilized on horizontal surfaces at failure in a simple shear test, conducted under appropriate normal stress conditions.

Adoption of the residual angle of friction in equation (4.19) (based on arguments of slip-planes formed during the augering process) does not appear consistent with the assumption of failure *within* the clay (due to the roughness of the pile shaft). However, residual friction angles maybe appropriate where the augered shaft is relatively smooth, and failure occurs at the concrete-soil interface. Overall, a reasonable design approach seems to be to take the horizontal effective stress as $0.5(1 + K_0)\sigma_v'$, and the friction angle as that mobilized at failure in a simple shear test. Where the horizontal effective stress is greater than the vertical effective stress (i.e. in heavily overconsolidated clays), the friction angle will lie in the range 15 to 20° (Randolph and Wroth, 1981).

For the design of bored piles in soil deposits where the profile of K_0 with depth is not known, either the approach for driven piles may be used with an appropriate factor to allow for stress relief on excavation of the pile shaft, or a direct 'total stress' approach may be used. Assuming that little change in the in-situ effective stress state takes place during construction of the pile, Meyerhof's (1976) observation of a K value close to $1.5K_0$ would indicate that the shaft friction for a bored pile may be estimated as about 70% of that for a driven displacement pile. This figure may be conservative for piles in soil with low in-situ values of K_0, since some increase in lateral stress may occur during the process of casting the pile.

For bored piles in stiff, heavily overconsolidated clay, an α value of 0.45 was suggested in order to estimate the shaft capacity (Skempton, 1959). This approach is based largely on experience in London Clay, with shear strengths measured on small (38 mm diameter) samples. In fissured clays, such as London clay, the shear strength is affected by sample size and the modern use of 100 mm diameter samples leads to lower shear strengths, and thus a higher value of α will be appropriate. Recent reviews suggest α values of 0.5 for conventional bored piles, and 0.6 for CFA piles, based on shear strengths determined from unconsolidated undrained triaxial tests on 100 mm diameter samples. Care should be taken when applying the same value of α to clays where the strength ratio, c_u/σ'_v, is higher than for London Clay, since an effective stress approach may well indicate substantially lower values of shaft friction than the total stress approach based on α values in the range 0.5 to 0.6.

Weltman and Healy (1978) analyzed a number of pile tests in boulder clays and other glacial tills, and suggested values of α varying with the undrained shear strength of the soil as shown in Figure 4.14. The ratio of values for bored and driven piles is just over 80%, but both curves are derived from data showing considerable scatter. The figure also shows a separate curve for very short, stubby piles driven into till overlain by soft clay. For such piles, the ratio of c_u/σ'_v will be high and the resulting low α values are consistent with the effective stress approach outlined earlier.

4.1.2.6 Correlations with field tests

Comparisons of measured pile capacities with results of cone penetration or standard penetration tests show a much greater scatter than for non-cohesive soils. The drainage conditions during a penetration test will be rather different from those existing during a pile test, and this may account for some of the variation. The end-bearing pressure

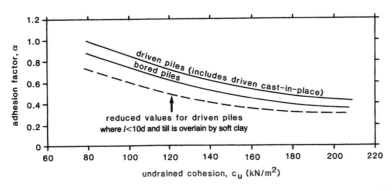

Figure 4.14 Variation of α with shear strength of glacial till (Weltman and Healy, 1978).

mobilized by a driven pile should, in principle, be comparable with that measured during a cone penetration test. However, effects of scale and of different rates of testing combine to yield variations in the ratio q_b/q_c ranging from under 0.5 to over 2. Unless there is particular evidence to the contrary, it is suggested that the end-bearing pressure for driven closed-ended piles may be taken equal to that measured in a cone penetration test, with a reduction of 30% for open-ended or bored piles.

Estimates of the shaft friction for driven piles are often made from friction sleeve measurements, taking $\tau_s = f_c$. However, caution should be exercised using this approach, owing to the sensitivity of the value of friction measured to the degree of consolidation in the soil (a) around the pile and (b) around the friction sleeve. Pile tests show that the value of shaft friction measured during installation (a similar process to the continuous penetration of the cone and friction sleeve) may be as low as 20% of the long-term shaft friction (see section 4.1.5). Estimates of shaft friction directly from the cone resistance, q_c, offer a better guide. However, recommended correlations range from $\tau_s = q_c/10$ (Fleming and Thorburn, 1983) down to $\tau_s = q_c/40$ (Thorburn and McVicar, 1971). It is probably more reliable to deduce profiles of undrained shear strength from the cone penetration tests and then to use the expressions given earlier, relating shaft friction to the undrained shear strength and to the effective overburden stress.

As discussed in Chapter 2, results from standard penetration tests in medium to firm clays and soft rocks may be used to estimate the shear strength of the soil, taking a ratio for c_u/N of 4 to 5 kPa for clays of medium plasticity, rising to 6 to 7 kPa for plasticity indices less than 20 (see Figure 2.4). The values of shear strength may then be used to provide estimates of end-bearing pressure and of shaft friction in the normal way.

4.1.3 Capacity of piles in rock

Piles driven to refusal on firm or intact rock are often designed as purely end-bearing piles ignoring any shaft capacity the pile might have, although, in practice, considerable shear transfer may occur in the upper layers of soil, at working loads. In many cases, the maximum design load for such a pile will be determined by the stresses in the pile material itself rather than the permitted bearing pressure on the rock.

Great care should be exercised where steel piles are driven to bedrock, to ensure that good sound rock contact is achieved. Pile driving may shatter the rock, and indeed it is good practice to ensure that the piles penetrate a small distance into the rock. This will help to prevent the piles heaving and coming away from the bearing stratum during driving of subsequent piles. Alternatively, where access permits, each pile should be re-tamped at the end of installing the neighbouring piles, to ensure good contact with the bearing stratum.

In weathered rock, careful monitoring of driving records is essential owing to the wide variability in strength exhibited by soft rocks such as chalk and marl in a weathered state. Refusal of the piles may not be achieved and the contribution of the pile shaft to its overall capacity may need to be relied upon. Accurate estimation of the shaft friction around a pile driven into soft rock is difficult owing to the disruption of the structure of the rock as the pile penetrates. The shaft friction will depend on the magnitude of the normal effective stress, which is liable to vary in a haphazard fashion.

Recorded values of shaft friction for piles driven into low-grade chalk vary from as low as 8 kPa (Hodges and Pink, 1971) up to 26 kPa (Hobbs and Robins, 1976) for similar pile penetrations.

4.1.3.1 End-bearing pressure

It has been customary for design codes to stipulate allowable bearing pressures on rocks of different types. Tomlinson (1986) presents an extensive list of allowable pressures, in terms of type, quality and joint spacing of the rock. In view of the great variability in strength (and allowable end-bearing pressure) of individual rock types, it appears best to express allowable bearing pressures for pile design in terms of the unconfined compressive strength of the rock. Theoretical (and experimental, where failure has actually been achieved) values of ultimate end-bearing pressure are generally in excess of $10q_u$ (where q_u is the unconfined compressive strength). Large settlements would be required to mobilize such values of end-bearing, and, in practice, a rather lower 'ultimate' end-bearing pressure of about $3q_u$ is often adopted (Rowe and Armitage, 1987).

A rational design approach is to limit allowable bearing stresses to below that which would initiate crushing of the rock beneath the pile. Even where the joints are relatively open, this value will not fall below the unconfined compressive strength of the rock. Thus, allowable end-bearing stress, q_{ba}, may be expressed as (Rowe and Armitage, 1987)

$$q_{ba} = q_u \tag{4.23}$$

The above value is rather higher than that of $0.3q_u$ suggested by, for example, Poulos and Davis (1980). However, the higher value is supported by experimental evidence and is consistent with a factor of safety of 3 applied to the (conservative) ultimate bearing pressure of $3q_u$ discussed above.

A more extensive database has been presented by Zhang and Einstein (1998), where end-bearing pressures mobilized at a displacement of 10% of the pile diameter were found to be proportional to the square root of the unconfined compression strength, according to

$$\frac{q_b}{p_a} \approx 15 \sqrt{\frac{q_u}{p_a}} \tag{4.24}$$

where p_a is atmospheric pressure. This expression has been compared with results from cone penetration tests and model pile tests in calcarenite by Randolph et al. (1998), as shown in Figure 4.15.

4.1.3.2 Shaft friction

Shear transfer along the shaft of a rock socket is a complex phenomenon that depends on the frictional characteristics of the interface, the degree of roughness of the socket and the strength properties of the host rock. Values of peak friction will depend strongly on the normal stress along the shaft, and the amount of dilation that occurs

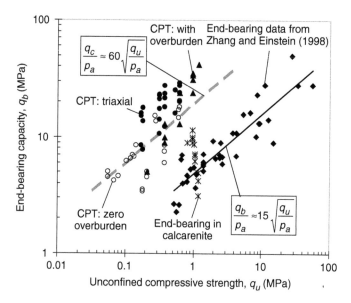

Figure 4.15 Correlations of end-bearing capacity for piles in rock and cone resistance in calcarenite (Randolph et al., 1998).

during shearing (Johnston and Lam, 1989). In practice, it has been customary to correlate measured values of shaft friction with the unconfined compressive strength of the rock (Horvath *et al.*, 1980; Williams and Pells, 1981; Rowe and Armitage, 1987). While there are minor differences, most of these correlations show a trend of shaft friction which is proportional to the square root of the compressive strength. However, traditional approaches show a major discontinuity in the transition zone between hard soils and soft rocks, as indicated in Figure 4.16 (based on Kulhawy and Phoon, 1993). The trend lines through the data are given by

$$\frac{\tau_s}{p_a} = \psi \left(\frac{c_u}{p_a}\right)^{0.5} \quad \text{or} \quad \psi \left(\frac{q_u/2}{p_a}\right)^{0.5} \tag{4.25}$$

where $\psi = 0.5$ for the clay data and 2 for the main soft rock data. Also shown on the figure are data from laboratory grouted driven pile data, and laboratory and field test data related to a soft limestone site at Overland Corner in South Australia (Randolph *et al.*, 1998).

The discontinuity in the data, and the high shaft friction values found for soft rocks, relative to the unconfined compression strengths, is attributable to the high roughness of bored sockets for the strength range between about 300 and 3000 kPa. Seidel and Haberfield (1995) have presented a detailed model for the shearing of rock-sockets, where combined shearing and bearing resistance of asperities at the borehole edge determine the limiting shaft friction. Their model is able to simulate the trends in the data, provided appropriate assumptions (based on field measurements) are made in relation to the asperity height and angles, (see Figure 4.17). Of particular note is the effect of pile diameter, with shaft friction reducing as the diameter increases. For large

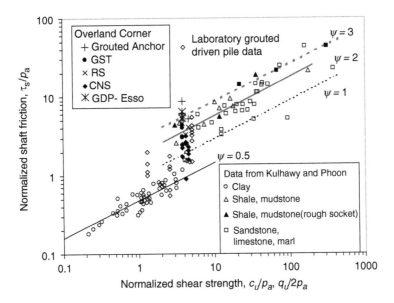

Figure 4.16 Shaft friction data from piles in cemented materials (extended from Kulhawy and Phoon (1993) by Randolph *et al.* (1998)).

piles, the value of ψ is close to unity, which also gives a conservative bound to the data in Figure 4.16. A check should also be made in any design, that the assumed shaft friction is not greater than 5% of the concrete strength.

4.1.3.3 Correlations with SPT in chalk

Hobbs and Healy (1979) have presented an extensive review of the performance of piles in chalk, and have correlated shaft friction and end-bearing pressures for different pile types with the SPT value of the chalk. For large displacement driven piles, they recommend using an effective stress approach for estimating the shaft capacity, taking a value for K of 0.06 N, and interface friction angles of 20° (for steel piles) to 24° (for concrete piles). Higher values of K (up to 0.15 N) were recommended for driven cast-in-situ piles where high-slump concrete was compacted against the chalk, while lower values of K (0.03 N) were suggested for low-displacement piles. End-bearing pressures were shown to correlate with N values with ratios of q_b/N lying between 0.2 and 0.25 MPa. Figure 4.18 shows summaries of measured values of shaft friction and end-bearing pressure, plotted against SPT values, taken from Hobbs and Healy (1979).

For bored piles in chalk, where interlocking and possible chemical bonding occurs between the pile and the chalk, Hobbs and Healy found that the shaft friction correlated well with the compressive strength of the chalk, according to the relationship given in equation (4.25) with $\psi \sim 1$. The shaft friction of CFA piles has been reviewed by Lord *et al.* (2003) who suggest an effective stress approach, although the correlation of shaft friction with vertical effective stress is very poor. A direct correlation of

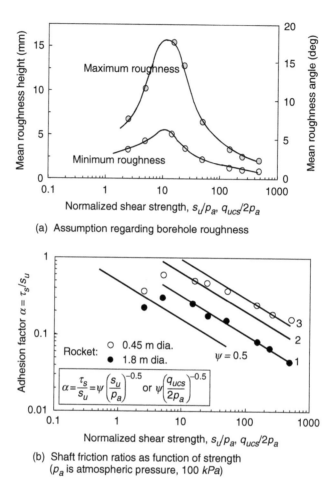

(a) Assumption regarding borehole roughness

(b) Shaft friction ratios as function of strength
(p_a is atmospheric pressure, 100 kPa)

Figure 4.17 Predictions of rock-socket shaft friction from ROCKET (Seidel and Haberfield, 1995).

shaft friction and the SPT values reported at the level of the pile base shows a clearer trend, with the average shaft friction around $3N_{base}$ kPa.

4.1.4 Driving formulae and dynamic estimates of pile capacity

In many instances, particularly in heavily stratified soil or rock of varying degrees of weathering, the capacity of each pile installed needs to be checked from the driving record. The capacity of the pile at the time of installation should represent a conservative value. With time after installation, dissipation of excess pore pressures generated during the driving process leads to an increase in effective stresses around the pile, and a resulting increase in the pile capacity. For piles driven into clay, this increase may be as much as a factor of 5 to 6 (Vesic, 1977). Thus pile load tests aimed at assessing the long-term capacity of a working pile should allow sufficient time for this capacity to be achieved often a matter of 3 to 4 weeks (see section 4.1.5).

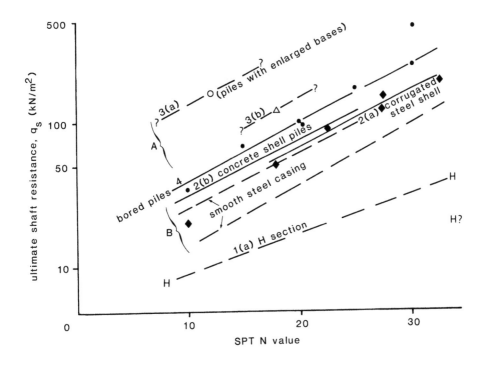

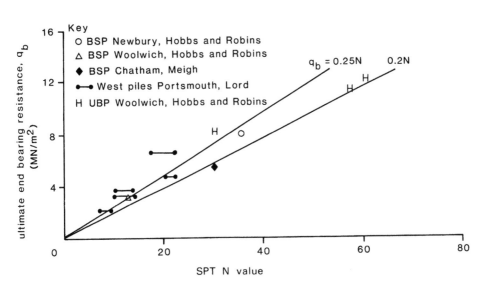

Figure 4.18 Correlations of skin friction and end-bearing pressures with SPT value, for piles in chalk (Hobbs and Healy, 1979). (Top) Ultimate shaft resistance against N value for all piles. (Bottom) Ultimate and bearing resistance for some driven piles. A, Driven cast-in-place piles (fresh concrete in contact with the chalk). B, displacement piles (including preformed and cast-in-place types).

4.1.4.1 Pile driving formulae

The most widespread method of estimating the dynamic capacity of piles is the use of some form of pile driving 'formula', relating the measured permanent displacement (or 'set') of the pile at each blow of the hammer, to the pile capacity. Driving formulae are based on an energy balance between the (dynamic) input energy of the hammer, and the (static) work required to move the pile permanently a small distance, s. As such, they offer a compromise (often rather a poor one) between the static and dynamic capacities of a pile.

There is a variety of pile driving formulae in common use, many of which have been reviewed by Whitaker (1975), who comments that, in some situations, pile capacities predicted by the different formulae may differ by a factor of 3. Rather than describe all the different approaches here, the basic approach will be outlined and some of the main variables discussed.

A simplified picture of the driving process, and variation of pile resistance with displacement is shown in Figure 4.19. This picture gives rise to the fundamental pile driving formula

$$R = \frac{\eta W h}{(s + c/2)} \tag{4.26}$$

where R is the pile resistance.
        η is the efficiency of the hammer (allowing for energy loss on impact).
        W is the weight of the hammer.

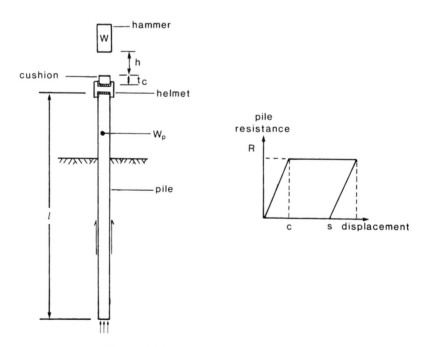

Figure 4.19 Schematic diagram of pile driving.

h is the drop height.

s is the permanent set of the pile.

c is the elastic, or recoverable, movement of the pile.

This expression is widely used in the American continent, adopting a value of 0.8 to 1 for η, and 0.1 in (2.5 mm) for c (giving the so-called *Engineering News* formula). In Europe, both η and c are chosen with regard to the type of hammer, the material used as a cushion at the head of the pile, and the physical properties of the pile. Probably the best known variant is that due to Hiley (1925), where the overall hammer efficiency is given by

$$\eta = \frac{k\left(W + e^2 W_p\right)}{W + W_p} \tag{4.27}$$

where k is the output efficiency of the hammer (ratio of power delivered at the cushion, to rated power).

W_p is the weight of the pile.

e is the coefficient of restitution between the hammer and the pile cushion (or top of pile if there is no cushion).

The recoverable movement, c, is taken as the sum of the cushion compression, $c_c = R t_c/(AE_c)$ (where t_c is the thickness and (AE_c) the cross-sectional rigidity of the cushion), the elastic shortening of the pile (considered as a column), $c_p = Rl/(AE)_p$, and the recoverable movement (or 'quake') of the ground, c_g. The last quantity may be taken as 0.5% of the pile diameter. Table 4.2 shows commonly adopted values for the quantities k, e and E depending on hammer type and cushion material.

It should be emphasized that the form of equation (4.24) is inherently inaccurate for assessing, by means of the measured set, s, whether a required pile resistance, R, has been achieved. Uncertainties in the values of k and the various components of c can lead to wide variations in the deduced pile resistance. Over a period of a few hours driving, the characteristics of both the hammer and the cushion may alter significantly; thus

Table 4.2 Values of parameters for pile driving formulae

Type of hammer		Power efficiency, k
Drop hammer (triggered fall)		1
Steam or compressed air hammer		0.9
Drop hammer (winch operated)		0.8
Diesel hammer		0.6–0.8
Cushion type	Coefficient of restitution, e	Young's modulus, E_c (MN/m²)
Micarta plastic	0.8	3×10^3
Greenheart oak	0.5	3×10^2
Other timber	0.3	2×10^2

piles driven to a given set at the start and end of the period may have widely varying capacities.

4.1.4.2 Dynamic analysis of pile driving

A more rigorous assessment of the driving characteristics of a pile may be achieved by means of a proper dynamic analysis of the pile–soil system. The inertial effects of the soil around the pile, and the viscous nature of the soil, may both be taken into account by appropriate numerical techniques, such as the finite element method (Smith and Chow, 1982). A common simplification of the soil is as a massless medium providing frictional resistance alone, while the pile is modelled as a discrete assembly of mass elements, interconnected by springs. This 'one-dimensional' model is discussed in detail in section 9.3 (Dynamic testing of piles).

At the design stage, the main objective of dynamic analysis of pile driving is to assess the 'drivability' of a pile in given soil conditions, and with a particular type and size of hammer. This is commonly accomplished using one-dimensional wave equation programs, based on the method described in section 9.3. Probably the most widely used program is the WEAP program developed originally by Goble and Rausche (1976) for the Federal Highway Administration in North America. The program, now marketed as GRLWEAP has been updated over the years with improved user interfaces, and other technical improvements (Goble and Rausche, 1986; Rausche et al., 1988).

The main outcome of a drivability study is a series of curves that give the penetration rate (or blow count) as a function of the assumed 'static' resistance of the pile. These curves may be used to choose an appropriate hammer, to assess the time and cost of installation of each pile and to provide a quality control on pile installation (in terms of required 'set' per blow for a given capacity). In addition, a drivability study will include assessment of the maximum tensile and compressive stresses in the pile during driving, maximum acceleration levels, and range of required (or permitted) hammer stroke. Limiting the range of hammer stroke may be particularly necessary during the early driving of concrete piles, in order to avoid excessive tensile stresses in the pile.

4.1.5 Increase in capacity of piles with time

It is important to consider the effect of time on the axial capacity of piles, particularly driven piles in clay, and also the effect of loading rate on the measured capacity. The design methods that have been described are based on observed performance of piles which have been load tested over a relatively short period (of tens of minutes or hours), usually a matter of a few days or weeks since installation. Except in particular circumstances involving dynamic loading, piles must generally withstand their design loads for very long periods of time, and the question arises whether creep, or stress relaxation, in the soil around the pile, may lead to reduced capacity. In soils which are known to creep, some allowance may need to be made for these effects, although it should be borne in mind that the stress level in the soil will generally be relatively low. It is unlikely that creep would lead to more than 5 to 10% reduction in the capacity of a pile.

4.1.5.1 Pore pressure dissipation in clay

The question of the effect of time since installation is of much more consequence, particularly for displacement piles installed in soils of low permeability. Piles driven into cohesive soil generate high excess pore pressures close to the pile. The pore pressures arise partly as a result of a decrease in effective stress, as the soil is sheared and remoulded, and partly due to the increase in total stress, as the pile forces soil out of its path. Typically, the excess pore pressures may be as high as the effective overburden stress, and may extend out over a zone up to 10 times the diameter of the pile. Randolph *et al.* (1979) suggest that the excess pore pressure, Δu, close to a driven pile may be estimated from the expression

$$\Delta u = 4cu - \Delta p' \tag{4.28}$$

where $\Delta p'$ is the change in mean effective stress due to shearing and remoulding the soil. In normally or lightly overconsolidated clay, $\Delta p'$ will be negative and may be as high as 2 to $3c_u$ for sensitive clays. In more heavily overconsolidated clay, $\Delta p'$ will become positive as the clay attempts to dilate on shearing. Figure 4.20 shows typical field measurements of excess pore pressures due to pile installation. The values reflect the trend implied by equation (4.28).

After the pile is installed, the excess pore pressures will dissipate, primarily by radial flow of pore water away from the pile, and the soil will consolidate. During this process, the water content of the soil will decrease, and its shear strength increase. Seed and Reese (1955) report a decrease in water content of 7% close to a pile driven into San Francisco Bay mud. This was accompanied by a threefold increase in the remoulded strength of the soil (measured from samples taken from close to the pile). A similar figure for the decrease in water content adjacent to a displacement pile has been reported by Francescon (1983), from tests on model piles installed in soil of varying overconsolidation ratio.

The axial capacity of a driven pile will vary with time after installation, because of this consolidation process. Seed and Reese report a sixfold increase in pile capacity, over a period of 30 days. Similar increases in capacity have been discussed by Vesic (1977) and by Thorburn and Rigden (1980). Analytical studies of the stress changes caused by pile installation, and of the consolidation process, have shown that the shear strength of the soil may increase by between 30 and 100% close to the pile, over a time

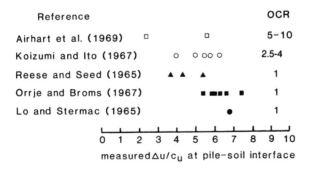

Figure 4.20 Field measurements of excess pore pressure.

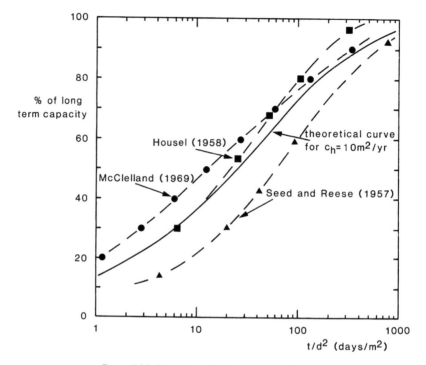

Figure 4.21 Variation of pile capacity with time.

period which is governed by the dimensionless group $c_h t/d^2$, where c_h is the coefficient of consolidation in a horizontal plane, t is the time since installation of the pile, and d is the diameter of the pile (Randolph *et al.*, 1979; Randolph and Wroth, 1979).

Figure 4.21 shows the variation of pile capacity with time, for three piles driven into soft silty clay. The time scale has been normalized by the square of the pile diameter. Other factors which will affect the rate of dissipation of excess pore pressures are (a) the consolidation coefficient, c_h, and (b) the extent of the zone of excess pore pressures. The latter will be a function of the rigidity index, G/c_u, of the soil (where G is the shear modulus), and the sensitivity of the soil. Also shown on Figure 4.21 is a theoretical curve of the change in excess pore pressure, normalized by the maximum value, with time. The curve is taken from Randolph and Wroth (1979), assuming that the excess pore pressures extend out to a distance of 10 pile radii from the axis of the pile. A value of 10 m^2/year has been taken for c_h, which is reasonable for silty clays under conditions of horizontal drainage. The measured variations of pile capacity follow the theoretical curve reasonably well. When piles are driven in a group, the magnitude of the excess pore pressures generated will be similar to that for a single pile. However, the zone over which they extend will be considerably larger, and the resulting consolidation process will be more protracted.

The dissipation time will depend on the magnitude and extent of the excess pore pressure field generated around the pile, and will thus be much shorter for open-ended piles than for closed-ended piles. Dissipation curves for open-ended piles of different wall

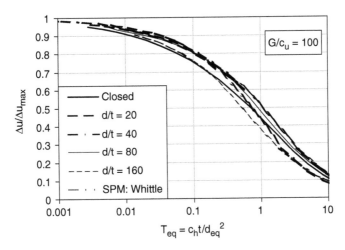

Figure 4.22 Pore pressure dissipation for open-ended and closed-ended piles.

thicknesses are shown in Figure 4.22 for lightly overconsolidated conditions, modelled as elastic-perfectly plastic material with $G/c_u = 100$, corresponding to $\Delta u_{max} = 4.6c_u$ (Randolph, 2003). The time axis has been non-dimensionalized as $T_{eq} = c_h t / d_{eq}^2$, where d_{eq} is the equivalent diameter of a pile with the same solid volume as the open-ended pile in question. The equivalent diameter may be expressed in terms of the actual diameter, d, area ratio, ρ, and wall thickness, t, as

$$d_{eq} = d\sqrt{\rho} \approx 2\sqrt{dt} \qquad (4.29)$$

A comparison with a dissipation based on the strain path method (SPM) is also shown (Whittle, 1992). Note that the average dissipation curve in Figure 4.22 may be applied to situations where the maximum excess pore pressure is greater or less than $4.6c_u$, by adjusting d_{eq} accordingly, on the basis of equations (4.22) and (4.29).

Pile installation in soft clay may temporarily reduce the stability of nearby buildings or slopes. The remoulding of soil during pile driving may decrease the strength of a sensitive clay sufficiently to cause a bearing capacity failure or slope instability (see Figure 4.23(a)). Dissipation of the excess pore pressures generated during pile driving may allow pore pressures to rise in the vicinity of a slope, thus leading to failure of the slope, even where the soil has not been remoulded (see Figure 4.23(b)). However, in general, the effects of pile installation will not extend beyond 10 to 15 pile diameters. Where there is danger of instability, consideration should be given to the use of non-displacement piling techniques, or to preboring a slightly undersize hole before driving piles – a technique commonly adopted in Scandinavia.

4.1.5.2 Effect of time for driven piles in sand

Recent studies have suggested that the shaft friction of steel piles driven in sand shows significant increase with time (Chow *et al.*, 1998), with gains of 50 to 100% possible.

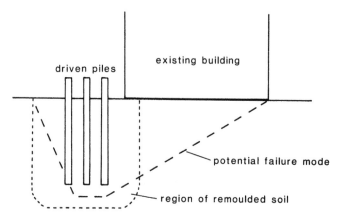

(a) Piles driven into sensitive clay close to an existing building

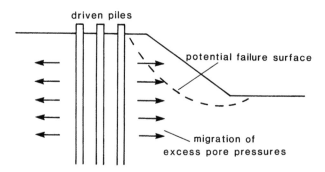

(b) Piles driven close to an existing slope

Figure 4.23 Potential problems due to pile driving.

Alternative explanations ranging from chemical changes (essentially corrosion) at the pile–soil interface to relaxation of the arch of highly stressed sand surrounding the pile have been put forward to explain these increases. Further work is needed before relying on increased shaft friction in design, particularly as the resulting shaft capacity may prove somewhat brittle, reducing rapidly as the pile slips relative to the surrounding sand.

4.2 Axial deformation of piles

The load deformation response of piles under axial load has been examined extensively, using numerical methods, particularly integral equation or boundary element methods (Poulos and Davis, 1968; Mattes and Poulos, 1969; Butterfield and Banerjee, 1971a, b; Banerjee and Davies, 1977). Such methods enable charts to be developed showing how the settlement of a pile depends on the various parameters of pile geometry and stiffness, and soil stiffness. Poulos and Davis (1980) have compiled an extensive collection of such charts, enabling the load settlement response of any given pile to be estimated readily. As an aid to understanding the manner in which a pile transfers load to the soil under working conditions, an approximate solution is developed here for the load settlement response of a pile in elastic soil. The solution leads to an expression for the pile stiffness (that is applied load divided by settlement) in closed form, which may be used as an alternative to charts.

In developing this solution, the manner in which load is transferred to the soil from the pile shaft, and from the pile base, will be examined separately, before combining the two to give the response of the complete pile. In the first instance, the soil will be treated as an elastic material, characterized by an appropriate secant value of elastic modulus, varying in some fashion over the depth of interest. Choice of values of elastic modulus for different soil types is discussed in Chapter 5.

Some discussion is appropriate concerning the geometric modelling of real piles. Analytical solutions for the response of piles have tended to treat the pile as a circular cylinder of diameter d (or radius r_0) and length l. In practice, non-circular piles, such as square or hexagonal precast concrete piles or H-piles, are common. Such piles may be adequately represented as circular piles by taking a pile of equivalent (gross) cross-sectional area. Thus for an H-section pile of outer dimensions $b \times w$ (see Figure 4.24), a circular pile of diameter d may be taken, where $\pi d^2 = 4bw$. It is also convenient to consider the pile as solid, with an appropriate Young's modulus E_p, such that the rigidity is equivalent to that of the original pile. Thus, when considering axial deformation, the equivalent pile modulus is

$$E_p = \frac{4(EA)_p}{\pi d^2} \tag{4.30}$$

where $(EA)_p$ is the cross-sectional rigidity of the actual pile.

4.2.1 Basic solution for a rigid pile

4.2.1.1 Pile shaft

Finite element and boundary element analyses of the response of friction piles (Frank, 1974; Randolph, 1977) have shown that load is transferred from the pile shaft by shear stresses generated in the soil on vertical and horizontal planes, with little change in the vertical normal stress (except near the base of the pile). Schematically, a pile

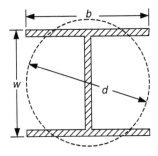

Figure 4.24 Idealization of H section pile.

may be considered as surrounded by concentric cylinders of soil, with shear stresses on each cylinder. For vertical equilibrium, the magnitude of the shear stress on each cylinder must decrease inversely with the surface area of the cylinder (Cooke, 1974; Frank, 1974). Writing the shear stress on the pile shaft as τ_0, the shear stress at radius r (for a pile of radius, r_0) is given by

$$\tau = \frac{\tau_0 r_0}{r} \tag{4.31}$$

Since the mode of deformation around a pile is primarily one of shear, it is more natural to develop the solution in terms of the shear modulus G and Poisson's ratio, v, rather than Young's modulus, E (noting that the shear modulus may be related to Young's modulus by $G = E/2(1+v)$). The shear strain in the soil, γ, will thus be given by $\gamma = \tau/G$. Since the main deformation in the soil will be vertical, the shear strain may be written approximately as

$$\gamma \sim \frac{dw}{dr} \tag{4.32}$$

where w is the vertical deflection.

These relationships may be assembled and integrated to give

$$w = \int_{r}^{r_m} \frac{\tau_0 r_0}{Gr} dr = \frac{\tau_0 r_0}{G} \ln\left(r_m/r\right) \tag{4.33}$$

A maximum radius, r_m, has been introduced at which the deflections in the soil are assumed to become vanishingly small. Empirically, this radius has been found to be of the order of the length of the pile (Randolph and Wroth, 1978). The deflection of the pile shaft, w_s, is given by

$$w_s = \zeta \frac{\tau_0 r_0}{G} = \zeta \frac{\tau_0 d}{2G} \quad \text{where } \zeta = \ln\left(r_m/r_0\right) = \ln\left(2r_m/d\right) \tag{4.34}$$

This approximate analytical derivation of the deflection of the pile shaft in terms of the average applied shear stress τ_0 shows a number of important features of the response of a pile to axial load.

1 Stress changes induced in the soil are primarily shear stresses, decreasing inversely with distance from the pile axis; thus, only soil very close to the pile is ever highly stressed.
2 The resulting deflections decrease with the logarithm of distance from the pile axis; thus, significant deflections extend some distance away from the pile (up to about one pile length).
3 The deflection of the pile shaft, w_s, normalized by the pile radius r_0, is ζ times the local shear strain, $\gamma_0 = \tau_0/G$, in the soil: the parameter ζ is found to vary between 3 and 5, with an average value of about 4 (Baguelin and Frank, 1979).

4   A load transfer stiffness, relating the load transferred at a given depth to the local
    displacement may be expressed as

$$k = \frac{\pi d \tau_0}{w_s} = \frac{2\pi}{\zeta} G \qquad (4.35)$$

Equation (4.34) is of particular importance in deducing the magnitude of deflection
necessary to mobilize full shaft friction (see section 4.2.4). The theoretical basis for
the load transfer approach, and analytical expressions for the ratio k/G have been
explored by Mylonakis (2001).

The overall load taken by the pile shaft is $P_s = \pi d / \overline{\tau}_0$, where $\overline{\tau}_0$ is the average shear
stress mobilized at the pile shaft. Thus the load settlement ratio (or stiffness of the
pile-soil system) is

$$\frac{P_s}{w_s} = \frac{2\pi L \overline{G}}{\zeta} \qquad (4.36)$$

where $\overline{G}$ is the average shear modulus of the soil over the embedment depth, L, of
the pile.

4.2.1.2 Pile base

At the pile base, it is sufficient to ignore the pile shaft and surrounding soil, and treat
the base as a rigid punch acting at the surface of a soil medium (that, in reality, starts at
a depth, $z = L$). The base stiffness is obtained from the standard solution (Timoshenko
and Goodier, 1970) as

$$\frac{P_b}{w_b} = \frac{2d_b G_b}{(1 - v)} \qquad (4.37)$$

where the subscript b refers to the pile base.

4.2.1.3 Combining shaft and base

For a stiff pile, the base settlement and shaft settlement will be similar to the settlement
of the pile head, w_t. The total load, P_t, may thus be written as

$$P_t = P_b + P_t = w_t \left(\frac{P_b}{w_b} + \frac{P_s}{w_s} \right) \qquad (4.38)$$

In developing a general solution for the axial response of a pile, it is convenient to
introduce a dimensionless load settlement ratio for the pile. The pile stiffness is P_t/w_t
and this may be made dimensionless by dividing by the diameter of the pile and an
appropriate soil modulus. It has been customary to use the value of the soil modulus at

the level of the pile base for this purpose, written as G_L. Thus equation (4.36) becomes (making use of equations (4.34) and (4.35))

$$\frac{P_t}{w_t d G_L} = \frac{2}{(1-v)} \frac{d_b}{d} \frac{G_b}{G_L} + \frac{2\pi}{\zeta} \frac{\overline{G}}{G_L} \frac{L}{d} \qquad (4.39)$$

In order to obtain a feel for the relative importance of the terms in equation (4.39), the shear modulus variation with depth may be idealized as linear, according to $G = G_0 + mz$ (where z is the depth), with the possibility of a sharp rise to G_b below the level of the pile base (see Figure 4.25). Defining parameters $\rho = \overline{G}/G_L$ and $\xi = G_L/G_b$, the constant ζ has been found to fit the expressions

$$\zeta = \ln\left\{\left[0.25 + (2.5\rho(1-v) - 0.25)\xi\right]\right\}2L/d \qquad (4.40)$$

$$\zeta = \ln\left[5\rho(1-v)L/d\right] \quad \text{for } \xi = 1 \qquad (4.41)$$

Figure 4.26 shows how the load settlement ratio varies with slenderness ratio L/d for a range of values of ρ and ξ, taking $\eta = d_b/d = 1$. It should be noted that the dependence on v is small and the results in Figure 4.26 are all for $v = 0.3$. It may be seen that, as might be expected, the stiffness of the pile increases nearly linearly with pile length, holding other factors constant. However, it must be borne in mind that this result is only true of piles that may be considered rigid.

4.2.2 Pile compression

Most piles exhibit some shaft compression at working loads, and this should be allowed for in estimating pile deflection. The axial strain at any level down the pile is

$$\varepsilon_z = -\frac{dw}{dz} = \frac{4P}{\pi d^2 E_p} \qquad (4.42)$$

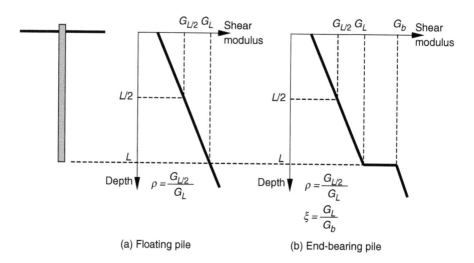

(a) Floating pile   (b) End-bearing pile

Figure 4.25 Assumed variation of soil shear modulus with depth.

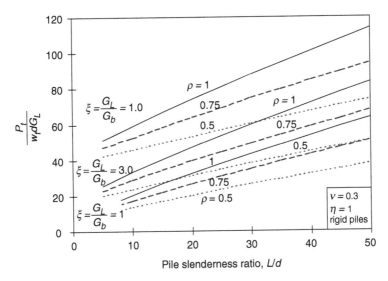

Figure 4.26 Load settlement ratios for rigid piles.

taking E_p as the Young's modulus of an equivalent solid pile (see equation (4.30)). The load P will vary down the pile as load is shed into the surrounding soil, so that

$$\frac{dP}{dz} = -\pi d\tau_0 \tag{4.43}$$

Finally, w_s and τ_0 may be related by equation (4.33) to give

$$\frac{d^2w}{dz^2} = \frac{8G}{\zeta E_p d^2} w \tag{4.44}$$

This differential equation may be solved to yield w_s in terms of hyperbolic cosine and sine functions of z. Substituting in the appropriate boundary conditions at the pile base yields an expression for the load settlement ratio of the pile head of

$$\frac{P_t}{w_t d G_L} = \frac{\frac{2\eta}{(1-\nu)\xi} + \frac{2\pi\rho}{\zeta}\frac{\tanh(\mu L)}{\mu L}\frac{L}{d}}{1 + \frac{8\eta}{\pi\lambda(1-\nu)\xi}\frac{\tanh(\mu L)}{\mu L}\frac{L}{d}} \tag{4.45}$$

where, summarizing the various dimensionless parameters

$\eta = d_b/d$ (ratio of underream for underreamed piles)
$\xi = G_L/G_b$ (ratio of end-bearing for end-bearing piles)
$\rho = \bar{G}/G_L$ (variation of soil modulus with depth)

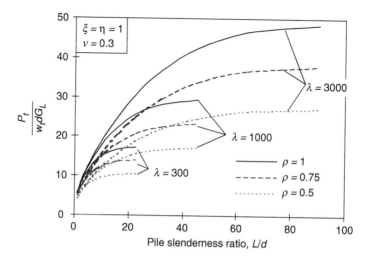

Figure 4.27 Load settlement ratios for compressible piles.

$\lambda = E_p/G_L$ (pile–soil stiffness ratio)
$\zeta = \ln(2r_m/d)$ (measure of radius of influence of pile)
$\mu L = 2\sqrt{2/\zeta\lambda}(L/d)$ (measure of pile compressibility)

Figure 4.27 shows the variation of the load settlement ratio with slenderness ratio, L/d, for $\eta = \zeta = 1, v = 0.3$.

The load at the pile base is given by

$$\frac{P_b}{P_t} = \frac{\dfrac{\eta}{(1-v)\xi}\dfrac{1}{\cosh(\mu L)}}{\dfrac{\eta}{(1-v)\xi} + \dfrac{\pi\rho}{\zeta}\dfrac{\tanh(\mu L)}{\mu L}\dfrac{L}{d}} \tag{4.46}$$

Figure 4.28 shows this ratio as a function of L/d for a range of values of ρ and λ (again for $\eta = \xi = 1$, $v = 0.3$). It is clear that, under working conditions, very little load reaches the base of long piles. The situation would be different for underreamed or for end-bearing piles and also for groups of piles, where additional load is transferred to the base of the piles.

From Figures 4.27 and 4.28, it may be seen that there are combinations of slenderness ratio, L/d, and stiffness ratio, λ, beyond which very little load is transmitted to the pile base. Further increase in pile length yields no corresponding increase in the load settlement ratio of the pile. This limiting behaviour is the converse of a stiff, rigid pile, and corresponds to the case where the pile starts behaving as if it were infinitely long, with no load reaching the lower region.

The two limits may be quantified. Piles may be taken as essentially rigid where L/d is less than about $0.25\sqrt{E_p/G_L}$ and equation (4.45) then reduces to equation (4.39). At the other extreme, for piles where L/d is greater than about $1.5\sqrt{E_p/G_L}$, the value

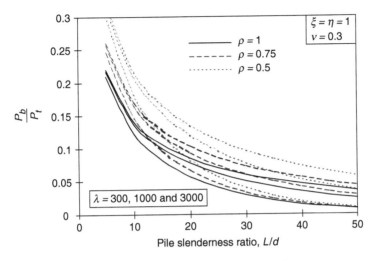

Figure 4.28 Proportion of load taken by pile base.

of $\tanh(\mu L)$ approaches unity and equation (4.45) reduces approximately (exactly for $\rho = 1$) to

$$\frac{P_t}{w_t dG_L} = \pi\rho\sqrt{\frac{\lambda}{2\zeta}} \qquad (4.47)$$

As expected, the load settlement ratio is now independent of the length of the pile (since no load reaches the lower end). The modulus G_L should be interpreted as the soil shear modulus at the bottom of the active part of the pile, that is at a depth that corresponds to $z/d = 1.5\sqrt{E_p/G_L}$, rather than at $z = L$.

The solution developed above is based on a linear variation of soil modulus with depth, although it can be extended to layered profiles. An alternative solution for soil modulus distributions expressed as $G = mz^n$, where $0 \leq n \leq 1$, has been presented by Guo and Randolph (1997).

In deriving the solution in equation (4.45), various simplifying assumptions had to be made and more rigorous evaluation of the pile stiffness requires numerical analysis such as described by Poulos and Davis (1980). Poulos and Davis write the load settlement ratio of the pile in terms of a coefficient, I, where

$$I = \frac{w_t dE_L}{P_t} \qquad (4.48)$$

and E_L is the Young's modulus of the soil at pile tip level. The coefficient I is obtained from the product of a number of other coefficients that reflect features such as the pile-soil stiffness ratio, the value of Poisson's ratio for the soil and so forth. Thus

$$I = I_0 R_k R_h R_v \qquad (4.49)$$

where I_0 is the value of I for a rigid pile in an infinitely deep homogeneous
deposit of incompressible soil ($v = 0.5$).
R_k allows for the compressibility of the pile.
R_h allows for the finite depth of a layer of soil.
R_v allows for values of Poisson's ratio less than 0.5.

Comparison between the two approaches is generally reasonably good, considering the simplifying assumptions adopted in the solution developed here. For long compressible piles, the results from Poulos and Davis give higher values of pile stiffness than obtained using equation (4.45), but this may be partly due relatively coarse discretization of the very long piles, leading to numerical inaccuracies in the boundary element solutions.

Two examples are included here to illustrate the approach described above.

Example 1

Estimation of the working settlement of bored pile, 460 mm in diameter, founded at a depth of 16 m in stiff clay; the upper 2 m of soil is made ground, with no shear transfer assumed. The strength of the clay increases uniformly from 70 kPa just below the made ground up to 200 kPa at the pile base, with an average value of 135 kPa. The design load for the pile is 800 kN. Taking a shear modulus for the soil of $G = 150c_u$ (see section 5.2.4), with $v = 0.2$, and assuming the Young's modulus for the pile is $E_p = 25\,000$ MPa, we obtain

$$G_L = 30 \text{ MPa}; \quad \rho = 0.675; \quad \lambda = 833$$

For the lower 14 m of pile, $L/d = 30.5$, thus

$$\zeta = \ln(2r_m/d) = 4.4; \quad \mu L = 2\sqrt{\frac{2}{\zeta\lambda}\frac{L}{d}} = 1.42; \quad \frac{\tanh(\mu L)}{\mu L}\frac{L}{d} = 19.1$$

Thus

$$\frac{P_t}{w_t} = dG_L \times \frac{\dfrac{2}{0.8} + \dfrac{2\pi \times 0.675}{4.4} \times 19.1}{1 + \dfrac{1}{\pi\lambda}\dfrac{8}{0.8} \times 19.1} = 269 \text{ kN/mm}$$

The deflection at depth of 2 m, for the working load of 800 kN, is thus $800/269 = 3.0$ mm. In addition, 0.4 mm of compression will occur in the upper 2 m of the pile, giving an overall working settlement of 3.4 mm.

Example 2

Estimation of the working settlement of a driven cast-in-situ pile, 510 mm in diameter and driven through low-grade chalk to bear on grade I/II chalk at a depth of 12 m.

Back analyses of earlier pile tests have indicated that the shear modulus for the upper chalk may be taken as increasing at a rate of 1 MPa for every metre of depth, taking

$v = 0.24$. The shear modulus of the bearing chalk, allowing for disturbance due to pile installation, may be taken as $G = 36$ MPa (Lord, 1976). Young's modulus of the piles is $E_p = 28\,000$ MPa.

The various parameters may be calculated as

$$G_L = 12 \text{ MPa}; \quad \rho = 0.5; \quad \xi = G_L/G_b = 0.33; \quad \lambda = E_p/G_L = 2333$$

$$L/d = 23.5; \quad \zeta = 3.1 \text{(from equation (4.40))}; \quad \mu L = 0.78; \quad \frac{\tanh(\mu L)}{\mu L}\frac{L}{d} = 19.6$$

Thus

$$\frac{P_t}{w_t} = dG_L \times \frac{\dfrac{2}{0.76 \times 0.33} + \dfrac{2\pi \times 0.5}{3.1} \times 19.6}{1 + \dfrac{1}{\pi\lambda}\dfrac{8}{0.76 \times 0.33} \times 19.6} = 157 \text{ kN/mm}$$

For a working load of 1000 kN, the settlement will therefore be $1000/157 = 6.4$ mm.

4.2.3 Layered soil profiles

The solution described above may be extended or adapted to different conditions, such as to deal with step-tapered piles or piles that penetrate more than one soil type (with widely differing soil stiffnesses). Essentially each section of the pile shaft may be treated independently, retaining compatibility of displacement of the pile shaft between each section (see Figure 4.29). Effectively, the load settlement ratio for the pile base

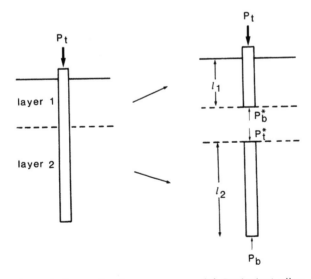

(a) Pile in layered soil   (b) Equivalent piles

Figure 4.29 Modelling of pile in layered soil.

(the term $2/(1-v)$ in equation (4.45) is replaced by the load settlement ratio for the section of pile below that which is currently being considered.

An example calculation will illustrate this. Consider a straight-shafted bored pile, 0.6 m in diameter and 20 m long, with Young's modulus of 25×10^3 MPa, which penetrates

(i)   20 m of stiff clay, with shear modulus reasonably constant at 40 MPa;
(ii)   the same stiff clay, but now overlaid by 8 m of a softer deposit with G varying from 5 MPa at the surface up to 15 MPa at 8 m depth.

In the first case, straightforward substitution of the pile and soil parameters into equation (4.45) results in $\eta = \xi = \rho = 1$, $E_p/G_L = 625$, $\zeta = 4.76$ (taking $v = 0.3$). Thus the overall load settlement ratio is $P_t/(w_t dG_L) = 24.2$, or $P_t/w_t = 581$ kN/mm. In the second case, the lower 12 m of pile maybe treated in the normal way to give $P_t/w_t = 540$ kN/mm (surprisingly high considering how much shorter this section of pile is than before). Now, the top 8 m of pile may be analyzed, substituting the appropriate value of the load settlement ratio at the lower end instead of the terms $2/(1-v)$. In this example

$$P_b/w_t dG_L = 540/(0.6 \times 15) = 60$$

and, taking $\rho = 0.67$, $\zeta = 3.44$, the calculated stiffness is $P_t/w_t = 431$ kN/mm.

It is interesting to note that, in the second case, the average shear modulus over the full depth of penetration of the pile is 28 MPa. If the pile were to be analyzed in a single step, then the parameter, ρ, would be $28/40 = 0.7$ and the stiffness from equation (4.45) for the full 20 m pile is calculated as $P_t/w_t = 444$ kN/mm, only 3% higher than that given in the more precise calculation.

An alternative and elegant expression for the pile stiffness in homogeneous soil has been given by Mylonakis and Gazetas (1998), who exploit the elastic load transfer approach for vertically loaded piles to evaluate interaction effects (see Chapter 5). The expression for the single pile stiffness is given here for completeness. Adopting a load transfer (or Winkler) stiffness, k, as defined previously in equation (4.35), the pile head stiffness is expressed as

$$K = (EA)_p \mu \frac{\Omega + \tanh(\mu L)}{1 + \Omega \tanh(\mu L)} \tag{4.50}$$

where

$$\Omega = \frac{P_b}{w_b (EA)_p \mu} \quad \text{and} \quad \mu L = \sqrt{\frac{k}{(EA)_p}} L \tag{4.51}$$

(noting this definition of μ is actually identical to that used previously). As in the above worked example, this solution may be used for segmented piles, or piles in multiple soil layers, by taking P_b/w_b as the 'base' stiffness of the part of the pile currently under consideration, although that stiffness may be derived from consideration of deeper sections of the pile.

4.2.4 Behaviour of long slender piles

It was shown earlier that as the length of a pile is increased, the stiffness of the pile response approaches a limiting value. If piles are being used to reduce settlements, it would clearly be illogical to design piles of such a geometry that this limit was reached or even approached. However, there are many cases where piles are used mainly to provide sufficient bearing capacity, for example, to support storage tanks on soft ground, or for offshore oil production platforms. In these situations, it may prove economically advantageous to use very long slender piles, and, in many cases, compression of the pile may be sufficient to cause relative slip between pile and soil in the upper part of the pile, under working conditions.

The load level at which slip starts to occur between pile and soil close to the ground surface may be estimated by combining equations (4.34) and (4.47). Thus, the local movement of the pile to mobilize full shaft friction is obtained by substituting τ_s for τ_0 in equation (4.34) to give

$$(w_s)_{\text{slip}} = \zeta d \tau_s / 2G \tag{4.52}$$

The ratio G/τ_s varies from between 100 to 400 for piles in clay, to over 1000 for piles in sand. Thus the movement for slip to occur will be 0.5 to 2% of the pile diameter in clay, and as low as 0.2% of the diameter in sand. Substitution of equation (4.52) into equation (4.47) leads to an expression for the pile load, P_{slip} at which slip starts to occur at ground level (where the pile movement is greatest), given by

$$P_{\text{slip}}/Q_s = 1/\mu L = \sqrt{\frac{\zeta \lambda}{2}} \frac{d}{2L} \tag{4.53}$$

where Q_s is the ultimate shaft capacity of the pile, which equals $\pi d L \bar{\tau}_s$. It is not uncommon for this ratio to be as low as 0.2, particularly for piles used offshore, in which case considerable slip will occur at working loads, where P/Q_s may be of the order of 0.5 to 0.6.

It is clear that an elastic solution for the pile load settlement ratio becomes inappropriate at load levels greater than that given by the expression above. Estimation of the pile settlement must then allow for the region of slip between pile and soil. This may be accomplished in the same manner as for a pile in a layered soil profile, essentially treating the upper part of the pile (where full shaft friction has been mobilized) independently from the lower part. Some iteration will be required, since, for a given applied load at the pile head, the transition point between the two regions must be estimated initially. This procedure will be illustrated by an example, taken from pile load tests reported by Thorburn *et al.* (1983).

The two piles in question were precast concrete, 250 mm square, driven into silty clay to penetrations of 29 m and 32 m. The shear strength varied with depth according to $c_u = 6 + 1.8z$ kPa (where z is the depth in m). Taking $\alpha = 0.9$ for the clay, which was lightly overconsolidated, the shaft capacities of the two piles may be estimated as 0.84 MN and 1.00 MN respectively. In calculating the load settlement ratio for the piles, a cylindrical pile of radius 141 mm (giving the same cross-sectional area) will be taken. The short-term Young's modulus of the piles was $E_p = 26\,500$ MPa.

The response of the piles to constant rate of penetration tests may be matched by adopting a shear modulus of $G = 250c_u = 1.5 + 0.45z$ MPa, taking Poisson's ratio equal to 0.5. Equation (4.47) then gives the pile stiffness as $P_t/w_t = 90$ kN/mm. The local movement to mobilize full shaft friction may be estimated from equation (4.52) as 2.5 mm. Thus, slip between pile and soil will start at the top of the pile when the applied load reaches $90 \times 2.5 = 225$ kN. Figure 4.30 gives the measured pile responses, showing both piles having the same initial stiffness, with an essentially linear response up to loads of about 500 kN.

The complete load settlement curve may be estimated by calculating the load and corresponding settlement for a given depth of slip between pile and soil. Thus, consider the stage at which full shaft friction has been mobilized over the top 10 m of the pile. The load shed over the top 10 m may be calculated from the given profile of shear strength, as 150 kN. For the bottom part of the pile, the load settlement ratio may be calculated from equation (4.45) as $P/w = 112$ kN/mm. Thus the load level at the transition point (where the settlement is 2.5 mm) is 280 kN. The overall load at the pile head is then $280 + 150 = 430$ kN. The pile compression over the top 10 m of the pile may now be calculated as 2.3 mm, giving an overall settlement at the pile head of 4.8 mm.

The above process may be repeated for different depths of slip between pile and soil to give the two estimated load settlement curves shown in Figure 4.30. In practice, adequate estimates of these curves may be obtained merely from the initial slopes of the load settlement curves, and estimates of the pile settlements when full shaft capacity has been mobilized. Thus for the shorter pile, the compression of the pile for the shaft

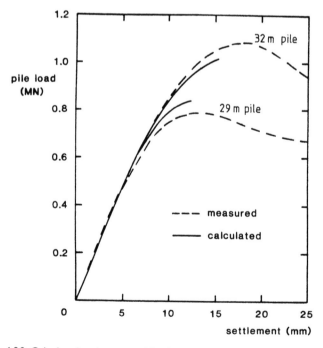

Figure 4.30 Calculated and measured load settlement curves for two long piles.

capacity of 838 kN is 9.3 mm. Together with 2.5 mm of settlement at the bottom of the pile (in order to mobilize full shaft friction there), the pile head movement is 11.8 mm. This corresponds reasonably well with the measured movement to peak load.

It should be noted that, for simplification, no allowance has been made in the above calculations for the small amount of load taken by the pile base. In addition, the measured pile responses show clear evidence of strain softening, which will affect the distribution of shaft friction down the pile at the point of failure of the pile (see following discussion). The working loads for these piles were 440 kN and 550 kN, respectively, for the short and long piles. Maintained load tests at these load levels showed some creep movement, with final settlements of about 8 mm, of which probably about half reflected slip between the upper part of the pile and the soil.

As may be seen from the pile tests just discussed, large relative movement between the pile shaft and the soil may lead to a reduction in shear transfer from a peak value of shaft friction to a residual value. Such strain softening may be significant (a) in loose compressible sands (particularly calcareous sediments) and (b) in soft, lightly overconsolidated clays. Model pile tests in soft clay reported by Francescon (1983) and by Chandler and Martins (1982) indicate that the residual value, $(\tau_s)_{res}$, may be estimated from

$$(\tau_s)_{res} = K_0 \sigma'_v \tan \delta_{res} \tag{4.54}$$

where δ_{res} is the residual interface friction angle. The model pile tests indicated that shear transfer from the pile shaft may reduce by a factor of two or more after as little as 30 mm of movement of the pile, particularly for cyclic relative displacements. Similar findings have been reported from ring shear tests (Bishop $et\ al.$, 1971; Lupini $et\ al.$, 1981; Jardine $et\ al.$, 2005). The consequences of this in the design of long slender piles to anchor offshore platforms have been discussed by Randolph (1983).

The strain softening behaviour can also have consequences for the ultimate capacity of long piles, where a form of progressive failure becomes possible. At failure, the displacements at the head of a long pile may have been sufficient to reduce the shaft friction to a residual value (see Figure 4.31). Thus the measured capacity will be lower than would be calculated from the profile of peak values of shaft friction. In order to quantify this effect, a flexibility ratio, K, may be introduced (Randolph, 1983) given by

$$K = \frac{\pi d \tau_s L^2}{(EA)_p \Delta w_{res}} \tag{4.55}$$

where Δw_{res} is the relative movement needed between pile and soil for the value of shaft friction to reduce from peak to residual. The flexibility ratio is similar to the parameter π_3, introduced by Murff (1980), but based on Δw_{res} rather than the movement w_{slip} to mobilize peak shaft friction initially. The flexibility ratio, K, may be thought of as the ratio of elastic shortening of the pile (treated as a column under the full shaft capacity) to the relative movement Δw_{res} to go from peak to residual conditions.

Figure 4.32 shows values of reduction factor, R_f, which should be applied to pile shaft capacities based on peak values of shaft friction, in order to allow for progressive failure of the pile. The different curves reflect different ratios of residual to peak

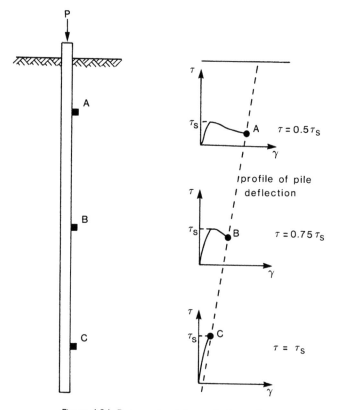

Figure 4.31 Progressive failure of a long pile.

shaft friction. In general, land-based piles will have values of K which are greater than unity, and the effect of strain softening will be small. However, piles used offshore frequently have values of K in the range 2 to 4, and the effect may be significant. For a particular situation, estimation of the reduction factor, R_f, will require an estimate of the ratio of $(\tau_s)_{res}$ to $(\tau_s)_{peak}$, and also of the movement Δw_{res} to cause the degradation. Ring shear, or direct shear box tests on the soil in question, will enable the value of δ_{res} to be assessed, while Δw_{res} may be taken, conservatively, as 30 mm. This value is likely to be largely independent of pile size, since the degradation is concentrated in a thin band close to the pile shaft, with re-orientation of soil particles responsible for most of the decrease in the angle of friction.

4.2.5 Rock sockets

The axial deformation of rock sockets deserves special consideration. Although rock sockets are generally relatively short, with typical length-to-diameter ratios of less than 10, the high stiffness of the host rock compared with the pile stiffness results in the rock socket behaving in a compressible fashion. As an example, consider a rock socket of length 10 m and diameter 1.4 m, embedded in soft rock with a shear modulus of

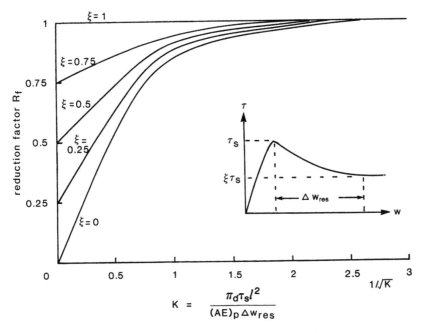

Figure 4.32 Variation of reduction factor with pile stiffness ratio.

$$K = \frac{\pi_d \tau_s l^2}{(AE)_p \Delta w_{res}}$$

700 MPa. Assuming a Young's modulus for the concrete of 25 000 MPa, the quantity that controls the compressibility, μL, may be calculated as 1.9 (see equation (4.45)). The initial stiffness may be estimated approximately from equation (4.45) as $P/w = 7.3$ MN/mm.

The relatively compressible nature of rock sockets, together with high base capacity, results in partial (or complete) slip along the pile shaft generally occurring under working conditions. Analytical and numerical solutions for estimating settlement and load transfer in rock sockets under conditions of partial slip have been presented by Rowe and Armitage (1987) and Carter and Kulhawy (1988). As slip occurs, the shaft friction will often decrease, as the cohesive or dilational component of the shaft friction will reduce to zero (Johnston and Lam, 1989). Since the shaft capacity is mobilized at much smaller displacements than the base capacity, the overall load-displacement response of the rock socket may then show a decrease in capacity at intermediate displacements (due to strain softening of the shaft) prior to increasing again at larger displacements as the base capacity is mobilized. Figure 4.33 shows schematically the different types of rock socket response, depending on the rapidity of strain softening and the rate of mobilization of the base capacity.

4.3 Lateral capacity of piles

Almost all piled foundations are subjected to at least some degree of horizontal loading. In many cases, the magnitude of the loads in relation to the applied vertical loading is small, and no additional design calculations are necessary. For example, the piled

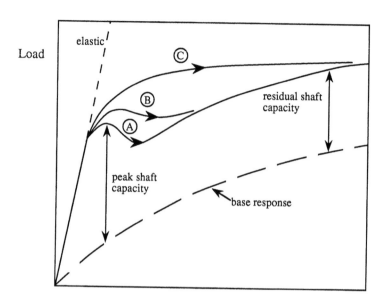

Figure 4.33 Different types of rock socket responses.

foundation of a building of moderate height will easily withstand the small wind shear loading that the building will be subjected to. In other cases, such as a piled earth retaining wall or a jetty, horizontal loading may prove critical in the design. Traditionally, piles have been installed at an angle to the vertical in such cases, providing sufficient horizontal resistance by virtue of the component of the axial capacity of the pile that acts horizontally. However, it is very conservative to ignore the capacity of a pile to withstand lateral loading (i.e. loading applied normal to the pile axis). This may typically be an order of magnitude less than the axial capacity of a pile, but may well be sufficient to obviate the need for so-called 'raking' or 'battered' piles, which are more expensive to install.

Lateral loading of a pile may be divided broadly into the two categories of 'active' loading, where external loads are applied to the pile, with the soil resisting the load, and 'passive' loading, where movement of the soil subjects the pile to bending stresses. This, and the subsequent section, are concerned primarily with the first category of loading, although some of the features discussed (for example, the limiting pressure between pile and soil) are applicable to both categories. Passive loading of piles by consolidating soil is discussed further in section 4.5.

As a pile is loaded laterally, normal stresses will increase in front of, and decrease behind the pile. Displacements in the soil will tend to be radially away from the pile in front of the pile, and radially towards the pile behind it. At some stage, near the ground surface, a gap will probably open up between the back of the pile and the soil, with soil in front of the pile failing in a wedge type of mechanism, as shown in Figure 4.34. Further down the pile shaft, soil will eventually fail by flowing around the pile, with no gap present. In calculating the distribution of limiting pressure which

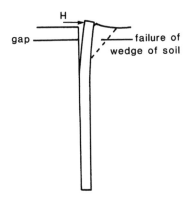

Figure 4.34 Deformation of pile under lateral load.

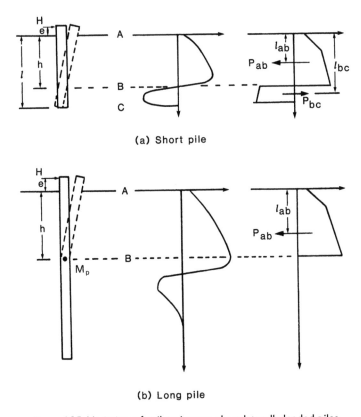

(a) Short pile

(b) Long pile

Figure 4.35 Variation of soil resistance along laterally loaded piles.

may be mobilized by the pile, these different types of failure mechanism in the soil must be taken into account.

Collapse of the pile itself may occur in one of two modes. For short piles, or piles with a large plastic moment, the pile will rotate essentially as a rigid body. The resulting profile of resistance offered by the soil will be as shown schematically in Figure 4.35(a).

Above the centre of rotation, passive pressures will develop in front of the pile, while below the centre of rotation, passive pressures will develop behind the pile. For longer piles, a plastic hinge will develop at some depth down the pile, and only the upper part of the pile will undergo significant displacement (Figure 4.35(b)). Although the soil will be subjected to loading below the plastic hinge, calculation of the failure load will only require knowledge of the limiting pressures acting over the upper part of the pile, above the plastic hinge.

Where a pile is embedded in a pile cap which is restrained from rotation, three different collapse modes are possible, as shown in Figure 4.36. Short piles will translate as a rigid body with the pile cap, while progressively longer piles will first form a plastic hinge at the level of the pile cap, and then a plastic hinge at some distance down the pile. In practice, the majority of piles encountered behave as long piles as regards lateral capacity. Failure will thus be by the mode shown in Figure 4.36(c).

In calculating failure loads, it is customary to treat the soil as a rigid plastic material, idealizing the profiles of limiting pressure as shown in Figure 4.35, with sharp transitions from full limiting pressure acting in front of the pile just above the hinge point, down to zero at the hinge point and, for short piles, full limiting pressure acting behind the pile just below the hinge point. The relevant equivalent forces, P, and their appropriate lever arms, l, are shown in Figure 4.35 for an unrestrained pile. For short piles, the moment at point B is $P_{bc}(l_{bc} - h)$, where h is the depth of the hinge point, and this must be less than the plastic moment for the pile section. Horizontal equilibrium then gives

$$H_f = P_{ab} - P_{bc} \tag{4.56}$$

while moment equilibrium entails that

$$H_f(e + h) = P_{ab}(h - l_{ab}) + P_{bc}(l_{bc} - h) \tag{4.57}$$

For a given profile of limiting pressure acting on the pile, P_{ab} and P_{bc} may be evaluated and these two equations solved to find H_f.

If the pile fails by forming a plastic hinge, this will occur at the point of maximum bending moment, and thus of zero shear force. In Figure 4.35(b), the moment at B

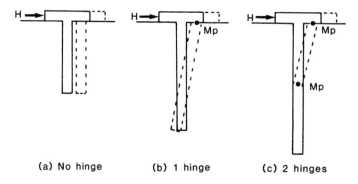

(a) No hinge   (b) 1 hinge   (c) 2 hinges

Figure 4.36 Failure modes for laterally loaded piles.

may be set equal to the plastic moment, M_p, for the pile section, and the pressures acting on the pile below point B may be ignored. For zero shear force at the hinge point, H_f must equal P_{ab}. Taking moments about B then gives

$$H_f[e + h - (h - l_{ab}] = H_f(e + l_{ab}) = M_p \tag{4.58}$$

For the case of a pile where the head is restrained from rotating, the above equations may be used where one or two plastic hinges form, merely adding an extra term M_p to the right-hand sides of equations (4.57) and (4.58) to allow for the hinge at the top of the pile. For short piles, where the whole pile translates laterally (Figure 4.36(a)), the failure load may be calculated in a straightforward manner, by integrating the limiting pressure over the length of the pile. It should be noted that the bending moments at the top of piles built in to a pile cap must be taken into account in the design of the pile cap itself.

The following sections discuss values of limiting pressure appropriate for different soil types.

4.3.1 Limiting lateral pressures in non-cohesive soils

At shallow depths (less than one diameter) a circular or square pile will appear to soil elements in front of the pile as a long retaining wall. At failure, a wedge of soil will be pushed ahead of the pile, and the limiting pressure acting on the pile (for non-cohesive soil) will be close to K_p times the local vertical effective stress, where K_p is the passive earth pressure coefficient, equal to $(1 + \sin \phi')/(1 - \sin \phi')$. At greater depths, however, much larger limiting pressures may develop, owing to the three-dimensional nature of the problem. In front of the pile, the soil will be deformed much as in a pressuremeter test, while behind the pile, the contact stresses will reduce until, in the limit, the pile breaks away from the soil. From the results of pressuremeter tests in sand (Hughes *et al.*, 1977; Fahey and Randolph, 1984), pressures greater than 10 times the effective overburden stress are typically required to increase the radius of the pressuremeter by 10%. No limiting pressure is normally reached, but rather the pressure increases proportionally with the cavity strain to some exponent. The value of the exponent depends on the friction angle and rate of dilation of the soil (Hughes *et al.*, 1977).

In practice, of course, the pile will fail at some stage, normally by the formation of a plastic hinge at some point down the pile. Lateral movement of the pile to cause such failure will generally be in excess of 10% of the pile diameter, at which stage the pressuremeter analogy may no longer be accurate.

Although no analytical solution has been proposed, Broms (1964) suggested a net limiting force per unit length of pile, P_u given by

$$P_u = 3K_p \sigma_v' d \tag{4.59}$$

where d is the diameter or width of the pile. This expression furnishes average limiting pressures which will be slightly greater than the figure of $10\sigma_v'$ mentioned above, for practical values of ϕ'. However, comparisons with field test results show a tendency for the measured capacities to be underestimated by about 30% using this expression (Poulos and Davis, 1980). Reese *et al.* (1974) postulated a wedge failure mechanism in front of the pile, which leads to a variation of P_u, with depth that is initially proportional to K_p, but at greater depths becomes proportional to K_p^3. This relationship has been incorporated in the API (1993) guidelines for offshore pile design.

Brinch Hansen (1961) and Meyerhof (1995) also developed charts describing quite complex variations of the limiting lateral resistance with depth, while Barton (1982) found that her data from lateral pile tests could be matched sufficiently by the simple variation given by

$$P_u = K_p^2 \sigma_v' d \tag{4.60}$$

For almost all naturally occurring sand, K_p will be greater than 3, so that this expression will give a better fit to results of load tests than did equation (4.59). For the dense sand used by Barton (1982) the friction angle was $\phi' = 43°$, giving $K_p = 5.3$, and thus lateral resistances some 75% greater than suggested by Broms (1964). These various relationships are compared in Figure 4.37.

Equations (4.59) and (4.60) are particular cases of the more general case where the soil resistance is assumed to vary proportionally with depth according to $P_u = ndz$, where n is the gradient of the average ultimate pressure across the width of a pile of diameter, d. Thus for these two expressions, n would equal $3K_p\gamma'$ or $K_p^2\gamma'$ respectively. The general relationship may be used to derive charts giving the lateral capacity of short and long piles embedded in uniform soil with a given resistance gradient, as

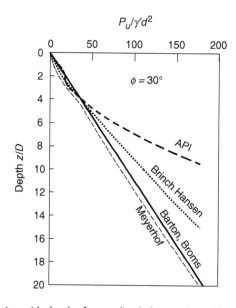

Figure 4.37 Variation with depth of normalized ultimate lateral force per unit length.

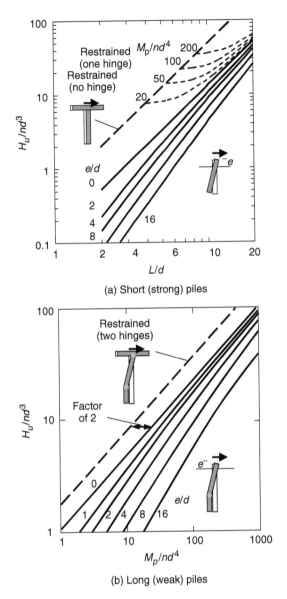

(a) Short (strong) piles

(b) Long (weak) piles

Figure 4.38 Design chart for laterally loaded piles in soil with resistance proportional to depth.

shown in Figure 4.38. For short piles, the curved dashed lines correspond to the failure mechanism in Figure 4.36(b).

4.3.2 Limiting lateral pressures in cohesive soils

For piles in cohesive soil, the limiting pressure exerted on the front of the pile by the wedge of soil near the ground surface may be taken as $2c_u$, where c_u is the undrained shear strength of the soil. Reese (1958) and Matlock (1970) suggested taking a net limiting pressure at shallow depths, given by

$$p_u = \frac{P_u}{d} = 2c_u + \sigma_v' + \alpha c_u \frac{z}{d} \qquad (4.61)$$

where the factor α lies between 0.5 and 3. Randolph and Houlsby (1984) suggest taking $\alpha = 1.5$. (Note that σ_v' should be replaced by σ_v where no free water is available to flow in behind the pile.)

A more general failure mechanism has been explored by Murff and Hamilton (1993), comprising a one-sided or two-sided conical wedge where the radial velocities in the soil vary in proportion to $\cos\theta$ (see Figure 4.39). Apart from very transient loading, the conservative assumption is to assume a one-sided mechanism, with a gap opening behind the pile.

The mechanism may be used as the basis of upper bound estimates of the limiting lateral resistance for a pile. Murff and Hamilton (1993) show that the resulting distribution of limiting resistance may be expressed as

$$p_u = \frac{P_u}{d} = N_p c_u + \sigma_v' \qquad (4.62)$$

where N_p is expressed as a function of depth, allowing for the variation of strength with depth. For the case of soil strength varying linearly with depth z according to $c_u = c_{u0} + kz$, the variation of N_p is

$$N_p = N_{pl} - (N_{pl} - 2)\,e^{-\xi z/d} \qquad (4.63)$$

where

$$\xi = 0.25 + 0.05\frac{c_{u0}}{kd} \leq 0.55 \qquad (4.64)$$

The quantity N_{pl} is the limiting value of N_p at depth, and can be taken either as 9, following Broms (1964), or using the solutions described below.

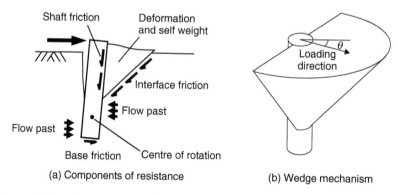

Figure 4.39 Collapse mechanism for upper bound calculation. (After Stewart, 1999.)

4.3.2.1 Plasticity solution

Randolph and Houlsby (1984) describe an analytical solution for the flow of soil around a cylindrical object. The solution was originally considered exact, within the confines of the rigid plastic model of the soil, although subsequently it was discovered that the best upper bound lies slightly above the lower bound. The solution allows for limiting friction, f_s, at the pile–soil interface which is less than or equal to the shear strength of the soil. Defining $\Delta = \sin^{-1}(f_s/c_u)$, the lower bound for the limiting bearing factor, N_{pl}, is given by

$$N_{pl} = \frac{P_u}{c_u d} = \pi + 2\Delta + 4\cos\left(\frac{\pi - \Delta}{4}\right)\left(\sqrt{2} + \sin\left(\frac{\pi - \Delta}{4}\right)\right) \tag{4.65}$$

The variation of the limiting pressure with friction ratio f_s/c_u is shown in Figure 4.40. The value of N_{pl} varies from 9.14 for a perfectly smooth pile, up to 11.94 for a perfectly rough pile, with a mean value of about 10.5.

Figure 4.41 shows the failure mechanism for the case of $f_s/c_u = 0.5$.

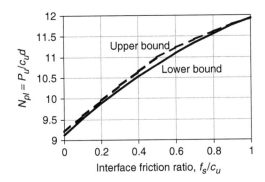

Figure 4.40 Variation of limiting lateral resistance with interface friction ratio (Randolph and Houlsby, 1984).

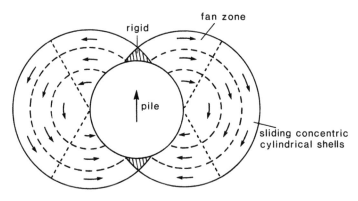

Figure 4.41 Flow mechanism for soil around laterally loaded pile.

4.3.2.2 Pressuremeter analogy

Solutions based on rigid-plastic response of the soil may overestimate the true capacity where the plastic flow is confined. An alternative approach uses the analogy of the deformation field around a pressuremeter, such as noted by Swain (1976) from tests on deeply-embedded strip anchors. The geometry of the strip anchor is similar to that of a pile loaded laterally, and the experimental results indicated a pattern of outward radial movement of the soil in front of the strip anchor, and inward radial movement behind the anchor.

For this pattern of deformation, it is reasonable to assume that the pressure exerted by soil in front of a laterally-loaded pile will approach the limiting pressure measured in a pressuremeter test. Behind the pile, the lowest normal stress attainable will be a suction of the order of 100 kPa. If a gap forms, then this stress will increase to zero (in a dry hole) or to the ambient head of water if free water is available (such as for offshore piles). In addition, at the sides of the pile, some friction may be present between the pile and the soil. Typically, this friction may be estimated as $c_u d$, per unit length of pile.

The limit pressure, p_l, obtained from a pressuremeter test may be written

$$p_l = \sigma_h + c_u \left[\ln(G/c_u) + 1 \right] \tag{4.66}$$

where σ_h is the (total) in-situ horizontal stress in the soil, and G is the shear modulus of the soil. In practice, for typical values of G/c_u, this expression may be approximated (Marsland and Randolph, 1977) by $p_l \sim \sigma_h + 6c_u$. Thus the net force per unit length acting on the pile lies between the limits

$$(\sigma_h' + 7c_u)d \leq P_u \leq (\sigma_h + p_a + 7c_u)d \tag{4.67}$$

where the lower limit corresponds to the case of the ambient head of water acting behind the pile, and the upper limit is where a suction develops behind the pile (p_a denoting atmospheric pressure).

For normally or lightly overconsolidated clay, σ_h'/c_u will be about 2, and the lower limit in equation (4.67) corresponds to the result for a smooth pile obtained from plasticity theory. For short-term loading, some suction will probably develop behind the pile, and the plasticity solution will give the lowest estimate of limiting force per unit length acting on the pile. For stiff, overconsolidated clay, the ratio σ_h'/c_u will be considerably lower and may be as low as 0.5 at shallow depths. Even if suctions develop behind the pile, the limiting force per unit length may fall below $9c_u d$.

4.3.2.3 General profile of limiting pressure

The discussion provides a theoretical background for choosing particular profiles of limiting pressure for laterally loaded piles. In many cases, however, the suggested values for P_u lead to a profile which increases from $2c_u d$ at the ground surface up to a value of about $9c_u d$ at a depth of 3 pile diameters, much as was originally suggested by Broms (1964). (Broms conservatively took P_u equal to zero at the ground surface, to allow for any gap which might be present around the head

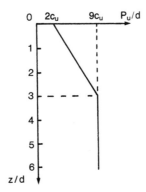

Figure 4.42 Idealized variation of limiting resistance with depth.

of the pile.) This profile of P_u is shown in Figure 4.42. The limit of $9c_u d$ at depth is likely to be conservative, except for the case of piles in stiff overconsoldiated clay, particularly if there is free water readily available to flow into any gap behind the pile.

Adopting the profile of limiting force per unit length shown in Figure 4.42, charts may be prepared giving the lateral capacity of a pile in terms of its geometry and limiting plastic moment. The charts, shown in Figure 4.43, are similar to those presented by Broms (1964), giving marginally higher failure loads as a result of taking P_u equal to $2c_u d$ at the ground surface. Stewart (1999) has shown that the resulting pile capacities are also very close to those obtained using the approach of Murff and Hamilton (1993), with the conservative assumption of weightless soil.

For other calculated variations of P_u with depth, the expressions derived in section 4.3 may be used to produce similar sets of charts to those shown in Figure 4.43. For layered profiles, the procedure may involve some iteration choosing a value for the depth of plastic hinge, following the principles described in section 4.3 to calculate a value of H_f, and then checking that the position of the hinge corresponds to the point of zero shear force in the pile.

4.4 Deformation of single piles under lateral loading

The deformation of a pile under lateral load is generally confined to the upper part of the pile, seldom extending beyond about 10 pile diameters below the ground surface. This has led to a common – though imprecise – idealization of laterally loaded piles in terms of equivalent cantilevers. The pile is replaced by a cantilever, fixed at some depth determined by folklore and ignoring the soil support above that depth. The limitations of this idealization are discussed further in section 4.4.3. More accurate modelling of the response of a pile to lateral load is obtained by allowing for the soil support over the upper part of the pile. Two alternative approaches are possible, one where the soil is represented by discrete springs down the length of the pile, and one where a more accurate, continuum model of the soil is adopted. The two approaches are discussed in the following two sections.

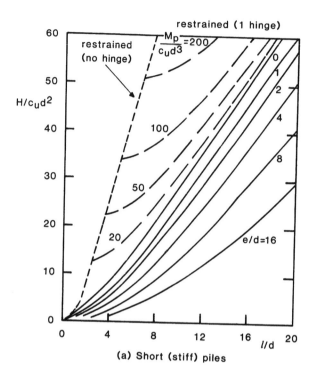

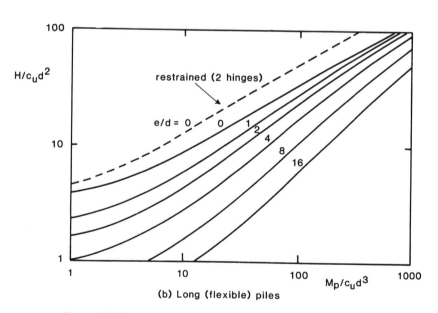

Figure 4.43 Design chart for laterally loaded piles in cohesive soil.

4.4.1 Subgrade reaction approach

Early analysis of the response of piles to lateral loading treated the soil as a series of springs down the length of the pile – the so-called Winkler idealization of soil (see Figure 4.44). The spring stiffness, k, giving the load per unit length of pile induced for unit lateral deflection of the pile, is generally referred to as the coefficient of subgrade reaction. If k is assumed constant down the length of the pile, then analytical solutions are possible, giving the deflected shape of the pile, and the shear force and bending moment distribution down the pile (Matlock and Reese, 1960). For a pile of a given bending rigidity, EI_p, in soil with a coefficient of subgrade reaction, k, there is a critical length beyond which the pile behaves as if it were infinitely long.

This critical length is given by

$$L_c = 4\left[(EI)p/k\right]^{1/4} \tag{4.68}$$

Thus, as for the case of long slender piles subjected to axial loading, the effects of load applied at the top of the pile die out at some depth down the pile. In fact the large majority of piles encountered in practice behave as 'flexible' piles (i.e. longer than their critical length). For such piles, the deflection, u, and rotation, θ, at ground level due to applied load, H, and moment, M, are given by

$$u = \sqrt{2}\frac{H}{k}\left(\frac{L_c}{4}\right)^{-1} + \frac{M}{k}\left(\frac{L_c}{4}\right)^{-2}$$
$$\theta = \frac{H}{k}\left(\frac{L_c}{4}\right)^{-2} + \sqrt{2}\frac{M}{k}\left(\frac{L_c}{4}\right)^{-3} \tag{4.69}$$

Similar expressions for the critical length and for the load deformation response may be obtained for cases where the value of the coefficient of subgrade reaction varies with depth, in particular where it is proportional with depth (Reese and Matlock, 1956).

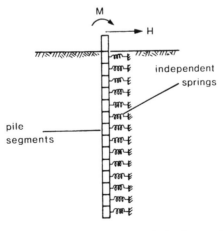

Figure 4.44 Subgrade reaction model of soil around pile.

For the coefficient of subgrade reaction varying proportionally with depth according to $k = nz$, the critical length may be estimated from

$$L_c = 4\left[(EI)_p/n\right]^{1/5}$$ (4.70)

and the corresponding ground level deformations are given by

$$u = 2.43\frac{H}{n}\left(\frac{L_c}{4}\right)^{-2} + 1.62\frac{M}{n}\left(\frac{L_c}{4}\right)^{-3}$$

$$\theta = 1.62\frac{H}{n}\left(\frac{L_c}{4}\right)^{-3} + 1.73\frac{M}{n}\left(\frac{L_c}{4}\right)^{-4}$$ (4.71)

Modern computer techniques have permitted the extension of this approach to include non-linear springs for representing the soil. In place of a coefficient of subgrade reaction, a complete load transfer (or p (load per unit length of pile) – y (deflection)) curve is specified. Finite difference modelling of the bending of the pile then allows a complete load deflection curve to be obtained for the pile (Reese, 1977). Typical forms for the p–y curves have been presented by Reese *et al.* (1974) for piles in sand and by Matlock (1970) and Dunnavant and O'Neill (1989) for piles in clay. In particular situations, notably for offshore pile design, load transfer curve analysis of piles can play an essential role in design, mainly with a view to assessing the effects of cyclic loading. Software such as LPILE (Reese and Wang, 1993) or PYGMY (Stewart, 1999b) incorporate standard algorithms for generating p–y curves and can be used for non-linear analysis of single piles. However, for onshore applications simple linear analysis is often sufficient, particularly where piles are part of a group and interaction effects need to be allowed for.

4.4.2 Elastic continuum approach

The main limitation in using subgrade reaction theory for lateral pile analysis lies in assessing an appropriate value of subgrade reaction coefficient, k, for the soil. Comparisons of the subgrade reaction approach with rigorous elastic analysis have shown that the correct choice for k depends not only on the soil properties, but also on the pile stiffness and the form of loading (that is the eccentricity of the applied lateral load). In addition, there is no rational way in which interaction effects can be quantified when a group of piles is loaded laterally.

In order to overcome these limitations, solutions based on finite element and boundary element modelling of the pile and soil continuum have been presented (Poulos, 1971; Kuhlemeyer, 1979; Randolph, 1981). These solutions allow expressions similar in form to those in equations (4.68) and (4.69), to be written for the critical pile length and the ground level deformations. Dealing initially with piles that are longer than their critical length, it is first necessary to consider the average soil stiffness over the active length of the pile. In order to avoid needing different solutions for different values of Poisson's ratio, Randolph (1981) introduced a modulus G^* defined by

$$G^* = G(1 + 3v/4)$$ (4.72)

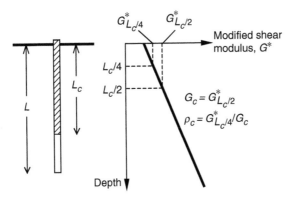

Figure 4.45 Definition of ρ_c and G_c.

Referring to Figure 4.45, a characteristic modulus, G_c, is defined as the average value of G^* over the active length of the pile. In addition a parameter, ρ_c, is introduced reflecting the degree of homogeneity in the soil stiffness. For the linear variation of modulus shown in Figure 4.45, ρ_c may conveniently be defined by

$$\rho_c = \frac{G^*_{L_c/4}}{G^*_{L_c/2}} = \frac{G^*_{L_c/4}}{G_c} \tag{4.73}$$

It is helpful to cast the solution in terms of an equivalent solid pile of the same cross-sectional area and the same bending rigidity as the real pile. For an equivalent pile of diameter d, the appropriate Young's modulus may be calculated as

$$E_p = \frac{(EI)_p}{\pi d^4/64} \tag{4.74}$$

where $(EI)_p$ is the bending rigidity of the actual pile.

The critical pile length is now defined by

$$L_c = d(E_p/G_c)^{2/7} \tag{4.75}$$

The form of this equation is similar to equation (4.68), with an exponent of 2/7 instead of 1/4. Also, since E_p for an equivalent pile is used instead of the bending rigidity $(EI)_p$, it is possible to think in terms of a critical slenderness ratio, L_c/d, for the pile which is deduced directly from the stiffness ratio, E_p/G_c.

It may seen from Figure 4.45 that the definition of G_c requires the knowledge of the critical length, L_c, which is in turn defined in terms of G_c. Thus some iteration is required except for the extreme cases of homogeneous soil ($\rho_c = 1$) and soil where G is proportional to depth ($\rho_c = 0.5$ and the critical length reduces to $L_c = d(2E_p/m^*d)^{2/9}$, where m^* is the rate of increase of G^* with depth). For example, suppose that the variation of shear modulus with depth, z, is approximated as

$$G = 10 + 4z \text{ MPa } (z \text{ being the depth in m})$$
(4.76)

For a pile of diameter 0.6 m, with equivalent modulus $E_p = 30 \times 10^3$ MPa, a first guess for the critical pile length might be 6 m (that is 10 pile diameters). Taking Poisson's ratio as 0.3, the characteristic modulus, G_c, is calculated as

$$G_c = (10 + 4 \times 6/2)(1 + 3 \times 0.3/4) = 27.0 \text{ MPa}$$
(4.77)

This gives a revised estimate of the critical pile length of

$$L_c = 0.6 \times (3 \times 10^4/27.0)^{2/7} = 4.45 \text{ m.}$$
(4.78)

Further iteration yields final values of $L_c = 4.63$ m, $G_c = 23.6$ MPa.

The concept of a characteristic shear modulus, G_c, and a critical pile length, L_c, may be used to write expressions for the ground level deformation of the pile. Thus the lateral deflection, u, and rotation, θ, are given by

$$u = \frac{\left(E_p/G_c\right)^{1/7}}{\rho_c G_c} \left[0.27 \frac{H}{L_c/2} + 0.30 \frac{M}{\left(L_c/2\right)^2} \right]$$

$$\theta = \frac{\left(E_p/G_c\right)^{1/7}}{\rho_c G_c} \left[0.30 \frac{H}{\left(L_c/2\right)^2} + 0.80\sqrt{\rho_c} \frac{M}{\left(L_c/2\right)^3} \right]$$
(4.79)

These expressions have been derived by synthesizing a number of finite element analyses (Randolph, 1981). The similarity with those obtained from the subgrade reaction approach (equations (4.69)) is evident.

Generalized profiles of deflection and bending moment down the pile may be drawn for piles subjected to shear force loading, H, or moment loading, M. These profiles are shown in Figures 4.46 and 4.47. The maximum moment for a pile under a lateral load of H, occurs at a depth between $L_c/4$ (for homogeneous soil) and $L_c/3$ (for soil with stiffness proportional to depth). The value of the maximum moment may be estimated as

$$M_{\max} = (0.1/\rho_c)HL_c$$
(4.80)

For piles within a group, the pile cap may prevent rotation of the head of the pile. For such 'fixed-headed' piles, equations (4.79) may be used to find the fixing moment, M_f. Setting $\theta = 0$, the fixing moment is given by

$$M_{\max} = -\left[0.1875/(\rho_c)^{1/2} \right] HL_c$$
(4.81)

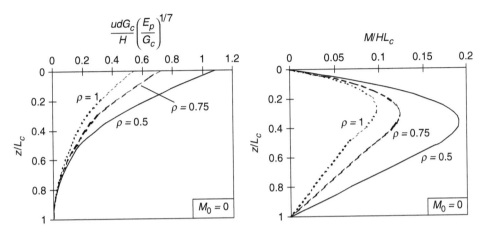

Figure 4.46 Generalized curves of lateral deflection and bending moment profile for force loading.

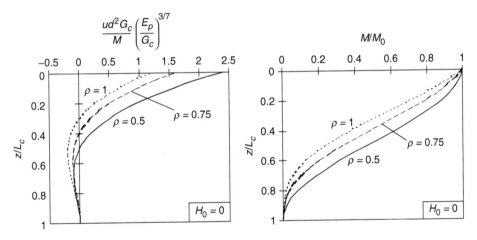

Figure 4.47 Generalized curves of lateral deflection and bending moment profile for moment loading.

The resulting deflection of the pile head may then be calculated as

$$u_f = \frac{\left(E_p/G_c\right)^{1/7}}{\rho_c G_c}\left(0.27 - \frac{0.11}{\sqrt{\rho_c}}\right)\frac{H}{L_c/2} \tag{4.82}$$

This is approximately half what it would have been for a free-headed pile. Figure 4.48 shows the generalized deflected profile and variation of bending moment down the pile for this case.

To illustrate the use of these solutions in design, consider the response of a steel pipe pile, 1.5 m in diameter, with 50 mm wall thickness, embedded in soft, normally consolidated clay with a shear strength that increases at a rate of 2.5 kPa/m. The

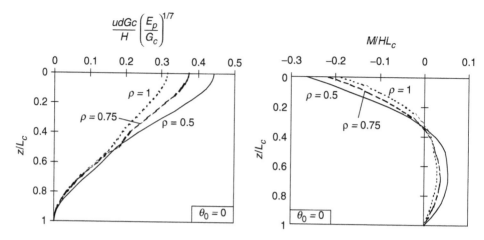

Figure 4.48 Generalized curves of lateral deflection and bending moment profile for force loading of fixed head pile.

equivalent modulus of the pile may be calculated as

$$E_p = E_{steel}\left[1 - \left(\frac{d_i}{d}\right)^4\right] = 50\ 600 \text{ MPa} \tag{4.83}$$

where d_i is the inner diameter of the pile and E_{steel} has been taken as 210 GPa. Taking a shear modulus for the soil of $G = 100c_u = 0.25z$ MPa, and Poisson's ratio as 0.3, the critical pile length may be calculated as

$$L_c = d(2E_p/m^*d)^{2/9} = 23.1 \text{ m} \tag{4.84}$$

where $m^* = 0.25(1 + 0.75 \times 0.3) = 0.306$ MPa/m. The value of G_c is then $0.5 \times 23.1 \times 0.306 = 3.53$ MPa, and $\rho_c = 0.5$.

Under a lateral load of 1 MN, with no rotation allowed at ground level, the maximum bending moment and ground level deflection may be calculated from equations (4.81) and (4.82) respectively as

$$M_f = -6.1 \text{ MNm}; \quad u_f = 21.3 \text{ mm} \tag{4.85}$$

Some comment is appropriate concerning the applicability of solutions for the response of piles to lateral load, based on elastic soil properties. Clearly, the large strains that occur locally around the head of the pile will reduce the relevant secant modulus to a low value. It is probable that idealization of the soil as a material with stiffness proportional to depth is a better idealization than as a homogenous material. Even with this idealization, however, choice of soil modulus will depend on the magnitude of movement expected. Centrifuge model tests on piles embedded in dense

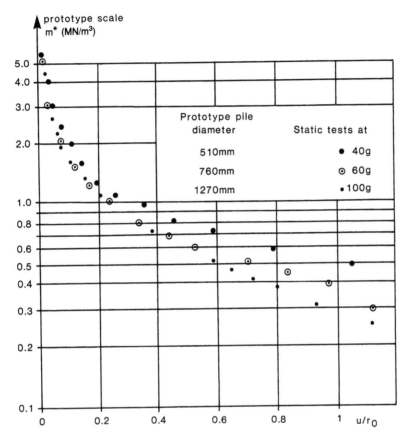

Figure 4.49 Variation of deduced soil modulus with pile deflection (Barton, 1982).

sand, reported by Barton (1982), show that the deduced stiffness of the soil decreases sharply as the head deflection of the pile increases. Taking a shear modulus for the sand which increases proportionally with depth, z, according to $G = mz$, the variation of the deduced value of $m^* = m(1 + 3v/4)$ with head deflection was as shown in Figure 4.49.

In practice, the critical design criterion will usually be the maximum bending moment down the pile rather than the head deflection. Equation (4.80) shows that this maximum moment will be relatively insensitive to the value of soil modulus adopted. Indeed, Barton's centrifuge tests showed that the elastic solution gave a reasonably good estimate of the position and magnitude of the maximum bending moment.

4.4.2.1 Short piles

For piles that are shorter than their critical length, the head deformation will be larger than given by equations (4.79). The increase in deflection is small until the pile length falls below about $0.8L_c$. Carter and Kulhawy (1988) have presented a solution for essentially rigid piles in homogeneous soil ($\rho_c = 1$). For piles where $L/d \leq 0.05(E_p/G_c)^{1/2}$, the ground level deflection and rotation are given by

$$u = 0.32\frac{H}{dG_c}\left(\frac{L}{d}\right)^{-1/3} + 0.16\frac{M}{d^2G_c}\left(\frac{L}{d}\right)^{-7/8}$$

$$\theta = 0.16\frac{H}{d^2G_c}\left(\frac{L}{d}\right)^{-7/8} + 0.25\frac{M}{d^3G_c}\left(\frac{L}{d}\right)^{-5/3}$$

(4.86)

With such short piles, the design will normally hinge on the possibility of the pile failing by rigid body movement through the soil. Analysis of this type of behaviour has been discussed in section 4.3. Further charts giving the response of short piles to lateral loading may be found in Poulos and Davis (1980).

4.4.3 Cantilever idealization

The main shortcoming of the idealization of laterally loaded piles as cantilevers, fixed at some depth below ground surface, has been the lack of consideration of the role of the relative pile and soil stiffnesses in determining an appropriate depth of fixity. Equations (4.79) may be rewritten in a form more familiar to engineers with a structural background, by substituting for G_c in terms of E_p and the critical length, L_c (see equation (4.73)) and then replacing E_p by $(EI)_p/(\pi d^4/64)$. In most cases, it will be appropriate to take ρ_c equal to 0.5, allowing for very low soil stiffness near the ground surface, where the strain level in the soil is high. For this case, equations (4.79) transform to

$$u = 0.424\frac{H}{EI}\left(\frac{L_c}{2}\right)^3 + 0.472\frac{M}{EI}\left(\frac{L_c}{2}\right)^2$$

$$\theta = 0.472\frac{H}{EI}\left(\frac{L_c}{2}\right)^2 + 0.887\frac{M}{EI}\left(\frac{L_c}{2}\right)$$

(4.87)

These equations are similar to those that would be obtained from a cantilever idealization of the pile, with a depth of fixity of $L_c/2$. In that case the three independent coefficients on the right-hand sides would be 1/3, 1/2 and 1 respectively, instead of the values of 0.424, 0.472 and 0.887. Thus the errors involved in estimating pile deformations from a cantilever idealization will be small, providing the appropriate depth of fixity of $L_c/2$ is adopted.

The cantilever idealization is not suitable for estimating the profile of induced bending moments. For example, in the simple case of a pile under force loading, H, only, equation (4.80) gives a maximum bending moment of $0.2HL_c$ (taking $\rho_c = 0.5$), occurring at a depth of $L_c/3$. The cantilever approach would give a maximum bending moment at the base of the cantilever (a depth of $L_c/2$) of magnitude $0.5HL_c$.

4.4.4 Layered soil profiles

The response of piles to lateral loading is greatly affected by the soil properties close to ground level. The presence of a desiccated crust or a shallow layer of granular material overlying softer cohesive soil, can have a significant beneficial effect, reducing both deflections and bending moments down the pile. General solutions for the response of piles in a layered soil profile have been presented by Davisson and Gill (1963) using a subgrade reaction approach, and by Pise (1982) for an elastic soil continuum. The two approaches yield similar results which are best presented in the form of modifications to the solution given in section 4.4.2. Thus, consider a layer of soil of thickness, t_l, whose stiffness is a factor, f_l, greater than that of the underlying soil (see Figure 4.50). For given values of t_l and f_l, the deformation of the pile at ground level may be compared with that for $f_l = 1$ (that is, a single layer). Similarly, the maximum bending moment induced in the pile will be affected.

Figure 4.51 shows the effect of soil layering on the ground level deflection and bending moment for a fixed-headed pile subjected to lateral loading. The ratios u_f/u_{f1} and M_f/u_{f1} (where the subscript 'f' denotes fixed-head, and the subscript '1' denotes the value for $f_l = 1$) are plotted against the ratio t_l/f_l for different values of f_l. These curves may also be used with acceptable accuracy for free-headed piles. It is interesting to note that soil conditions over the upper 10% of the active length of pile are critical in determining the lateral response of piles.

An alternative approach for analyzing the response of a pile in a layered soil profile may be used where ground level deflections are high (10% or more of the pile diameter). For such cases, it is sufficient to assume that the limiting resistance is mobilized in the layer. The pile deformation and bending moment distribution may then be obtained by treating the pile as a cantilever in that region, while the lower part of the pile is analyzed using the expressions presented in section 4.4.2.

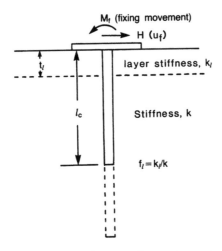

Figure 4.50 Definition of notation for pile in two-layer soil.

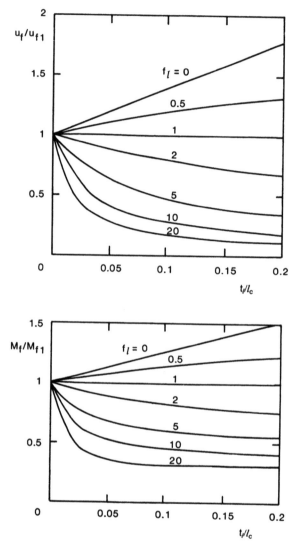

Figure 4.51 Effect of upper soil layer on pile deflection and bending moment (fixed-head piles).

4.4.5 Raking piles and effect of lateral loading on axial capacity

The response of piles to inclined loading (i.e. combined axial and lateral loading) has been discussed by various authors (Broms, 1965; Meyerhof and Ranjan, 1972; Poulos and Davis, 1980). In non-cohesive soil, it has been argued that the axial capacity of a pile may be increased by small amounts of lateral load, owing to the increase in normal stress between pile and soil. This argument has only limited support from experimental data, except for very stubby, caisson-type piles. For most piles, the effects of lateral loading are confined to the upper few diameters of the pile. For small amounts of lateral load, the increase in normal stress in front of the pile will be at least partially offset by a reduction in normal stress behind the pile.

In practical terms, the axial and lateral response of a pile may be treated independently, although Hanna and Nguyen (2003) found that the shaft capacity of raking piles in sand decreases slightly with increasing rake angle. The response of a raking pile to load applied at a particular angle may then be assessed by taking the components of load parallel and normal to the axis of the pile. The overall capacity of a pile will be determined by taking the axial and lateral components of the applied load and comparing them with the respective capacities. Thus for the general case shown in Figure 4.52, where a pile raking at an angle ψ to the vertical is subjected to a load inclined at θ to the vertical, axial failure will occur when

$$Q \cos(\theta - \psi) = Q_a \quad \text{(axial failure)} \tag{4.88}$$

or when

$$Q \sin(\theta - \psi) = Q_l \quad \text{(lateral failure)} \tag{4.89}$$

It may be seen that the maximum capacity will be achieved when both axial and lateral pile capacities are mobilized simultaneously. This occurs for

$$\tan(\theta - \psi) = Q_l/Q_a \tag{4.90}$$

when the overall capacity will be

$$Q = (Q_a^2 + Q_l^2)^{1/2} \tag{4.91}$$

Evangelista and Viggiani (1976) showed that the axial and lateral deformation response of a pile is virtually independent of the angle that the pile makes with the ground surface, for angles of rake up to about 30°. Thus, as for the calculation of the pile capacity, the deformation response may be assessed by considering the components of the applied load parallel and normal to the pile axis. For vertical piles, the

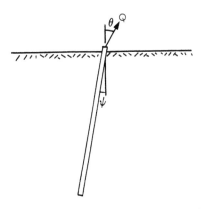

Figure 4.52 Oblique loading of a ranking pile.

vertical and horizontal deformations may be written in terms of a flexibility matrix, F, and the applied load as

$$
\begin{bmatrix} w \\ u \\ \theta \end{bmatrix} = \begin{bmatrix} F_{wP} & 0 & 0 \\ 0 & F_{uH} & F_{uM} \\ 0 & F_{\theta H} & F_{\theta M} \end{bmatrix} \begin{bmatrix} P \\ H \\ M \end{bmatrix}
\tag{4.92}
$$

For a pile battered at an angle ψ to the vertical, the matrix F must be replaced by a transformed matrix, F', where

$$
F' = T^T F T
\tag{4.93}
$$

and

$$
T = \begin{bmatrix} \cos\psi & -\sin\psi & 0 \\ \sin\psi & \cos\psi & 0 \\ 0 & 0 & 1 \end{bmatrix}
\tag{4.94}
$$

One aspect of the effect of lateral loading on axial capacity that deserves particular attention is the possibility of a gap opening up around the pile near ground level – a so-called 'post-holing' effect. Piles subjected to axial loading together with cyclic lateral loading generally fail at a point where the lateral component of load is zero, and the pile is then close to its neutral position. In the design of piles for offshore structures, it is generally recognized that load transfer into the soil close to the sea-floor (under axial loading) should be ignored because of the risk of post-holing. The question is: to what depth will such post-holing occur?

Some estimate of the extent of any gap forming around the pile (particularly in fine-grained soils) may be made by considering the lateral deflection at which yield in the soil close to the pile will occur. Baguelin *et al.* (1977) have considered the case of a circular disc moving sideways through elastic, perfectly plastic material, with shear modulus, G, and shear strength, c_u. They showed that the lateral movement is related to the normal stress change, $\Delta\sigma_n$, in front of and behind the disc, by

$$
u/d \sim \Delta\sigma_n/G
\tag{4.95}
$$

The load deflection curve for the disc was essentially linear up to a normal stress change of about $2c_u$. This would indicate that non-linear effects, and the possibility of gapping around the pile, would start to occur at a normalized deflection of $u/d \sim 2c_u/G$. For a typical value of G/c_u of 100, this would be about 2% of the pile diameter.

Thus, in assessing the extent of any post-holing, and loss of shaft friction, it is necessary to estimate the depth down the pile at which the cyclic lateral movements are less than about 2% of the pile diameter. Above this level, it would be imprudent to rely on significant shaft friction under axial loading.

4.5 Piles in deforming soil

There are many situations in which loads are induced in a pile by virtue of the deformation of the surrounding soil. The most common example is where piles are installed into, or through, a layer of soil that is consolidating, usually due to the recent placement of surface loading. Where the downward movement of the surrounding soil exceeds the settlement of the pile, the pile resists the movement and load in excess of the external applied load will be transferred into the pile. The shear stresses on the pile shaft will act in the reverse sense by comparison with the normal situation for a pile loaded in compression; accordingly, the term 'negative friction', or 'downdrag' is used (see Figure 4.53(a)).

Piles may also be loaded laterally by soil movement, for example on a steeply sloping site, where creep of the upper layers of soil may occur. Eccentric (vertical) loading of the ground surface adjacent to a piled foundation may also lead to significant lateral pressures on the piles, as the soil moves away from the loaded area (see Figure 4.53(b)).

4.5.1 Negative friction

Negative friction or downdrag of piles may occur wherever piles are driven through, or adjacent to, recently placed fill (Johanessen and Bjerrum, 1965). Such fill may merely be to raise the existing ground level, or may be part of an embankment, for

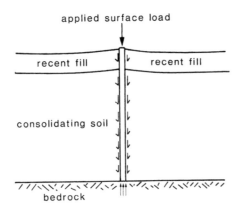

(a) Pile subjected to negative friction

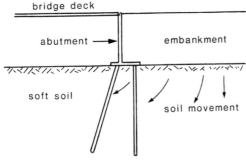

(b) Piles subjected to lateral thrust from soil

Figure 4.53 Piles in consolidating soil.

example, for a bridge approach. The problem is normally associated with soft, lightly overconsolidated deposits of clay. In some sensitive clays, remoulding of the soil during driving may lead to downdrag of the piles, even where no fill is placed (Fellenius, 1972). The settlement of the clay is then associated not with any long-term rise in the vertical effective stress, but rather with a reduction in voids ratio and water content as the remoulded soil consolidates.

The starting point for any estimation of the effects of negative friction is an assessment of the likely free-field vertical movement of the soil. Where this movement exceeds the estimated settlement of the pile, shear stresses acting in a reverse sense will be imposed on the pile shaft. The magnitude of the shear stresses may be determined from the amount of relative movement between soil and pile, with a limiting value reached where downward slip of the soil occurs. The magnitude of relative movement necessary to cause slip has been discussed in section 4.2.4 (see equation (4.52)). For relative movements greater than this limit, the mobilized shear stress may be estimated by

$$\tau_s = K\sigma'_v \tan \phi' \tag{4.96}$$

where, since the soil is consolidating, K should be taken as the appropriate value of K_0, the at-rest earth pressure coefficient corresponding to the current overconsolidation ratio. In most cases of practical interest, the consolidating soil will be normally consolidated, and K_0 may be approximated by $(1 - \sin \phi')$.

For relative movements less than that to cause slip, the mobilized shear stress (positive or negative) may be taken as a linear proportion of the limiting shaft friction. Thus, for small relative movements, the local shear stress on the pile shaft is

$$\tau_0 = \frac{2G}{\zeta d} (w_0 - w_{\text{soil}}) \tag{4.97}$$

where w_0 is the local settlement of the pile, and w_{soil} is the background consolidation settlement of the soil (see section 4.2.1).

Assessment of negative friction for a given distribution of soil settlement necessitates some iteration, since the profile of pile movement will depend on the distribution of shear stress, τ_0 down the length of the pile. The axial load at any depth, z, down the pile is given by

$$P = P_t - \int_0^z \pi \, d\tau_0 dz \tag{4.98}$$

where P_t is the applied (structural) load. The axial strain in the pile may then be integrated to give the settlement of the pile at any depth, z, as

$$w_0 = w_b + \int_z^L \frac{P}{(EA)_p} dz \tag{4.99}$$

where w_b is the settlement of the pile base.

From the calculated pile settlement profile, and the estimated profile of soil settlement, the distribution of shaft friction may then be calculated. This approach

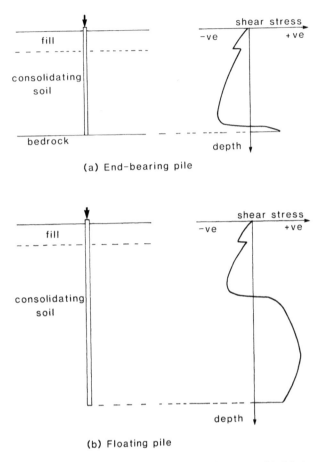

(a) End-bearing pile

(b) Floating pile

Figure 4.54 Typical variations with depth of negative skin friction.

is similar to that described by Fellenius (1989). For end-bearing piles, the distribution of negative friction will generally increase steadily with depth, until a point close to the pile base, where the soil movement is insufficient to mobilize full shaft friction (see Figure 4.54(a)). For a long friction pile, however, there will be a neutral point, generally some way above the mid-point of the pile, where pile and soil settlement are equal, and below which positive friction is mobilized to carry the load applied to the pile (see Figure 4.54(b)).

For single piles, the load transfer method may be used to assess pile response under combinations of loads applied at the pile head and external soil movements (for example, Randolph (2003)). A graphical method that illustrates the principles of how to establish the level of the neutral plane is shown in Figure 4.55. The external applied load, P_0, and the load induced by negative friction, P_n, is balanced by the end-bearing resistance, Q_b, and shaft capacity in the lower part of the pile, Q_s. The neutral plane occurs where pile and soil settlement are equal, and the depth of the plane will rise as the external load increases. The ultimate pile capacity, P_u or Q_u, is unaffected

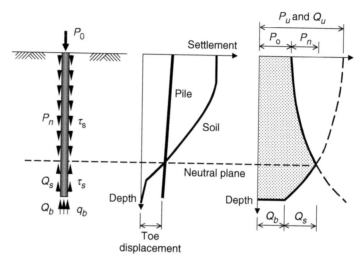

Figure 4.55 Establishing the neutral plane for a floating pile in consolidating soil.

by negative friction, since at that stage (by definition of pile failure) the pile settlement
will exceed the soil settlement.

For end-bearing piles, parametric studies for single piles have been presented by
Poulos and Davis (1980). The method of analysis has been extended to groups of
end-bearing piles by Kuwabara and Poulos (1989). This latter work has shown that
the internal piles within a group are subjected to much lower downdrag loads than
the peripheral piles, and that, for a given soil settlement, the average downdrag load
on the piles in a group is significantly lower than for a single isolated pile. Thus for
a surface settlement varying linearly from 5% of the pile diameter at the surface,
to zero at the bearing stratum, a typical single pile would attract over 90% of the
maximum available load due to negative friction, while a square group of 25 piles
would attract less than 50% of the potential load. Numerical analysis of pile groups
subjected to downdrag has been reported by Lee *et al.* (2002), who show that the
maximum downdrag loads are reduced by between 20 and 80% for edge piles and
central piles in a 5 × 5 group.

The appropriate design philosophy to be adopted in situations where negative friction may arise is a matter of some debate. If piles are designed so that they can take the
full negative friction (in addition to the load applied at ground level) without yielding,
then the structure founded on the piles will not settle significantly. In such cases, the
soil surrounding the structure will settle and may cause as great a problem (with services, and distortion of paved areas adjacent to the building) as in the more common
case of a building settling more than the surrounding soil. The problems can be particularly severe if there are adjacent, unpiled, structures. These will settle with the soil,
resulting in large differential settlements between the piled and unpiled structures.

In many cases, the optimum solution is for the piled structure to settle with the
surrounding soil. For this to be achieved, the piles should be designed to carry the loads
applied by the structure, with a relatively low factor of safety, and with no allowance

made for any negative friction. As the surrounding soil settles, negative friction will indeed drag the piles down, but, by its very nature, will not lead to bearing failure of the piles (provided they have been correctly designed for the given structural load). The low factor of safety ensures that the piles do not withstand too much negative friction before settling with the consolidating soil.

The above approach may certainly work well with floating piles, since these have a low base capacity. The shaft resistance relies on relative motion between pile and soil. Thus, if the soil settles around the piles, the pile must also settle in order to carry the applied load. The same approach is more sensitive to apply for end-bearing piles, owing to uncertainties in the load settlement response of the pile base. However, unless there is a danger of breaking through a competent stratum into softer soil beneath (which might lead to failure of the piles under the structural load alone) the same general philosophy of allowing the piles to settle with the consolidating soil should still provide a satisfactory foundation.

In situations where it is essential to restrict the absolute movement of the pile foundation, measures may be taken to relieve the piles of negative friction by coating the shafts or by sleeving them, particularly in the upper regions. Claessen and Horvat (1974) discuss the use of bitumen coatings, and recommend that, where such coatings are used, they should be omitted on the lower 10 diameters of the shaft, in order to ensure that the full end-bearing capacity may be relied on. Coatings as thin as 1.5 mm may be sufficient to eliminate nearly all negative friction (Walker and Darvall, 1973), provided that care is taken to ensure that the coating is not stripped off the pile during installation.

4.5.2 Lateral movement of soil

Before considering the case of vertical piles loaded laterally by horizontal soil movement, it is worth noting that a common cause of foundation failure is due to the use of raking piles in soil that is still consolidating. Vertical settlement of the soil around a raking pile can impose lateral pressures that are sufficient to break the pile. Raking piles should thus be avoided in situations where significant consolidation settlement of the soil may occur – including the case of piles being driven into sensitive clay, as mentioned at the start of section 4.5.1.

Eccentric loading of the ground around a piled structure, for example an embankment behind a piled bridge abutment, may generate sufficient horizontal soil movement to cause distress in the piles. This form of loading can also lead to significant movement of the structure. Cole (1980) reports the case of a piled bridge abutment, founded on a deep deposit of soft silty clay, which rotated away from the bridge (that is towards the retained embankment) due to lateral soil pressure below ground level.

A simple method of estimating the lateral pressure on piles close to a surcharged area has been presented by De Beer and Wallays (1972). The method differentiates between surcharge loads (particularly embankments) which have factors of safety against collapse (without the piles) which are greater or less than 1.4. For the less stable situation, it is assumed that the full limiting pressure is mobilized against the pile (see section 4.3) throughout the depth of the soft deposit. For cases with higher factors of safety, the lateral pressure on the pile is taken as the surcharge pressure, factored by a parameter which depends on the geometry of the loaded area relative to the piles. Figure 4.56

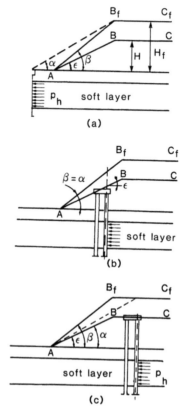

Figure 4.56 Piles subjected to lateral thrust near an embankment (De Beer and Wallays, 1972).

shows how the method would be applied for typical embankment geometries. The actual embankment, of height H, is replaced by a fictitious embankment of material of unit weight 18 kN/m³, of a height H_c, such that the vertical loading is the same. The lateral pressure on the piles is then taken as

$$\Delta p_h = \Delta p_v \frac{2\alpha - \phi'}{\pi - \phi'} \tag{4.100}$$

where Δp_v is the applied surcharge pressure, and ϕ' is the friction angle of the soft layer of soil. The angle α is defined in Figure 4.56.

In the above approach, no account is taken of the depth, or extent, of the soft layer, or even of its initial strength; the method is clearly very approximate. An approach based on elastic interaction of the moving soil and the pile, but with allowance made for the limiting pressure that the soil may exert on the pile, has been described by Poulos and Davis (1980). Their method relies on an initial estimate of the horizontal soil movement. A boundary element analysis of the problem then allows calculation of the resulting loading in the piles. Charts were presented for some common profiles of soil movement.

A design method for estimating the maximum bending moments and lateral deflection of piles adjacent to embankment loading has been presented by Stewart *et al.* (1994a, b). The method is based on a series of centrifuge model tests where the piles penetrated through a soft clay layer of height, h_s, to found in underlying dense sand. An embankment was then constructed adjacent to the piles, applying a surcharge, Δq. A distinct threshold was noted where the surcharge pressure exceeded about three times the shear strength of the clay, with a rapid increase in the rate of increase of bending moments and lateral deflections with increasing surcharge level.

Design charts are shown in Figure 4.57, with the pre-threshold and post-threshold rates of change of maximum bending moment and lateral pile head deflection non-dimensionalized as

$$M_q = \frac{\Delta M_{max}}{\Delta q d L_{eq}^2}; \quad y_q = \frac{\Delta y (EI)_p}{\Delta q d L_{eq}^4}; \quad K_R = \frac{(EI)_p}{E_s h_s^4} \tag{4.101}$$

where E_s and h_s are respectively the Young's modulus and thickness of the soft clay layer, and L_{eq} is an equivalent pile length allowing for the restraint condition at the pile head, with L_{eq} equal to L where the pile head is restrained from rotation, $1.3L$ for a free-headed pile, and $0.6L$ where the pile head is pinned (Stewart *et al.*, 1994b).

The data points have been extracted from a variety of field observations (triangular symbols) and centrifuge model tests, with the open symbols denoting pre-threshold rates of change (surcharge, $q < 3c_u$) and the solid symbols are best estimates of post-threshold rates of change (Stewart *et al.*, 1994a).

In cases where surcharge loading leads to excessive downdrag, or induced bending moments and lateral deflection in nearby piles, certain construction techniques may be

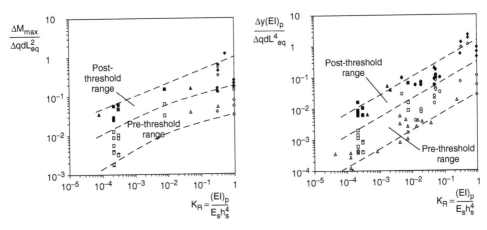

Figure 4.57 Experimental and calculated pile response due to surcharge loading: (a) bending moments (MNm), (b) lateral pressure (kPa), (c) lateral deflection (mm). —·—·— experimental (surcharge 53 k Pa); ——— SLAP (surcharge 53 kPa); ------ experimental (surcharge 93 kPa); ——— SLAP (surcharge 93 kPa).

used to circumvent the problem. One technique, which may be relatively costly, is to shield the upper part of the piles using cylindrical caissons. An alternative solution, particularly suitable for embankment approaches to bridge abutments, is the use of piles beneath part of the embankment. This technique, known variously as 'embankment piling' or 'bridge approach support piling' (Reid and Buchanan, 1983), is shown schematically in Figure 4.58. Widely-spaced piles (up to 10 diameters apart) are used, with small pile caps that typically occupy as little as 10% of the overall area. Arching action is relied on in order to shed most of the embankment load on to the piles, but Reid and Buchanan recommend the use of a latticed fabric membrane over the tops of the pile caps, in order to prevent cavities from forming. The membrane will carry a proportion of the embankment loading, and will thus aid the degree of arching between the pile caps.

The performance of the pile support may be expressed in terms of an 'efficacy', E, which gives the proportion of the total embankment load carried by the pile caps. Figure 4.59 shows a detail of reinforcing membranes spanning pile caps, of side, a, and centre-line spacing, s.

For granular fill of height, H, and friction angle, ϕ', overlying the pile caps, Hewlett and Randolph (1988) proposed a method of analysis based on the formation of arches spanning between the pile caps. Two alternative estimates of the average vertical pressure, p_r, acting on the ground between the pile caps were arrived at, considering the

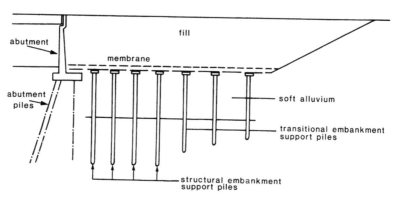

Figure 4.58 Typical section through embankment piling.

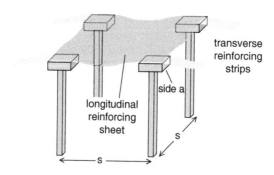

Figure 4.59 Detail showing reinforcing membranes overlying pile caps.

stress changes within the arch, and also limiting conditions at the pile caps. Defining $\delta = a/s$, so that the pile caps occupy δ^2 of the total area, the average pressure transmitted directly to the soil is estimated from the higher of either

$$p_r = \left[(1-\delta)^{2(K_p - 1)} \left(1 - \frac{s}{\sqrt{2}H} C \right) + \frac{s-a}{\sqrt{2}H} C \right] \gamma H \tag{4.102}$$

or

$$p_r = \frac{\gamma H}{\left(\dfrac{2K_p}{K_p + 1} \right) \left[(1-\delta)^{(1-K_p)} - (1-\delta)\left(1 + \delta K_p\right) \right] + (1 - \delta^2)} \tag{4.103}$$

where

$$C = \frac{2(K_p - 1)}{2K_p - 3} \quad \text{and} \quad K_p = \frac{1 + \sin \phi'}{1 - \sin \phi'} \tag{4.104}$$

The first of the expressions for p_r governs for large s/H (low embankment height relative to the pile spacing, while the second governs for low s/H (high embankments or closely spaced piles). This method has been found to give good agreement with measured data, such as the field studies reported by Reid and Buchanan (1983) where over 80% of the surcharge loading was shed onto pile caps spaced such as to occupy only 11% of the plan area. Reviews of this and other methods of analysis have been given by Russell and Pierpoint (1997) and Love and Milligan (2003).

The main advantages offered by embankment piling are:

(a) short-term stability of the embankment will be ensured, avoiding the need for staged construction;
(b) long-term settlement of the embankment close to the bridge will be reduced, giving lower maintenance costs for the roadway; and
(c) lateral pressure on the piles for the bridge abutment will be significantly reduced.

4.6 Buckling of piles

There are two main areas where the axial stability of piles may be critical in design. The first of these is during installation by driving, where instability can lead to the pile deviating significantly from its proposed alignment. The other area is under working conditions, where the axial load taken by a pile may be close to the critical buckling load for the pile. This latter case is relatively rare, but can occur for end-bearing piles driven through relatively soft alluvial deposits.

4.6.1 Instability during driving

Burgess (1976) has given a detailed analysis of the stability of piles driven or jacked into soil of uniform shear strength (see also Omar, 1978; Ly, 1980; Burgess, 1980). Two different forms of instability occur; one is a buckling phenomenon, and the other is known as 'flutter'. The latter may be viewed as directional instability at the advancing tip of the pile. Burgess' analysis for the buckling instability, which involves lateral movement of the pile, is based on a Winkler model for the response of the soil around the pile shaft. The flutter instability is independent of the lateral response of the pile.

The results of the analysis, based on the assumption that the pile is guided until just above ground level, indicate that the buckling mode of instability would rarely occur and, in any case, would always be preceded by flutter instability. The criterion for flutter to occur is shown in Figure 4.60. This figure is a dimensionless plot, essentially plotting the driving force, Q, against the proportion of that force that reaches the base of the pile, (Q_b/Q). The driving force is non-dimensionalized by the pile length, L, and the bending rigidity, EI, of the pile section, in the form

$$\Lambda = QL^2/EI \tag{4.105}$$

Burgess (1976) gives design curves for some typical pile cross-sections (see Figure 4.61), showing how the length of pile at which flutter would occur varies with the shear strength of the soil. It has been assumed that the full shear strength of the soil is mobilized in shaft friction down the pile shaft during installation, and the shear strength, c_u, should really be replaced by the estimated (average) shaft friction. The curves given in Figure 4.61 suggest that flutter may occur at relatively modest pile lengths.

4.6.2 Buckling under static loading

Buckling of piles under static loading will generally only be a problem where very soft soil overlies the founding material, giving rise to a depth in which the pile is only moderately supported laterally. The problem has been looked at by a number

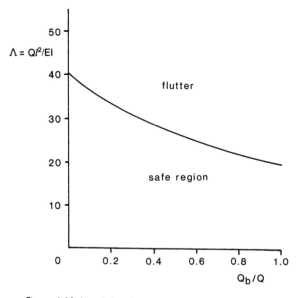

Figure 4.60 Instability due to flutter (Burgess, 1980).

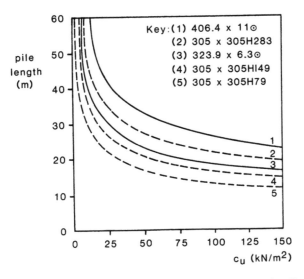

Figure 4.61 Typical design curves for limiting stable length of driven piles (Burgess, 1976).

of workers (Davisson, 1963; Davisson and Robinson, 1965; Reddy and Valsangkar, 1970). As might have been expected intuitively, analysis shows that buckling will be confined to the critical length of the pile under lateral loading (see section 4.4). This will be a region in the upper few diameters of the pile, and, for practical purposes, it is sufficient to ignore the variation of axial load in the pile over the length of interest. This will in any case furnish a conservative estimate of the load to cause buckling.

Figure 4.62 shows a general picture of the pile, with a length L_f of pile free-standing above the ground surface. In order to calculate the buckling load, the stiffness of the embedded section of the pile under lateral force and moment loading must be known. It is sufficient to replace the embedded section of the pile by a cantilever of such a length L_e, that provides an equivalent stiffness. This cantilever idealization was discussed in section 4.4, and it was shown that the equivalent length for the cantilever should be taken as half the critical length, that is

$$L_e \sim L_c/2 = (E_p/G_c)^{2/7}d/2 \sim 2(EI/k)^{1/4} \quad \text{or} \quad 2(EI/n)^{1/5} \tag{4.106}$$

where the parameters have been defined in section 4.4.

With the pile now replaced by an equivalent length, $L_f + L_e$, embedded at its lower end, the buckling load may be calculated using the standard Euler equations. Thus, for boundary conditions at the upper end which allow lateral movement and rotation (a free end), the buckling load is

$$P_{cr} = \frac{\pi^2 EI}{4\left(L_f + L_e\right)^2} \tag{4.107}$$

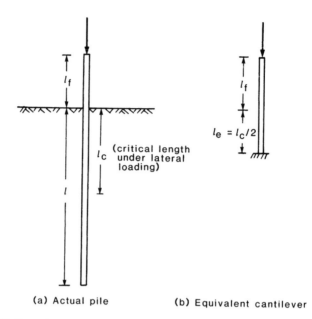

(a) Actual pile   (b) Equivalent cantilever

Figure 4.62 Equivalent cantilever for buckling calculation for a partially embedded pile.

while for conditions at the upper end that prevent rotation, but still allow lateral motion, the buckling load will be a factor of 4 times greater. This approach is consistent with that of Davisson and Robinson (1965), although restricted to the practical case of piles that are longer than their critical length under lateral loading.

It is interesting to compare the numerical values of the buckling load given by equation (4.107) with the condition for flutter to occur during installation shown in Figure 4.60. For a pile with no free-standing length above the ground ($L_f = 0$), the equivalent non-dimensional form of equation (4.107) would be

$$\Lambda = 2.5(L/L_e)^2 \tag{4.108}$$

Typically, the critical length of a pile will be about 10 to 15 diameters, giving an equivalent length of 5 to 7.5 diameters. From the example curves shown in Figure 4.61, the pile length at which flutter becomes a problem will be 50 to 100 diameters. Thus the non-dimensional parameter in equation (4.108) becomes about 250, well above that to cause flutter instability.

The conclusion, then, is that buckling of long piles driven into deep layers of soft soil will be a secondary problem to that of actually installing them. However, end-bearing piles, or piles with a significant length of free-standing section, which are installed through soft deposits, should be assessed for possible buckling under static load.

Chapter 5

Design of pile groups

The majority of piled foundations will consist not of a single pile, but of a group of piles, which act in the dual role of reinforcing the soil, and also of carrying the applied load down to deeper, stronger soil strata. As will be shown in this chapter, the effectiveness of a pile, in particular in respect of its stiffness, is generally reduced by the proximity of other piles. In recognition of this, current trends are towards the use of fewer, more widely spaced piles, where the reinforcement role of piles is emphasized. Modern analytical techniques allow more realistic estimates to be made of the response of pile groups under working loads, thus giving the designer more scope to optimize the foundation layout.

In particular, two areas of progress may be singled out. The first concerns the design of pile groups to withstand horizontal loading (Elson, 1984). The traditional view of piles as pin-ended columns, incapable of withstanding loads applied normal to their axis, entailed much unnecessary use of raking piles – often with two or more rows of piles aligned parallel to the resultant loading (Hambly, 1976) (see Figure 5.1(a)). The inefficiency of such a pile group as a means of reinforcing the soil is apparent. In addition, the horizontal component of load in most pile groups will generally contain a high time-varying component. Thus there will be periods when the maximum vertical load may be applied, but the horizontal load will be substantially less than the maximum designed for. Under such conditions, the raking piles are likely to be as severely stressed in bending, as they would have been as vertical piles under the full horizontal load.

While raking piles might still be used in a modern design, in order to minimize induced bending moments, awareness of the lateral load-carrying capacity of piles allows the piles to be spread more evenly through the soil (see Figure 5.1(b)). Computer analysis of the pile group, making proper allowance for the soil support around the piles, permits estimates of the resulting bending moments and deformations to be made with some confidence.

The other area of progress concerns the use of piles in conjunction with a raft foundation, rather than as an alternative. Where reasonably competent soil extends to the ground surface, the decision to include piles in a foundation may be made on the basis of excessive settlement of a raft foundation alone. The traditional approach has then been to design the foundation solely on the basis of the piles, ignoring any contribution of the pile cap (or raft) to the load-carrying capacity of the foundation. The illogicality of such an approach has been pointed out by Burland *et al.* (1977), who suggested the inclusion of a limited number of piles beneath a primarily raft

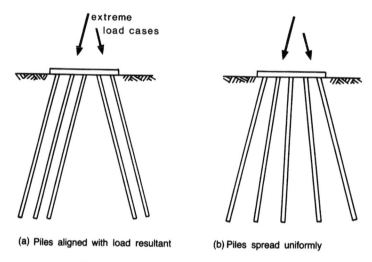

(a) Piles aligned with load resultant   (b) Piles spread uniformly

Figure 5.1 Piles as reinforcement for the soil.

foundation, in order to reduce settlements to an acceptable level. This approach is discussed further in section 5.3.

At the end of the chapter, there is a discussion of some of the main principles in the design of single piles and pile groups. The discussion includes comments on appropriate factors of safety to use for piles in different circumstances, and reviews the potential benefits and hazards of using raking piles.

5.1 Capacity of pile groups

A group of piles may be viewed as providing reinforcement to a particular body of soil. Failure of the group may occur either by failure of the individual reinforcing members (the piles) or as failure of the overall block of soil. When considering failure of the individual piles, it must be remembered that the capacity of each pile may be affected by the driving of subsequent piles in close proximity. This has been demonstrated experimentally by Vesic (1969) and Lo (1967), who showed that the shaft capacity of piles driven into sand could increase by a factor of 2 or more. Physically, this may be attributed to higher normal effective stresses acting on piles driven in a group, than on single piles. The beneficial effects of driving neighbouring piles may be particularly noticeable in loose sands, where compaction of the sand can lead to very low contact stresses around the pile shaft. Vibration and soil displacement due to driving of subsequent piles is likely to reduce the tendency for the soil to arch around each pile, leading to higher contact stresses, and thus shaft capacities, than for single piles.

In other situations, the capacity of a pile within a group may be reduced by comparison with a single, isolated pile. In particular, for piles driven into sensitive clays, the effective stress increase in the surrounding soil may be less for piles in a group, than for individual piles. This will result in lower shaft capacities.

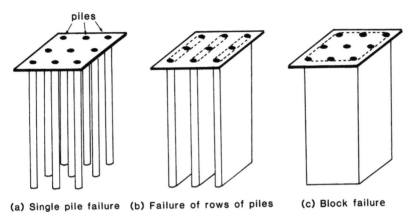

(a) Single pile failure   (b) Failure of rows of piles   (c) Block failure

Figure 5.2 Block failure of pile groups.

Instead of failure of the individual piles in a group, block failure may occur under axial or lateral load. Soil between the piles may move with the piles, resulting in failure planes which follow the periphery of the group (or parts of the group) as shown in Figure 5.2. In general, block failure will be associated with close spacing of the piles. However, in certain cases, such as the axial failure of a pile group with a ground-contacting pile cap, block failure may be enforced at quite large pile spacings.

In the past, the difference between the capacity of a group of piles, and the sum of the individual pile capacities, has been characterized by a group 'efficiency'. This concept has led to a certain amount of confusion. As stated above, the capacity of a pile within a group may be different from that of an equivalent isolated pile, and, in that sense, the pile may be thought of as more or less efficient when acting as part of a group. However, the block capacity of a group should not be confused with the sum of the individual pile capacities by the use of an 'efficiency' factor. Independent calculations should be made of both the block capacity and the individual pile capacities, to ensure that there is an adequate factor of safety against both modes of failure.

5.1.1 Axial capacity

The axial capacity of a pile group failing as a block may be calculated in a similar fashion to that for an individual pile, by means of equation (4.1), but now taking A_b as the base area of the block and A_s as the block surface area. In general, the enclosing block as shown in Figure 5.2(c) should be taken, but, where the pile spacing in one direction is much greater than that in the perpendicular direction, the capacity of the group failing as shown in Figure 5.2(b) should be assessed. It should be noted that the settlement needed to mobilize the base capacity of the block will usually be very large (for example, 5 to 10% of the width of the group).

Since the end-bearing pressure q_b is much greater than the average skin friction τ_s, block failure only becomes more likely than the failure of individual piles where the increase in base area, A_b, is offset by a much larger decrease in surface area, A_s.

Thus groups of a large number of long slender piles at a particular spacing are more likely to fail as a block than groups consisting of a few, short stubby piles at the same spacing. By a similar argument, group failure is less likely for piles in sand (where q_b/τ_s is 50 to 200) than for piles in clay (where q_b/τ_s is 10 to 20).

5.1.1.1 Non-cohesive soil

Calculation of the end-bearing pressure q_b for block failure of a pile group in non-cohesive soil follows a similar procedure to that described for single piles in section 4.1.1, except that the bearing capacity term due to the self-weight of the soil below bearing level may now be more significant. Thus the end-bearing pressure should be calculated from

$$q_b = N_q \sigma'_v + 0.4 \gamma B N_\gamma \tag{5.1}$$

where B is the width of the pile group (assumed approximately square). Values of the bearing capacity factor N_γ may be obtained from standard texts (Terzaghi and Peck, 1967).

The average frictional resistance along the sides of the block failure may be calculated on the basis of equation (4.7), taking K as the at-rest earth pressure coefficient, and δ equal to the angle of friction, ϕ'. Thus

$$\tau_s = K_0 \sigma'_v \tan \phi' \tag{5.2}$$

For most pile groups in sand, the calculated block capacity will be well in excess of the sum of the individual pile capacities. For displacement piles, the capacity of each pile in the group will itself tend to be larger than for isolated piles, owing to the compactive effect of pile installation. This increase in capacity has generally been interpreted in terms of a group efficiency, η, giving the ratio of group capacity to the sum of the individual pile capacities, taken as isolated piles (Vesic, 1969; Lo, 1976). Figure 5.3 shows group efficiencies reported by Vesic (1969). It may be seen that the base capacity of piles in a group is not affected by the neighbouring piles, but that the shaft capacities may increase by a factor of up to 3.

Accurate quantification of the increased capacity of piles in a group is difficult. As a conservative approach in design, the axial capacity of a pile group in sand is usually taken as the sum of individual pile capacities, calculated as in section 4.1.1. An exception to this is where the pile cap is cast directly on the ground, in which case the capacity of the pile group will correspond to that of the block. However, in these circumstances, the design of the pile group will probably hinge on the likely settlements. Factors of safety against overall failure will be high and accurate estimation of the group capacity will become less relevant.

5.1.1.2 Cohesive soil

For pile groups in cohesive soil, the block capacity may be calculated assuming end-bearing pressures given by equation (4.10), with an appropriate value of N_c, and

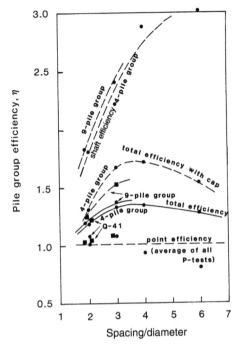

Figure 5.3 Pile group efficiencies in sand (Vesic, 1969).

assuming the full shear strength of the soil is mobilized round the periphery of the block. The factor N_c may be estimated approximately (Skempton, 1951) from the expression

$$N_c = 5(1 + 0.2B_g/L_g)[1 + 0.2(L/B_g)] \tag{5.3}$$

where B_g and L_g are the breadth and length of the pile group in plan ($B_g \leq L_g$) and L is the embedded pile length; a limiting value of 1.5 should be imposed on the second (square) bracketed correction factor.

In certain types of soil, particularly sensitive clays, the capacity of individual piles within a closely spaced group may be lower than for equivalent isolated piles. However, there is a paucity of field data allowing this to be quantified, and it is rarely taken into consideration in design. The effect is not likely to be large. A more common cause for concern has been that the block capacity of the group maybe less than the sum of the individual pile capacities, although evidence for this has largely been from small-scale model tests. De Mello (1969) summarized data for pile groups of up to 9 × 9 piles, showing that block failure of a pile group does not become pronounced until the pile spacing is less than 2 to 3 pile diameters.

With modern trends in pile group design for more widely spaced piles, the group capacity will rarely fall below that calculated from the sum of the individual pile capacities. Full-scale tests on a 3 × 3 pile group in stiff clay (O'Neill *et al.*, 1982) confirm that pile group capacity may be estimated with good accuracy as the sum of the individual pile capacities. For piled foundations with a ground-contacting cap, the group capacity will correspond to that of the block enclosing the piles. However, as for groups in sand, settlement considerations will generally dominate such a design and the overall capacity will be of little relevance.

In summary, design calculations for a pile group should include an assessment of the capacity both of the individual piles, and of the enclosing block. The former should generally be based on the capacity of equivalent isolated piles, only modified by the effect of installation of the surrounding piles where reliable data are available to quantify the effect. Adequate, and independent factors of safety should be ensured against either mode of failure, but the two modes should not be confused by the use of empirical group efficiency factors.

5.1.2 Lateral capacity

The lateral capacity of a pile group will rarely be critical in the design of the group. However, some discussion is appropriate concerning the limiting normal pressure that piles within a group may impose on the soil. The form of block failure relevant for laterally loaded pile groups will generally be that shown in Figure 5.2(b), where the applied load is parallel to the individual blocks. This mode of failure will govern when the shearing resistance of the soil between the piles is less than the limiting resistance of an isolated pile. Referring to Figure 5.4, the limiting force per unit length on the back pile will be the lesser of P_u calculated in section 4.3.1 for single piles, and $2s\tau_s$, where s is the pile spacing and τ_s is the friction on the sides of the block of soil between the two piles.

5.1.2.1 Non-cohesive soil

For non-cohesive soil, the limiting force per unit length for a single pile (at depth) may be taken as $P_u = K_p^2 \sigma_v' d$. The limiting shear stress on the sides of the block of soil shown in Figure 5.4 may be written

$$\tau_s = K\sigma_v' \tan \phi' \tag{5.4}$$

where the earth pressure coefficient K will lie between K_0 and K_p. Thus block failure will occur at pile spacings where

$$s/d < \frac{K_p^2}{2K \tan \phi'} \tag{5.5}$$

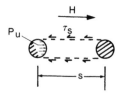

Figure 5.4 Plan view of block failure under lateral load.

For most practical pile spacings, this condition will be satisfied, and the limiting force per unit length, on piles that lie in the wake of another pile, may be estimated from

$$P_u = 2K\sigma'_v \tan \phi's \qquad (5.6)$$

The value to be assigned to the coefficient K remains open to question. The stress field will vary over the space between the piles, but an average value of around unity may be taken as a first approximation.

5.1.2.2 Cohesive soil

A similar approach may be adopted for cohesive soil, taking the limiting shear stress on the sides of the soil block as the undrained shear strength, c_u. Note that the failure mode depicted in Figure 5.4 is consistent with the Murff and Hamilton (1993) mechanism of a conical wedge failure ahead of the front pile, and either a gap or a collapsing wedge mechanism behind the trailing pile. Assuming a limiting force per unit length for an individual pile of $9c_u d$, block failure will occur at pile spacings less than 4.5 diameters. The limiting force per unit length for piles in the wake of another pile is then given by the lesser of $9c_u d$ and $2c_u s$.

5.1.2.3 Failure mode of pile group

Failure of a pile group under lateral load will involve rotation of the group as well as lateral translation. Referring to Figure 5.5, piles behind the axis of rotation will fail by uplift, while those in front will fail in compression. The overall stability of the group will be determined by the axial capacity of the piles in addition to their lateral capacity. The concept of group efficiency is thus even less applicable than for axial loading, and most pile groups will have efficiencies considerably greater than unity under lateral loading, due to the geometry of the group. The group capacity, and the mode of failure, may be determined from statical considerations, but taking due account of the lateral resistance of the soil.

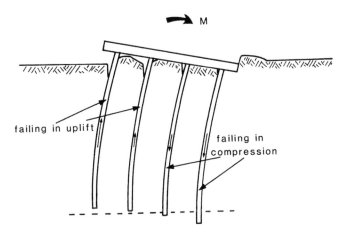

Figure 5.5 Failure of pile group under lateral load.

In practice, the lateral capacity of a pile group will seldom be a critical feature of a design. In situations where there is a large component of lateral load, it will be necessary to ensure that the maximum bending moments induced in the piles will not lead to failure of the piles. The maximum bending moments will generally be estimated on the basis of one of the methods described in section 5.2.3, without considering the lateral capacity of the pile group directly.

5.2 Deformation of pile groups

Over the last two decades, advances in analytical techniques and careful experimental research have combined to provide a much improved understanding of the manner in which pile groups deform under load. The traditional approach of replacing the pile group by an equivalent raft foundation in order to estimate settlements, has been replaced by techniques where the group stiffness may be calculated in terms of the combined stiffness of the individual piles, making due allowance for interaction between piles in the group.

Analysis of the response of pile groups became feasible with the growth of numerical methods of analysis, in particular the integral equation, or boundary element method (Banerjee and Butterfield, 1981). Application of the numerical techniques to pile group analysis has been described in detail by Poulos and Davis (1980), who provide useful design charts showing how the response of a pile group is affected by parameters such as the number and size of piles, the relative stiffness of the piles and the soil and so forth. Butterfield and Douglas (1981) have produced a similar collection of charts giving the stiffness of common forms of pile groups under vertical and horizontal loading.

In practical terms, one of the most useful concepts to emerge from the analytical work is the use of interaction factors. An interaction factor, α, is defined as the fractional increase in deformation (that is deflection or rotation at the pile head) of a pile due to the presence of a similarly loaded neighbouring pile. Thus, if the stiffness of a single pile under a given form of loading is k, then a load, P, will give rise to deformation δ, given by

$$\delta = P/k \tag{5.7}$$

If two identical piles are each subjected to a load, P, then each pile will deform by an amount, δ, given by

$$\delta = (1+\alpha)P/k \tag{5.8}$$

The value of α will of course depend on the type of loading (axial or lateral), and on the spacing of the two piles and the pile and soil properties.

The use of interaction factors may be regarded as equivalent to superimposing the separate deformation fields that each pile would give rise to by itself, in order to arrive at the overall deformation. This approach, illustrated in Figure 5.6 for axial loading, has been justified experimentally from carefully conducted tests on instrumented piles at field scale (Cooke et al., 1980), and also from model pile tests (Ghosh, 1975; Abdrabbo, 1976). As pointed out by Mylonakis and Gazetas (1998), the final

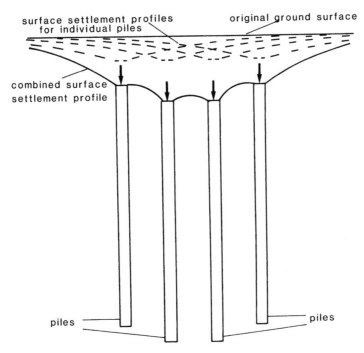

Figure 5.6 Superposition of settlement profiles for group of piles.

interaction factor must also take account of how the presence of the second pile will modify the deformation field imposed by the first pile.

The following sections describe how the stiffness of pile groups under axial and lateral loading may be evaluated from the stiffness of single piles and the use of appropriate interaction factors.

5.2.1 Axial loading

Randolph and Wroth (1979) describe how the solution for the axial load deformation response for single piles may be extended to deal with pile groups. Theoretical interaction factors are deduced from the form of the deformation field around a single pile, treating the pile shaft separately from the pile base. Under axial loading, the deformation field around the pile shaft varies approximately logarithmically with radius, r, according to

$$w = \frac{\tau_o d}{2G} \ln\left(r_m/r\right) \tag{5.9}$$

(see section 4.2.1). Thus the shaft interaction factor, α_s, for a pile a distance s from the pile in question may be estimated as

$$\alpha_s \approx \frac{\ln\left(r_m/s\right)}{\zeta} \tag{5.10}$$

where $\zeta = \ln(r_m/r_0)$ and $\alpha_s \geq 0$. At the level of the pile base, the deformation field due to the base load may be approximated by (Randolph and Wroth, 1979)

$$w = w_b \frac{d_b}{\pi r} \tag{5.11}$$

giving rise to a base interaction factor, α_b, for a pile at spacing s, of

$$\alpha_b \approx \frac{d_b}{\pi s} \tag{5.12}$$

For groups of piles symmetrically placed around a pitch circle, where each pile responds identically, the load settlement response of the ith pile may now be estimated directly from equation (4.43), but with the factor ζ replaced by ζ^* given by (Randolph, 1979)

$$\zeta^* = \zeta \sum_{j=1}^{n} (\alpha_s)_{ij} \tag{5.13}$$

where $(\alpha_s)_{ii}$ is taken as unity, and the factor η replaced by

$$\eta^* = \eta / \sum_{j=1}^{n} (\alpha_b)_{ij} \tag{5.14}$$

Interaction of the deformation fields around the pile shafts is much greater than around the pile bases, because of the logarithmic decay with pile spacing (compared with $1/s$ for the pile base). This results in a greater proportion of the applied load being transmitted to the base of piles in a group than for single piles, as confirmed by model pile tests (Ghosh, 1975).

For pile groups where the piles are not all subjected to the same load, for example under a stiff pile cap, the above approach must be modified by evaluating an overall interaction factor for each pair of piles in the group. The interaction factor must allow not only for the deformation field around the pile shaft, for example the logarithmic decay around the pile shaft given by equation (5.10), but also the stiffening effect of the neighbouring pile.

Mylonakis and Gazetas (1998) have shown that the interaction factor in respect of the pile head response may be expressed as the product of two terms representing the logarithmic decay and a 'diffraction factor', ξ, giving

$$\alpha = \frac{\ln(r_m/s)}{\zeta} \xi \tag{5.15}$$

For homogeneous soil conditions, ignoring the (relatively minor) interaction at the level of the pile bases, the diffraction factor, ξ, may be expressed in terms of the non-dimensional quantities Ω and μL (see equations (4.49)), as

$$\xi = \frac{2\lambda L + \sinh(2\lambda L) + \Omega^2 [\sinh(2\lambda L) - 2\lambda L] + 2\Omega[\cosh(2\lambda L) - 1]}{2\sinh(2\lambda L) + 2\Omega^2 \sinh(2\lambda L) + 4\Omega \cosh(2\lambda L)} \tag{5.16}$$

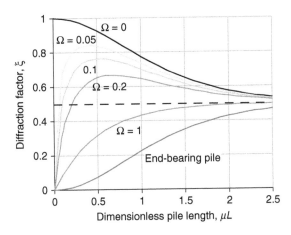

Figure 5.7 Diffraction coefficient, ξ (Mylonakis and Gazetas, 1998).

The form of equation (5.16) is illustrated in Figure 5.7, where it may be seen that ξ converges to 0.5 for long (or compressible) piles, is less than 0.5 for end-bearing piles (high Ω) and between 0.5 and 1 for most floating piles (low Ω).

Randolph (2003) has described how this approach may be extended to allow for piles of the same length, but different diameters, d_1 and d_2 (and corresponding values of ζ, Ω and μL) in soil where the modulus varies linearly with depth. Indeed, the general approach of Mylonakis and Gazetas (1998) may also be extended to pile groups where the piles are of differing length.

The above approach has been used to evaluate the stiffness of square groups of piles, from 2×2 up to 30×30, for $L/d = 25$, $E_p/G_L = 1000$, $\rho = 0.75$ and $v = 0.3$. The results are shown in Figure 5.8, where the pile group stiffness, K_g (ratio of total applied load to average settlement) has been normalized by $G_L B$, where B is the width of the pile group. Plotting the normalized stiffness against the normalized width, B/L, leads to an envelope of curves that tends to the stiffness of a surface raft as B/L becomes large. The stiffness envelope may also be matched closely by using an equivalent pier approximation of the pile group (Poulos and Davis, 1980; Randolph, 1994), demonstrating the robustness of calculations of pile group stiffness even with quite approximate models (Randolph, 2003).

From a practical viewpoint, it is also useful to link the average settlement of a pile group to the dimension, B, of the foundation and the factor of safety against ultimate capacity. This has been explored by Cooke (1986) and more recently by Mandolini (2003), who suggest that for fine-grained soils the average settlement is typically around 0.3 to 0.6% of the foundation width, B, for a factor of safety of 3.

For a given load per pile, piles in a group will settle more than a corresponding isolated pile. The increase in settlement may be quantified by a settlement ratio, R_s, giving the ratio of the flexibility of a pile in the group to that of an isolated pile (Poulos and Davis, 1980). An alternative way of quantifying the decreasing stiffness of each pile due to interaction effects is by means of an efficiency, η_w, which is the inverse of R_s (Butterfield and Douglas, 1981). The stiffness, K_g, of the pile group (total load

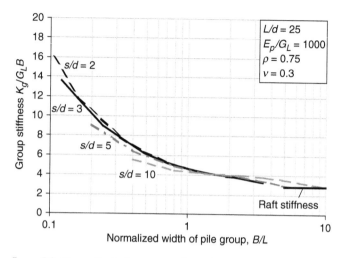

Figure 5.8 Normalized pile group stiffness for square groups of piles.

divided by average settlement) may be expressed as a fraction, η_w, of the sum of the individual pile stiffnesses, k. Thus for a group of n piles

$$K_g = \eta_w n k \qquad (5.17)$$

Butterfield and Douglas showed that plotting the efficiency, η_w, against the number of piles in a group gave essentially straight lines on logarithmic axes. The precise layout of the piles appeared to have little influence on the computed efficiency, rectangular groups of piles having the same efficiency as square groups at the same pile spacing.

The results from Figure 5.8 have been replotted as efficiencies in Figure 5.9. For typical pile spacing of $s/d \sim 3$ to 5, the resulting trends are linear in logarithmic space, and may be represented by

$$\eta_w \sim n^{-e} \quad \text{or} \quad K_g \sim n^{1-e} k \qquad (5.18)$$

with values of the exponent, e, typically in the range 0.5 to 0.6.

The results in Figures 5.8 and 5.9 have been obtained using the program PIGLET (see following section) for groups of up to 900 piles (30×30), based on the interaction approach of Mylonakis and Gazetas (1998). For large pile groups, the efficiency of each pile falls to 0.05 and lower, indicating that each pile is only 5% as stiff as it would be as an isolated pile. Thus a few widely spaced piles may be nearly as stiff as many more closely spaced piles. For example, a group of 300 piles at $s/d = 3$ has an efficiency of 0.03, or an overall stiffness equivalent to $400 \times 0.025 = 10$ isolated piles. A group of 36 piles at $s/d = 10$ (covering approximately the same ground area) has an efficiency of 0.1, or an overall stiffness equivalent to just under 4 isolated piles. Reducing the number of piles by an order of magnitude gives a group stiffness that is nearly 40% as high.

Natural soils exhibit non-linear response below yield, with the secant modulus reducing from a maximum value, G_0, at small strains to much lower values at strain

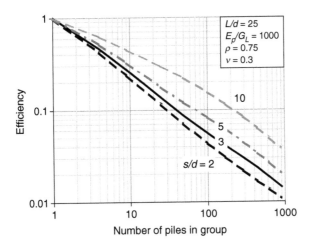

Figure 5.9 Efficiency η_ω for square pile groups under axial load.

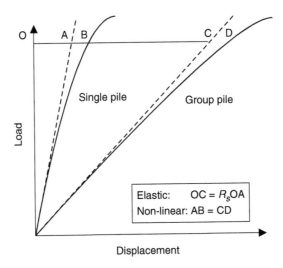

Figure 5.10 Calculation of interaction with non-linear load transfer response.

levels relevant to most foundation design. This may be taken into account for pile groups by basing elastic calculations of interaction on the initial (tangent) stiffness of the single pile response. The group stiffness is then evaluated using an appropriate settlement ratio, R_s, applied to the initial tangent stiffness, augmented by the (unfactored) non-linear displacement for the given load level (see Figure 5.10). This approach was first proposed by Caputo and Viggiani (1984) and has also been described by Randolph (1994). Mandolini and Viggiani (1997) back-analyzed the measured vertical response of several large pile groups and concluded that the response could be predicted with good accuracy by assuming elastic soil response, and adopting a shear modulus that is close to the small strain, or dynamic, modulus, G_0.

Their conclusion was that, at typical working loads, the contribution to settlement from non-linear effects is essentially negligible compared to the reduction in stiffness due to interaction.

5.2.1.1 Load distribution

A consequence of the interaction between piles is that, for uniformly loaded pile groups, the central piles will settle more than the piles at the edge of the group, while, for piles loaded though a stiff pile cap, the edge piles will take a higher proportion of the applied load than the central piles. This feature has been demonstrated experimentally by Cooke *et al.* (1980). Typical load distributions calculated for small pile groups (Poulos and Davis, 1980) show that corner piles may take up to 3 to 4 times as much load as piles near the centre. A case history reported by Cooke *et al.* (1981) showed that piles at the corner of a large pile group containing 351 piles took 2 to 3 times as much load as the central piles.

There has been some confusion among engineers as to how to react to these load distributions. There are a number of alternatives:

1 Lengthen the outer piles to ensure their stability under the higher loads.
2 Abandon the central piles since they contribute little to the stiffness of the pile group.
3 Lengthen the central piles in order to attract more load to them, thus evening out the load distribution among the piles.

The first alternative does not stand up to scrutiny since longer outer piles will attract even more load to the edge of the foundation, giving rise to higher bending moments in the pile cap or alternatively greater differential settlements. For the same reason, the second alternative is not attractive, unless the weight of the structure is concentrated at the edges of the foundation. In principle, the last alternative appears a possible option.

Assuming that the loading from the superstructure is reasonably uniformly distributed, the aim of the foundation must be to transmit the load as uniformly as possible to the soil. Ideally, bending moments in the pile cap and differential settlements should both be minimized. In general, this will require more support from piles towards the centre of the foundation than towards the outside (to counteract the natural tendency of a uniformly loaded region to dish in the centre). The approach becomes particularly attractive where the upper soil layers are reasonably competent, and the pile cap may transmit some load directly to the ground. Padfield and Sharrock (1983) discuss such a design, where differential settlements of a raft foundation could be virtually eliminated by the use of a few piles beneath the centre of the raft. This is discussed in more detail in section 5.3.

In many cases, the stiffness of the structure above the pile cap will be such as to obviate the need for concern over high loads in the peripheral piles. The simple approach of uniformly distributed piles will then be adequate, even if not the most efficient. The design should ensure that the pile material itself is not overstressed, and that due account is taken of possible high bending moments induced at the edge and corners of the pile cap.

5.2.1.2 Immediate and long-term settlements

The efficiency of piles in a group decreases as Poisson's ratio becomes smaller. This has consequences for the ratio of immediate to long-term settlement of pile groups. The solution for a single pile is relatively insensitive to the value of Poisson's ratio (for a given value of the shear modulus, G). Thus the immediate settlement of a single pile typically accounts for 80 to 90% of the long-term settlement (Poulos and Davis, 1980). However, for a group of piles, the immediate settlement is a smaller fraction of the long-term settlement. For wide pile groups (with overall width greater than the pile embedment), the stiffness becomes proportional to $G/(1-v)$ and thus the long-term stiffness for $v \sim 0.2$ to 0.3 would be only 60 to 70% of the short term stiffness $(v = 0.5)$.

5.2.1.3 Underlying soft layers

The effect of different layers of soil over the depth of penetration of the piles in a group may generally be dealt with adequately by adopting suitable values of the average shear modulus for the soil, and a value for the homogeneity factor, ρ, that reflects the general trend of stiffness variation with depth. However, particular attention needs to be paid to the case where a soft layer of soil occurs at some depth beneath the pile group, as shown in Figure 5.11. In assessing how much additional settlement may occur due to the presence of the soft layer, the average change in vertical stress caused by the pile group must be estimated.

Implicit in the solution for the load settlement response of a single pile is the idea of the transfer of the applied load, by means of induced shear stresses in the soil, over a region of radius r_m. The average vertical stress applied to soil at the level of the base of a group of piles may be estimated by taking the overall applied load and distributing it over the area of the group augmented by this amount, as shown in Figure 5.11. Below the level of the pile bases, the area over which the load is assumed to be spread may be estimated assuming a spreading gradient of 1:2 (Tomlinson, 1970).

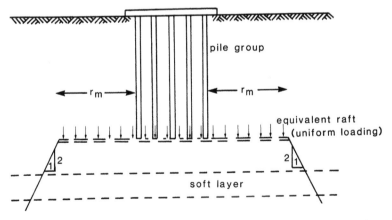

Figure 5.11 Use of equivalent raft for calculating effect of soft layer underlying pile group.

5.2.1.4 Long piles

It was noted in section 4.2.4 that long slender piles may mobilize their full skin friction over the upper part of each pile at working loads. In estimating the settlement of groups of such piles, allowance for slip between pile and soil must be made. As discussed in respect of non-linear response, the settlement ratio, R_s (inverse of the efficiency, η_w), should only be applied to the 'elastic' component of the single pile settlement. An example of this has been given by Thorburn *et al.* (1983) in relation to long piles (0.25 m square by 30 m long) used to support a molasses storage tank. A pile load test showed that a single pile would settle by 8 mm, at the working load. Of this, some 4 mm was estimated as relative slip between the upper part of the pile and the soil. Analysis of the pile group indicated a group settlement ratio of $R_s = 6.4$. Thus the working settlement of the foundation was estimated as 6.4 times the 'elastic' component of the single-pile settlement (4 mm) plus the 'plastic' settlement (4 mm), giving 30 mm, which agreed well with actual measured settlements. A straightforward factoring of the total single pile settlement of 8 mm would have given estimated settlements of over 50 mm, 60% greater than measured.

5.2.2 Lateral loading

The response of pile groups to lateral loading involves not only the lateral load deformation characteristics of the individual piles, but also the axial characteristics. As the pile group tries to rotate, piles at the edges of the group will be loaded in tension and compression, providing considerable rotational stiffness to the group. Computer-based methods such as those discussed in the following section are generally required for a complete analysis of the group under such loading, although an approach suitable for hand calculations has been described, in relation to offshore pile groups, by Randolph and Poulos (1982).

As discussed in Chapter 4, a distinction should be drawn between external lateral loads applied to a pile group, and loads induced in the piles by soil movement below ground level. The situation of a piled bridge abutment (see Figure 4.53(b)) is a common one, and due account must be taken of the effect of the embankment surcharge loading when estimating the deformation of the abutment and the load distribution in the piles.

For external load applied to the pile head, Poulos (1971) introduced four different kinds of interaction factor for piles under lateral load, depending on the loading at the pile head and the type of deformation. These were

α_{uH}: interaction for deflection of free-headed piles under force loading;
α_{uM}: interaction for deflection of free-headed piles under moment loading or for rotation of free-beaded piles under force loading;
$\alpha_{\theta M}$: interaction for rotation of free-headed piles under moment loading; and
α_{uf}: interaction for deflection of fixed-headed piles (restrained against rotation).

The last of these interaction factors is generally of most relevance in practice, since the rotational stiffness of pile groups is usually high.

Randolph (1981) has shown that interaction factors for fixed-headed piles may be estimated from the expression

$$\alpha_{uf} = 0.3\rho_c \left(\frac{E_p}{G_c}\right)^{1/7} \frac{d}{s}\left(1+\cos^2\beta\right) \tag{5.19}$$

where ρ_c and G_c are as defined in section 4.4.2, and β is the angle of departure that the piles make with the direction of loading (see Figure 5.12). At close pile spacings, this expression tends to overestimate the amount of interaction, and it is suggested that, where the calculated value of α_{uf} exceeds 0.33, the value should be replaced by

$$\alpha_{uf} = 1 - \frac{2}{\sqrt{27\alpha_{uf}}} \tag{5.20}$$

The form of equation (5.19) shows that interaction under lateral loading decreases much more rapidly with the spacing, s, than for axial loading. In addition, interaction of piles normal to the applied lateral load is only half that for piles in line with the load.

As discussed above, the lateral response of a pile group will depend on the axial stiffness of the piles in addition to their lateral stiffness, and general analysis of the group is necessary. Only if rotation of the pile cap (and thus axial movement of the piles) is prevented, do the piles deflect purely horizontally. For this case, it is possible to think in terms of a 'deflection ratio', R_u, defined in a similar way to the settlement ratio, R_s for vertical loading. Thus R_u, is the ratio of the average flexibility of piles in a group, to that of a single pile, for lateral deflection under conditions of zero rotation at ground level. Figure 5.13 shows this ratio for square pile groups containing up to 121 piles at $s/d = 3$, for a range of pile stiffness ratios (represented by different values of slenderness ratio L_c/d (see Chapter 4)). Typical values of deflection ratio are significantly lower than for vertical loading and decrease with decreasing stiffness (or slenderness) ratio. As for vertical loading, values for $\rho_c = 0.5$ are well below those for homogeneous soil ($\rho_c = 1$), reflecting the lower interaction between piles for soil where the stiffness is proportional to depth.

Limitations of the elastic approach to pile interaction under lateral loading have been revealed from model pile tests carried out on a centrifuge by Barton (1982). These tests showed that leading piles of groups embedded in sand carried a higher proportion of the overall applied load than the trailing piles. This arises out of the effects discussed in section 5.1.2, whereby the limiting pressure on a pile lying in the wake of another pile is reduced. For pairs of piles at close spacing ($s/d = 2$), Barton found that the applied load was shared 60% to the front pile and 40% to the back pile.

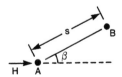

Figure 5.12 Definition of departure angle, β.

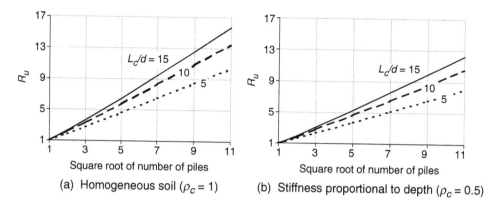

Figure 5.13 Deflection ratios for square fixed-head pile groups under lateral load.

In contrast, Matlock *et al.* (1980) report results of tests on pile groups in clay, where a uniform sharing of the applied load was found. Clearly, the adequacy of the elastic approach will depend on a number of factors such as the load level, soil type and pile spacing. However, at present there is little alternative for routine design work.

5.2.3 Computer programs for pile group analysis

The complexity of the problem of a pile-reinforced soil mass has necessitated the use of computer-based methods of analysis. In many cases, the results of the computations are amenable to presentation in graphical, or equation form, and examples of these have been given in the preceding sections and in Chapter 4. However, estimation of the deformations and bending moments induced in a group of piles under general loading conditions will usually require computer analysis. Even the relatively simple problem of the axial response of a pile group may still require analysis by computer, if the geometry of the group falls outside the range of charts such as those given in Poulos and Davis (1980).

It is important to allow for the axial and lateral support of the soil in the analysis of the pile group, even where the piles are primarily end bearing. The simplest approach is to use a standard (structural) frame analysis package, where the piles are replaced by equivalent cantilevers (Figure 5.14). Each cantilever is assumed to have the same bending rigidity as the pile it replaces, with the length chosen according to the criterion discussed in section 4.4.3 (approximately half the critical pile length under lateral loading). The cross-sectional stiffness of the cantilever is then chosen so as to give an appropriate axial stiffness (for example, by comparison with the axial stiffness calculated from equation (4.43)).

While the equivalent cantilever approach allows structural frame analysis packages to be utilized, its main limitation is that no allowance is made for interaction between piles. As discussed in section 4.4.3, the equivalent cantilever idealization will generally overestimate the maximum bending moments in the piles.

An alternative and preferable approach is to use purpose-written computer programs, where the pile and soil properties may be input directly. While there are a host

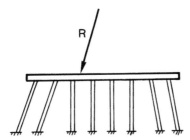

Figure 5.14 Frame analysis of pile group with piles treated as fixed-ended struts.

of individual computer programs written for pile group analysis, mention will be made here of three that are currently widely available, and for which some comparison of results has been undertaken. The three differ in the degree of rigour, and numerical complexity which is adopted and only a very brief description will be given here. Other piling software may be sourced through the Geotechnical and GeoEnvironmental Software Directory (www.ggsd.com).

1   *PGROUP*. Originally written by Banerjee and Driscoll (1976), this program is managed by the Transport Research Laboratory, UK. It is based on a full boundary element analysis of the complete pile group. The soil is modelled as an elastic material, and later versions of the program allow for variations of the soil modulus with depth, treating the soil as a two-layer continuum. The effect of a pile cap in contact with the ground may be analyzed and the piles may be raked in the plane of loading. This program is, in principle, the most rigorous of the three programs discussed, but needs considerable computing resources. The relatively coarse discretizing of the piles can lead to questionable accuracy in the results, particularly for laterally loaded piles.

2   *DEFPIG*. Written by Poulos (1980), the program is based on integral equation techniques for the analysis of single piles and the computation of interaction factors between pairs of piles. For pile groups, these interaction factors are used together with the single pile stiffnesses to yield an overall stiffness for the group. Variations in soil stiffness with depth, and allowance for yielding of the soil, are catered for in an approximate fashion. Pile cap elements may be included for vertical loading, and piles may be raked in the plane of loading. Initial calculation of the individual pile stiffnesses and interaction factors involves a large amount of computing. Thereafter, groups of similar piles may be analyzed relatively inexpensively.

3   *PIGLET*. Written by Randolph (1980), the program is based on the approximate solutions for single piles, and expressions for interaction between piles, described in this and the previous chapter. The soil is modelled as an elastic material, with a linear variation of stiffness with depth. Piles may be raked in any direction, and the effect of any type of loading analyzed, including torsion of the group. Since the program uses previously derived solutions, computing time is insignificant, and the program has recently been extended to allow load redistribution away from heavily loaded piles, by specifying maximum allowable axial loads on individual piles.

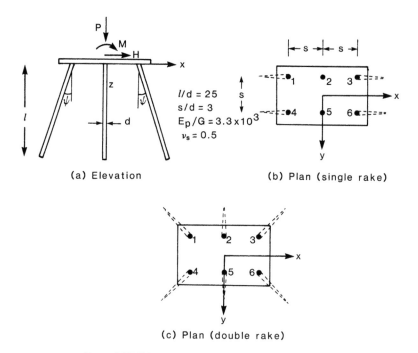

(a) Elevation

$l/d = 25$
$s/d = 3$
$E_p/G = 3.3 \times 10^3$
$\nu_s = 0.5$

(b) Plan (single rake)

(c) Plan (double rake)

Figure 5.15 Pile group analyzed in comparative study.

A detailed comparison of the results computed by the latter two programs has been presented by Poulos and Randolph (1983) (some results obtained using PGROUP are also included). As an illustration of the degree of agreement between the two programs, and also to show the type of parametric study which such programs may be used for, the effect of pile rake on a simple 3 × 2 pile group is presented here. The pile group details are given in Figure 5.15, the angle of rake of the piles being denoted by ψ. A flexibility matrix for the group may be written

$$\begin{bmatrix} w \\ u \\ \theta \end{bmatrix} = \begin{bmatrix} F_{wP} & F_{wH} & F_{wM} \\ F_{uP} & F_{uH} & F_{uM} \\ F_{\theta P} & F_{\theta H} & F_{\theta M} \end{bmatrix} \begin{bmatrix} P/Gr_0 \\ H/Gr_0 \\ M/Gr_0^2 \end{bmatrix} \tag{5.21}$$

Since the group is symmetric, the terms in F_{wH}, F_{wM} and $F_{\theta P}$ will be zero. Also, from the reciprocal theorem, F_{uM} must equal $F_{\theta H}$, giving only four independent flexibility coefficients.

The variation of these coefficients with the angle of rake, ψ, is shown in Figure 5.16. Both programs give very similar results, the largest discrepancy being 18%. The most striking effect of varying the angle of rake of the corner piles is the change in sign of the group rotation under horizontal loading; groups of vertical piles rotate in the direction of loading, but the inclusion of piles raked at even modest angles can reduce this rotation and even reverse the sense of it. The effect of double rake of the piles is relatively small in the plane of loading.

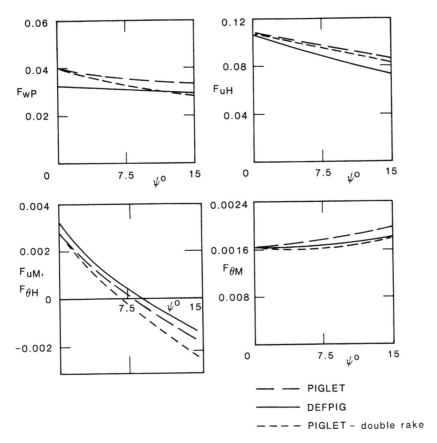

Figure 5.16 Variation of group deflection influence factors with angle of rake ψ.

5.2.3.1 Example application

An example application of pile group analysis is presented here, where the program, PIGLET, was used to evaluate the performance of piled foundations for the My Thuan Bridge in Vietnam (Chandler, 1998; Randolph, 2003). The pile foundations for the main bridge towers comprised 16 bored piles, 2.4 m in diameter and extending some 95 m below the pile cap (see Figure 5.17). The piles are grouped in two sets of 8 piles, beneath each of the legs of the tower. The critical design load cases were for ship impact, either parallel with the flow of the river (the x-axis) or at 45° to the x-axis, under conditions where a scour hole had developed to a depth of 47 m below the pile cap. Static load tests on the piles had led to a design geotechnical ultimate capacity for each pile under the scour conditions of 24.9 MN (Randolph, 2003).

Analysis of the pile group under the design load conditions leads to the distribution of loads among the piles shown in Figure 5.18. Assuming elastic response of the piles, 3 piles under load case 1 and 6 piles under load case 2 exceed the design capacity of 24.9 MN. In practice, non-linear response of the piles will reduce the extreme loads

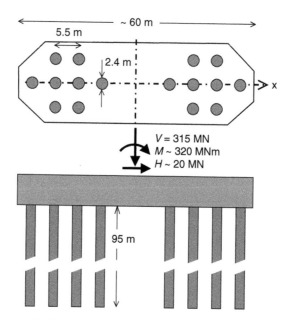

Figure 5.17 Pile layout for tower foundations of My Thuan bridge.

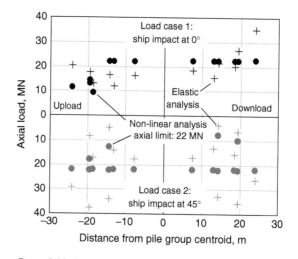

Figure 5.18 Computed axial loads for critical load cases.

predicted by elastic analysis in piles at the edges and corners of the pile group. Here, the analysis was repeated allowing redistribution of load away from the highly loaded piles, imposing a notional design limit of 22 MN on the axial 'capacity' of any pile. The load distributions for this non-linear analysis are shown in Figure 5.18. For each load case, only 4 piles remain with axial loads less than the imposed limit of 22 MN, suggesting that the pile group is close to failure in the sense of no longer being able to find a load distribution in equilibrium with the applied loading.

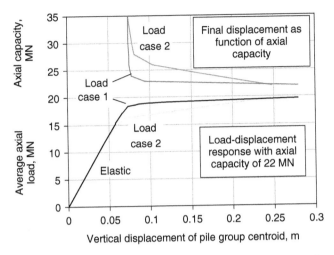

Figure 5.19 Non-linear response of pile group under design loads.

Figure 5.19 shows the development of vertical displacement at the pile group centroid for the non-linear analyses. The lower two curves are conventional load-displacement response for the two load cases, as the applied loads are factored up proportionally to their full values and with an imposed axial load limit of 22 MN on individual piles. Final displacements are 280 mm (load case 1) and 250 mm (load case 2), compared with the elastic value of 74 mm for both cases.

From a design perspective, more useful information is given by the upper two curves in Figure 5.19, which give the final vertical displacements for different magnitudes of the limiting axial load. This shows the more critical nature of load case 2, where displacements start to increase significantly once the axial load limit is reduced below 28 MN. For the actual design limit of 24.9 MN, the vertical displacement is 140 mm for this load case (and 76 mm for load case 1 – little more than the elastic value).

5.2.4 Elastic properties of soil

Elastic idealization of the non-linear stress-strain response of soil is a practical necessity for estimating the deformation of foundations under working load conditions. Provided appropriate secant values are chosen, and additional calculations are undertaken to ensure that failure, or relative slip between pile and soil, has not occurred, this approach should be satisfactory. The usefulness of the approach, however, rests on the degree of accuracy with which values of soil modulus may be chosen.

Modern site investigation techniques such as the pressuremeter and the screw plate test enable much better estimates to be made of the in-situ stiffness of the soil than those generally obtained from laboratory triaxial tests. Where such test results are not available, correlations must be relied upon, where the soil modulus is related to other properties or test results. However, the rapid development of geophysical methods, in particular the seismic cone, allows excellent evaluation of the small strain, or dynamic, shear modulus, G_0. This provides another route to estimate appropriate design values of shear modulus, by factoring down the small strain modulus according to strain levels expected in the soil. Typical shear strain levels around a single pile under working conditions will be less than 0.5% adjacent to the pile, with the shear strain level decaying inversely with distance from the pile. Within a pile group, however, interaction of displacement fields will lead to a reduction in the average shear strain level, and a working range of 0.05 to 0.1% may be assumed typically. The appropriate shear modulus for consideration of settlements will therefore be close to the small strain modulus, G_0.

For clays, an early correlation of soil modulus with the undrained shear strength was presented by Weltman and Healy (1978), who showed that the soil modulus applicable for pile loads corresponding to an average axial stress in the pile of 5 MPa was very high. A conservative fit to their data (expressed in terms of the shear modulus, G) gave

$$G \approx 0.6 \times 25^{(c_u/p_a)} \text{ MPa} \tag{5.22}$$

where p_a is atmospheric pressure (100 kPa). For c_u in the range 100 to 150 kPa, G/c_u would increase from 150 to 500.

Jardine et al. (1986) report ratios of small strain modulus to shear strength which range from $G_0/c_u = 500$ to 1000. This range is consistent with data presented by Kagawa (1992), who found that the initial shear modulus in normally and lightly overconsolidated clays was directly proportional to the consolidation (or yield) stress, with a typical ratio of $G_0/\sigma_y' = 160$. Assuming a strength ratio of $c_u/\sigma_y' = 0.25$, the implied rigidity index is $G_0/c_u = 640$. Other work has suggested a variation that varies with effective stress level to a power less than unity, for example (Viggiani, 1991)

$$\frac{G_0}{p_a} \approx A(OCR)^m \left(\frac{p'}{p_a}\right)^n \tag{5.23}$$

with typical values of m close to 0.25, n ranging from 0.65 for low plasticity clay to 0.85 for high plasticity clay, and corresponding modulus numbers, A, ranging from 500 down to 200.

For sands, Carriglio et al. (1990) presented data that are consistent with a relationship of the form

$$\frac{G_0}{p_a} \approx 400 \left(1 + 2I_D\right) \left(\frac{p'}{p_a}\right)^{0.5} \tag{5.24}$$

where I_D is the relative density, p' is the mean effective stress and p_a is atmospheric pressure (100 kPa). The value of the modulus number (400) and the exponent (0.5) will vary for different soils, with the exponent generally in the range 0.4 to 0.55, and the modulus number decreasing with increasing silt content of the soil.

Correlations of small strain shear modulus with cone resistance, q_c, and effective overburden stress, σ_{v0}', have been given by Baldi et al. (1989) and Robertson (1991), where the ratio G/q_c is taken as a function of $q_c/\sqrt{p_a\sigma_{v0}'}$ as shown in Figure 5.20.

Correlations of shear modulus with SPT value, N, have also been proposed. Wroth et al. (1979) have discussed a range of correlations, based on the work of

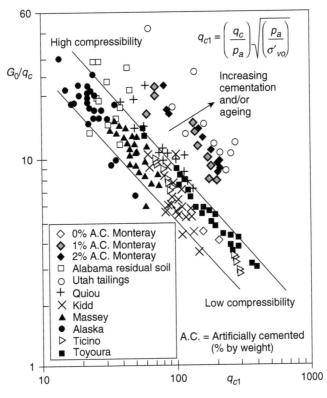

$$q_{c1} = \left(\frac{q_c}{p_a}\right)\sqrt{\left(\frac{p_a}{\sigma'_{vo}}\right)}$$

Figure 5.20 Correlation of G_0, q_c and σ'_{vo}. (After Fahey, 1998.)

Ohsaki and Iwasaki (1973) and other workers, that may be expressed as

$$\frac{G_0}{p_a} \approx 120N^{0.8} \qquad (5.25)$$

For vertical loading, strain levels are such that the small strain modulus may be used directly in estimating settlements (Mandolini and Viggiani, 1997), or possibly reduced by up to 50% where fewer piles are used (such as in piled rafts).

For horizontal loading, higher strains will occur in the soil and lower shear modulus values will be appropriate, perhaps a factor of 2 lower than for vertical loading. Randolph (1981) has suggested the simple, but conservative, correlation of

$$\frac{G}{p_a} \approx 10N \qquad (5.26)$$

at working load levels, which is 15 to 20% of the G_0 correlation in equation (5.25). Alternatively for laterally loaded piles, the soil shear modulus may be taken as varying from zero at the ground surface (where the strains are highest) up to the full value taken for vertical loading, at a depth equal to the critical pile length.

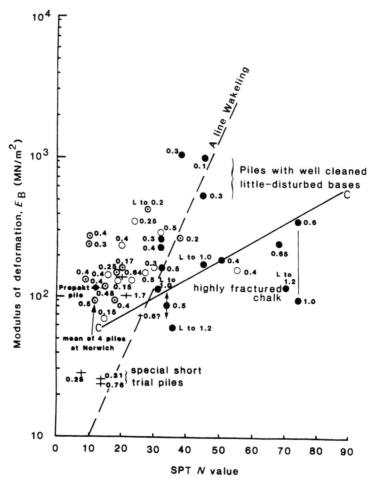

Figure 5.21 Correlation of Young's Modulus for chalk against SPT value: ● Borded piles category 4, ⊙ Driven piles category 3(a), O Driven piles category 3(b), + Driven piles category 2(b) (West Shell Piles, Portsmouth), C–C Slope derived from calculations to obtain a reasonable relationship between modulus, N value and depth.
Numerical values are stress on base (MPa) at the end of linear (L) portion of these curves. (Hobbs and Healy, 1979).

In soft rocks, the appropriate stiffness will be greatly affected by the degree of weathering, and by the method of pile installation. Hobbs and Healy (1979) have correlated values of soil modulus, back-analyzed from pile load tests in chalk, with the SPT number, N. Their correlation is shown in Figure 5.21. There is considerable scatter in the results, much of it reflecting varying amounts of disturbance due to pile installation. Even with good quality chalk, pile driving can loosen the block structure and reduce the stiffness by an order of magnitude. Randolph and Wroth (1978) quote values of shear modulus, deduced from load tests on driven cast-in-situ piles end bearing on good quality chalk, which are as low as 36 MPa for grade I–II chalk.

5.3 Piled rafts

As discussed at the start of this chapter, where competent soil exists close to the ground surface, a combined pile and raft foundation may provide the most economic form of foundation. Early analytical work by Butterfield and Banerjee (1971) suggested that, although the proportion of load taken by the piles would drop, inclusion of a ground-contacting pile cap would have little effect on the stiffness of a pile group. This finding has been substantiated by experimental work (Abdrabbo, 1976; Cooke *et al.*, 1979). It is also true that, in order to increase the stiffness of a raft foundation significantly, piles need to penetrate to depths greater than the width of the raft, or to reach strata that are substantially stiffer than the surface sediments.

Early numerical techniques for the detailed analysis of piled raft foundations, including assessment of bending moments and differential settlements, have been described by Hain and Lee (1978) and also by Padfield and Sharrock (1983) (although treatment of the piles in the latter approach is only approximate). Approximate treatments of mixed foundations are also available for rigid rafts using the programs PGROUP and DEFPIG, discussed in section 5.2.3. Over the last decade, however, a variety of analytical approaches of varying complexity have been developed specifically for piled rafts (Poulos, 1994, 2001; Clancy and Randolph, 1996; Franke *et al.*, 1994; Ta and Small, 1996; Russo, 1998; Katzenbach *et al.*, 2000; Reul and Randolph, 2003).

Design strategies for pile groups and piled rafts have been discussed by Viggiani (2001) and Mandolini (2003). Viggiani (2001) has suggested that they may be divided into two main categories:

1   'Small' pile groups, where the ratio or overall width B of the group to the pile length L is less than unity. Piles are needed to ensure adequate bearing capacity, and the pile cap (or raft) can easily be made sufficiently stiff to eliminate differential settlements. Even where the pile cap bears directly on the ground it will not contribute significantly to the overall performance of the foundation.

2   'Large' pile groups, with $B/L > 1$, where the pile cap alone will usually provide sufficient margin against bearing failure, and will contribute significantly in terms of transferring load directly to the ground. The design of such foundations hinges more on limiting the average and differential settlements to a acceptable level. Since for large rafts the flexural stiffness will be low, the location and length of any pile support should be chosen in order to minimize differential settlements.

A simple approach to evaluating the overall stiffness of a piled raft, and assessing the load sharing between pile group and cap, has been suggested by Randolph (1983). The approach is based on the use of average interaction factors, α_{rp} between the pile and pile cap (or raft). Writing the raft stiffness as k_r and the pile group stiffness as k_p, the overall foundation stiffness, k_f, is obtained from

$$k_f = \frac{k_p + k_r\left(1 - 2\alpha_{rp}\right)}{1 - \alpha_{rp}^2 k_r/k_p} \tag{5.27}$$

while the proportion of load carried by the raft, P_r and the pile group, P_p is given by

$$\frac{P_r}{P_r + P_p} = \frac{k_r\left(1 - \alpha_{rp}\right)}{k_p + k_r\left(1 - 2\alpha_{rp}\right)} \tag{5.28}$$

The stiffnesses of the pile cap and pile group may be evaluated conventionally, while the interaction factor, α_{rp}, may be obtained from the approximate expression (Randolph, 1983)

$$\alpha_{rp} = \frac{\ln\left(2r_m/d_r\right)}{\ln\left(2r_m/d\right)} = 1 - \frac{\ln\left(d_r/d\right)}{\zeta} \tag{5.29}$$

where r_m is the radius of influence of the piles (see section 4.2.1) and d_r is the effective diameter of the element of raft associated with each pile. This may be calculated so that, for a group of n piles, $n\pi d_r^2/4$ equals the actual area of the raft. More rigorous analyses reported by Clancy and Randolph (1993) show that, as the pile group size increases, the value of α_{rp} tends towards a constant value of about 0.8.

Figure 5.22 shows how the stiffness of a single pile under vertical load is increased by the presence of a pile cap in contact with the ground. The approximate method outlined here is compared with results from Poulos and Davis (1980) and shows good agreement. The results show that a small pile cap ($d_r/d \leq 4$) has very little effect on the stiffness of the foundation, although a significant proportion of the overall load may be carried by the pile cap (see Figure 5.23). Comparison of the approximate analysis with a more rigorous numerical analysis for the case of a 5×5 pile group

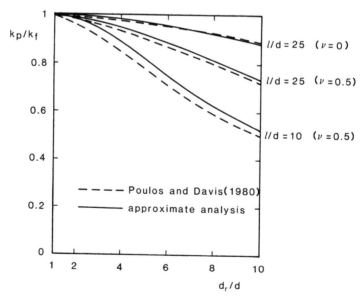

Figure 5.22 Effect of pile cap on single pile stiffness.

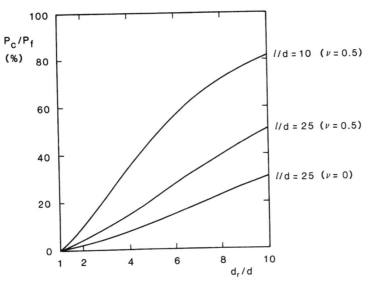

Figure 5.23 Proportion of load carried by pile cap.

shows good agreement in terms of overall stiffness (within 6%), but a tendency for the approximate analysis to underpredict the amount of load carried by the raft by up to 30%, especially where the pile-soil stiffness ratio is low (Griffiths *et al.*, 1991).

In recent years, there has been an increasing awareness that the pile support beneath (primarily) raft foundations may be optimized by the inclusion of a limited number of piles beneath the central region of the raft only (Padfield and Sharrock, 1983; Randolph, 1994; Horikoshi and Randolph, 1998) In principle, central pile support can minimize both differential settlements and bending moments within the raft. The key design decisions in this approach are the number and size of piles, and the raft area over which pile support is required.

5.3.1 Raft-soil stiffness ratio

For a raft foundation, it is important to assess initially what differential settlements and bending moments would occur without pile support. The raft-soil stiffness ratio is a key factor in that regard. Horikoshi and Randolph (1997) have proposed a consistent definition of raft-soil stiffness ratio, K_{rs}, defined as

$$K_{rs} = 5.57 \frac{E_r}{E_s} \frac{1 - v_s^2}{1 - v_r^2} \left(\frac{B_r}{L_r}\right)^{0.5} \left(\frac{t_r}{L_r}\right)^3 \qquad (5.30)$$

where E_r, v_r and E_s, v_s are Young's modulus and Poisson's ratio for raft and soil, respectively, B_r, L_r and t_r are the breadth, length $(B_r \leq L_r)$ and thickness of the raft. The factor 5.57 (or $\pi^{1.5}$) provides consistency between circular and square rafts of the same area, using the standard raft-soil stiffness ratio for circular rafts introduced by Brown (1969).

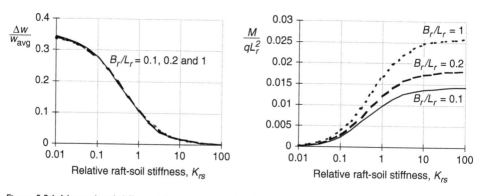

Figure 5.24 Normalized differential settlements, $\Delta w/w_{avg}$, and normalized central bending moment, M/qL_r^2.

The power of 0.5 in the B_r/L_r term in equation (5.30) ensures similar normalized differential settlements, regardless of the aspect ratio of the raft, as shown in Figure 5.24(a). In addition, the variation with raft-soil stiffness ratio of the maximum bending moment in the raft is qualitatively similar for rafts of different aspect ratios, even though the limiting maximum moments for very stiff rafts differ (Figure 5.24(b)).

5.3.2 Optimized central pile support

Where a primarily raft foundation is planned, a few piles may be incorporated beneath the central part of the raft, in order to reduce the differential settlements to an acceptable level (Burland *et al.*, 1977). Such 'settlement-reducing' piles provide a measure of reinforcement to the soil and help prevent dishing of the raft in the centre (see Figure 5.25). From a design point of view, they may be regarded as absorbing some part of the overall load applied to the raft. Under working conditions, the full shaft capacity of the piles may be mobilized over most of the length, excluding a region just beneath the raft where the shear transfer will be reduced (Combarieu and Morbois, 1982).

Horikoshi and Randolph (1996) demonstrated the effectiveness of central pile support by means of centrifuge model tests where a group of 9 piles was installed beneath the centre of an extremely flexible raft foundation. The piles were successful in largely eliminating differential settlements, although had relatively little effect on the average settlement. Since the raft alone will have sufficient bearing capacity in this type of foundation, the design question becomes not 'how many piles are needed to carry the weight of the building', but 'how many piles are needed to reduce the (differential) settlements to an acceptable level – and where should they be positioned'?

The various geometry and other parameters that need to be resolved are shown schematically in Figure 5.26. Horikoshi and Randolph (1998) investigated this problem in order to try and optimize the location and size of pile support, and indicated that under optimal conditions the absolute stiffness of the raft, k_r, by itself should

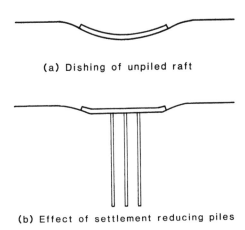

(a) Dishing of unpiled raft

(b) Effect of settlement reducing piles

Figure 5.25 Use of settlement-reducing piles to minimize differential settlement.

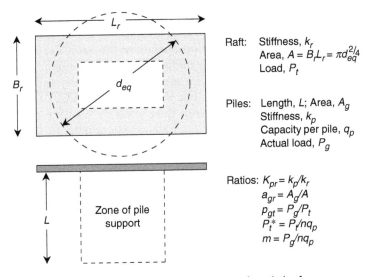

Raft: Stiffness, k_r
       Area, $A = B_r L_r = \pi d_{eq}^2 / 4$
       Load, P_t

Piles: Length, L; Area, A_g
        Stiffness, k_p
        Capacity per pile, q_p
        Actual load, P_g

Ratios: $K_{pr} = k_p / k_r$
         $a_{gr} = A_g / A$
         $p_{gt} = P_g / P_t$
         $P_t^* = P_t / n q_p$
         $m = P_g / n q_p$

Figure 5.26 Key geometric parameters for piled raft.

be similar to that of the pile group alone, k_p, hence $K_{pr} = k_p / k_r$ is close to unity. For moderate-sized rafts, they found that the pile group is most effective if the pile length is between 1 and 2 times the equivalent diameter of the rectangular raft ($L/d_{eq} \sim 1$ to 2).

Differential settlements were minimized by locating the pile support over the central 20 to 30% of the raft ($a_{gr} \sim 0.2$ to 0.3). For that arrangement, the optimum design conditions to give essentially zero differential settlement are where the total pile capacity, nq_p, is 40 to 50% of the total applied load P_t. As shown in Figure 5.27, for that load range, the piles are loaded to approximately 80% of their ultimate capacity ($P_g/nq_p \sim 0.8$) and the pile group as a whole carries some 30 to 40% of the total applied load (Horikoshi and Randolph, 1998).

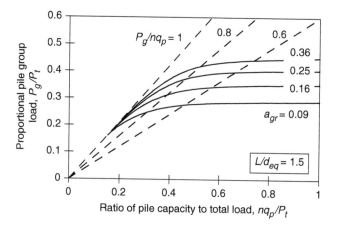

Figure 5.27 Load sharing between raft and pile group.

These results have been confirmed by other parametric studies, which have considered uniform loading of the raft (Prakoso and Kulhawy, 2001; Viggiani, 2001). For many buildings, however, a significant proportion of the load is applied at the edges of the raft through the outer walls of the building. Reul and Randolph (2004) (see also Randolph, 2003) have considered such a case, with a relatively extreme load distribution with 50% applied as a uniform line load around the edge of the raft, and the remaining load uniformly distributed over the central 25% of the raft (representing the load applied by the central core of the building containing lift shafts, etc.). Even for that case, it was found that differential settlements (for thin rafts), or bending moments (for thick rafts) could be minimized using central pile support.

As an example, application of the approach outlined for a complete raft, the foundations for a 16-storey block of flats built on London Clay are considered. The building, at Stonebridge Park in the London borough of Brent, was founded on a pile group consisting of 351 bored piles, each 0.45 m in diameter and 13 m long, capped by a pile cap in direct contact with ground (see Figure 5.28). The pile cap and piles were extensively instrumented to measure contact stresses between the raft and the ground, and the loads going into typical piles in the group. Results of the field study, and implications for the design of such foundations, have been presented by Cooke *et al.* (1981). A numerical analysis has also been reported by Padfield and Sharrock (1983).

The shear strength profile for the site may be approximated by the straight-line relationship

$$c_u = 100 + 7.2z \text{ kPa} \tag{5.31}$$

where z is the depth in m below foundation level (2.5 m below ground level). Adopting a ratio for G/c_u of 200 (Simpson *et al.*, 1979) for the London Clay, the shear modulus profile is given by

$$G = 20 + 1.44z \text{ MPa} \tag{5.32}$$

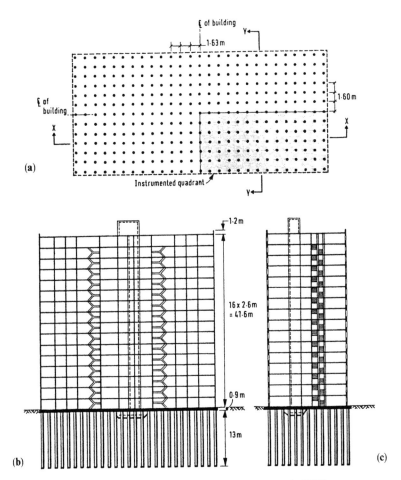

Figure 5.28 Plan and elevations of Stonebridge Park Flats (Cooke *et al.*, 1981):
(a) foundation plan (351 piles 450 mm diameter and 13 m long); (b) longitudinal
section $x - x$; (c) cross-section $y - y$.

The overall dimensions of the pile cap were 20.1 m by 43.3 m. Thus the average
stiffness of the cap, acting alone, may be estimated (Poulos and Davis, 1974) as

$$k_r = \frac{2G}{I(1-v)}\sqrt{20.1 \times 43.3}$$
(5.33)

where the influence factor I is 0.85 for a raft of this aspect ratio. The value of G may
be taken as the value at a depth of $B(1 - 0.5B/L)$, where B and L are the breadth and
length of the pile cap. This gives $G = 42$ MPa, and a cap stiffness (taking $v = 0.1$) of
$k_r = 3250$ kN/mm.

The stiffness of the pile group by itself may be estimated, using programs such
as PIGLET, as $k_p = 5300$ kN/mm (Randolph, 1983). The piles were laid out on a
rectangular grid, with spacing of 1.60 m by 1.63 m. Thus each pile had an area of

2.61 m^2 of cap associated with it, from which the effective diameter of each cap element may be calculated as 1.82 m. The radius of influence of each pile is $r_m = 22.2$ m (from equation (4.39), taking $v = 0.1$). Equation (5.29) then gives $\alpha_{rp} = 0.70$. The overall stiffness of the foundation may then be calculated as $k_f = 5720$ kN/mm with 24% of the load being carried by the pile cap (equations (5.27) and (5.28)).

Cooke *et al.* (1981) report an overall load on the foundation of 156 MN, from which the estimated average settlement may be calculated as $156/5.72 = 27$ mm. The field measurements indicate a long-term settlement of about 25 mm, with the proportion of load carried by the pile cap dropping from an initial value of about 50%, down to a long-term value of 23%. These values compare well with the estimated values obtained above.

Randolph (1983) commented that the piles in this foundation design were only contributing about 5% of their possible stiffness as individual piles. Halving the number of piles under the foundation would have reduced the pile group stiffness by only 13%, giving an estimated average settlement of 29 mm for the building, with some 31% of the load being taken by the pile cap.

Padfield and Sharrock (1983) discuss an alternative design of the foundation for Stonebridge Park (see Figure 5.28) which uses 40 piles, situated near the centre of the raft, at a spacing of 3.2 m (7.1 pile diameters). Horikoshi and Randolph (1999) suggest even fewer piles, with a 18 piles on a 3×6 grid over the central 25% of the raft. The proposed piles were 0.5 m in diameter, penetrating 28 m below the raft, at a spacing of 3.5 to 4 m. Both of these schemes appear to eliminate virtually all differential settlements, even for a flexible raft.

An estimate of the average settlement of the foundation proposed by Horikoshi and Randolph (1999) is based on their pile group stiffness of $k_p \sim 3000$ MN/m (estimated by two independent calculations) and an interaction factor α_{rp} of 0.66. Using the raft stiffness given above and the relationship in equation (5.27), the overall stiffness of the piled raft is calculated as $k_{pr} = 3710$ MN/m. The average settlement of the piled raft is therefore estimated as 42 mm under the design load of 156.6 MN.

This example serves to demonstrate that significant savings may be made in foundation design by eliminating piles that are not needed from a stability point of view. Relatively small increases in average settlement will occur due to the reduction in the number of piles, but this is offset by a significant reduction in the differential settlements.

There are of course many considerations that come into the design of a piled foundation. Combined pile and raft foundations are only viable where the surface soil layers are reasonably competent, and even then the number and spacing of piles may well be determined by the spacing of the columns carrying the structural load. The use of one pile beneath each column is a convenient way of minimizing bending moments in the pile cap. However, the piles may still be designed to take only a proportion of the column loads, the remaining load being shed into a pile cap cast directly on the ground. In other situations, the principle of concentrating the pile support in the central area of the raft may be followed by varying the pile length across the foundation. A good example of that has been reported recently by Liew *et al.* (2002), where piles of lengths varying from 36 m towards the centre down to 24 m at the edges were used to support a 20 m diameter oil tank on soft ground.

5.4 Discussion of design principles

Recent years have seen major advances in understanding the manner in which pile foundations interact with the surrounding soil. This and the preceding chapter have presented some of the modern analytical methods available, concentrating on those that may be used directly (that is without excessive computational effort) in design calculations.

The choice of piles of sufficient axial capacity is still generally the key issue in pile design. However, interpretation of the word 'sufficient' must be made in the light of the foundation requirements. In particular, settlement considerations now play a more dominant role in pile design. As will be discussed further below, the choice of what 'factor of safety' is appropriate for pile foundations will depend to a large extent on how much settlement may be tolerated.

Practical considerations, such as those discussed in Chapters 2 and 3, will play an important part in pile design, particularly as regards choice of pile type, installation method and even the overall layout of the pile group. The role of analysis is then to optimize the foundation and to assess, through parametric studies, the effect of uncertainties in assumed soil and other design parameters. It may be helpful to review here some of the main advances in analytical methods that have been made over the last decade or so.

1 In calculating the axial capacity of a pile simple 'total stress' methods have been augmented by a better understanding of the effective stress changes that occur during and after pile installation. No complex pile design would now be complete without an assessment of the effective stress state implied by a total stress calculation of capacity. Quantifying the gradual degradation in shaft friction along the length of piles driven into sand is a significant conceptual change from previous design methods based on a critical depth.

2 The concept of 'group efficiency', based on a comparison of the capacity of the individual piles in a group with that of the surrounding block, is generally considered inappropriate. Individual piles may be considered more or less efficient when part of a group, due to differences in the stress state around the pile. However, this is a separate consideration. The two modes, of individual pile and block failure, should be appraised independently.

3 The concept of 'efficiency' of a group is more appropriately used in respect of the stiffness of the foundation. Conventionally designed large pile groups will have efficiencies as low as a few percent, and this has highlighted the need to improve design methods for such groups, optimizing the quantity and location of the piles.

4 The lateral response of piles may be estimated simply using a consistent approach, treating the soil as a continuum. This approach avoids uncertainty in the choice of coefficient of sub-grade reaction, and enables group effects to be assessed within the same framework.

5 Computer methods of pile group analysis under general loading conditions are now widely available. The methods take proper account of the lateral support offered by soil around the piles and enable more realistic estimates to be made of the load distribution in the piles under working conditions. In particular, unnecessary use of raking piles to take any lateral component of load may be avoided.

5.4.1 Limit state design and Eurocode EC7

Modern design codes, such as Eurocode which is becoming mandatory in European countries, are generally based on limit states design (LSD) with factoring of the loads, soil parameters and calculated resistances with different values, i.e. the 'partial factors'. In this way they differ from traditional working stress design limits or the use of global factors of safety. With LSD there is scope for a variety of partial safety factors to be applied. In this way the system is somewhat complex because to provide flexibility whilst at the same time retaining a codified approach, there are a number of paths that can be followed. Under the EC7, the relevant Eurocode for Geotechnics, each path is designated and the permitted combinations are limited. The codes have as far as possible been tailored to suit the design approaches appropriate to a particular problem or reflect the methods that have been used successfully in the past, bearing in mind that design methods have evolved to suit the soil conditions and typical practices in each county.

Design loads and soil strengths are represented by 'characteristic values' and the partial factors are applied to these to obtain the 'design values'. The problem is then analyzed to ensure that the total effect of all the factored actions that are applied does not exceed the available resistance calculated from the factored values of soil strength, ie the calculations use the 'design values'. Partial factors of unity may be applied to given characteristic values depending on the particular Eurocode Design Approach that is adopted.

The effect of the EC7 approach on the outcome of typical calculations is somewhat variable but there is in fact no major change in the result of a typical foundation design compared with the traditional methods. Shallow foundations are not significantly different in size for a given load and ground conditions. Pile design is potentially more economical as there is scope for reducing partial safety factors in relation to increased pile testing, a feature that has become common in other codes around the world.

There are three permitted design approaches designated DA1, DA2 and DA3. The partial factors that are applied to the various elements involved in design (i.e. the actions, materials and the resistances) are stipulated for each of these methods and are given in the National Annexes to the EC7 Eurocodes that vary from country to country. There is a further tier of values of permitted partial factors within a given Design Approach. This is because the partial factors can be applied to the three elements of design in different combinations. For example, the approach under DA1 to be adopted in the United Kingdom stipulates that partial factors greater than 1 are applied to the actions or their effects and partial factors of unity are applied to the material strengths. Alternatively partial factors greater than 1 may be applied to the material strengths and factors of unity applied to the actions or their effects. A further check is then carried out to ensure that the serviceability limit state is satisfactory, and in this way deflections (especially settlement) are controlled. It is only where a piled solution relies on full mobilization of pile resistance that there may be some inconsistency in the use of LSD methods.

Although the codes are fairly prescriptive, the actual method used for design is still a free choice and there is no intention to stifle innovation. An engineer would usually apply judgement as to the soils parameters that might be used in a particular problem and also the loads that would be likely to be applied. Limit state design codes are in effect endeavouring to quantify this process, with separate consideration of each component.

5.4.2 Factors of safety for working stress design

The question of what factor of safety should be applied in pile design deserves particular consideration. Traditionally, it has been customary to design piles with ultimate axial capacities between 2 and 3 times the required working load. For example, for bored piles, Burland and Cooke (1974) recommend the use of a factor of safety of 2.0 on the combined shaft and base capacity, or, for underreamed piles, the use of partial factors of unity on the shaft capacity and 3.0 on the base capacity.*

In many areas of design, the size of the factor of safety reflects the confidence with which the ultimate capacity may be estimated. However, in many circumstances the allowable settlement of a pile foundation will be the overriding criterion in design, and the factor of safety against collapse will not be relevant. In general, it will be found that a tight settlement criterion will lead to factors of safety that are in excess of 3.

In situations where the settlement is not a key issue, for example where piles are used to support storage tanks or other structures that may undergo substantial settlement without damage, there are strong arguments for adopting relatively low factors of safety. Maintained load tests on piles generally show that the creep or consolidation settlement of the pile starts to increase significantly once the load reaches about 70% of the ultimate capacity. This finding may be used to advantage to control more precisely the load distribution within a group of piles. Thus, by designing piles with a factor of safety of 1.5, higher loads in piles at the edge or corners of a group may be avoided, since the stiffness of such piles will start to decrease at higher load levels. This approach relies on relatively uniform ground conditions.

In Sweden, there has been a move towards what is termed 'creep piling' (Hansbo and Jenderby, 1983). In essence, piles are designed at the load at which they will start to 'creep' significantly, which corresponds to a factor of safety of about 1.5. It has been found that the load distribution among the piles in a foundation may be controlled more precisely using this approach, giving resulting savings in the structural design of the pile cap.

In summary, then, it may be argued that the factor of safety should be determined by the required settlement characteristics of the structure. Low factors of safety may be used where large settlements may be tolerated, even to the extreme of a 'factor of safety' of close to unity, where piles are used as settlement reducers beneath a primarily raft foundation (see section 5.3). For large groups of piles, the settlement criterion will almost always dominate the design. Thus, provided there is a minimum factor of safety, which may be as low as 1.5, the design should be approached in terms of satisfying the settlement criterion. For another possible approach, see section 10.3.

---

*A factor of safety of 2.0 is often deemed sufficient where test piles have been loaded to failure. However, a higher factor of safety, 2.5, is recommended where only proof loads are applied to working piles.

5.4.3 Use of raking piles

Raking piles are often used to provide the necessary support to structures which are subjected to horizontal or inclined loads. The extent to which they are used is partly due to early analytical methods for estimating the load distribution in pile groups (see section 5.2.3), which ignored any lateral support offered by the pile. Modern analytical methods enable better estimates of the lateral performance of piles to be made, and more sophisticated analyses are now available for groups of piles. Such analyses indicate that, in certain circumstances, particularly where the vertical component of load is relatively small, raking piles may be used to advantage to reduce bending moments in the piles and to minimize horizontal deflections.

There are, however, certain circumstances where raking piles should be avoided.

1   Raking piles should not be used in soil that is consolidating, since vertical movement of the soil may lead to overstressing the pile in bending.
2   Raking piles should be avoided in any structure that may undergo significant vertical settlement. Again, where raking piles are forced to move vertically, the lateral component of movement may lead to overstressing of the pile.

Where raking piles *are* used, it should be borne in mind that the standards of alignment tolerance are generally lower than for vertical piles. Thus the centreline may deviate from the design by up to 1:25, compared with 1:75 for a vertical pile, and the position in plan of the pile will also be subject to a greater variation than for a vertical pile. The technical and practical difficulties in the use of raking piles are discussed further in section 10.3.2.

Chapter 6

Retaining walls

6.1 General

Although for many years sheet piles have provided the commonest way of supporting deep excavations, there are many instances in which for reasons of vibration, noise, the sheer difficulty of driving, or because the piles can not be withdrawn for re-use, the use of sheet piling is less attractive than it might otherwise be.

The practical alternative solutions are to use contiguous reinforced concrete bored pile, secant pile, or diaphragm walls; the last-mentioned might in practice be regarded as a particular form of non-circular secant pile wall. The use of bored piles for a retaining wall often goes hand in hand with the use of bored piles for the support of the main structure on a site.

Contiguous piles are constructed in a line with a clear spacing between the piles of 75 to 100 mm as a rule and in consequence cannot easily be used for water-retaining structures. Their main use is in clay soils where water inflows are not a problem, though they have also often been used to retain dry granular materials or fills. Clearly, where water is not a problem the spacing of the piles can be adjusted so long as the gap between piles is such as to prevent soil collapse between them.

Secant piles are constructed so that there is an intersection of one pile with another, the usual practice being to construct alternate piles along the line of the wall leaving a clear space of a little under the diameter of the required intermediate piles. The exact spacing is determined by the attainable construction tolerances. These initially placed piles do not necessarily have to be constructed to the same depth as the intermediate piles which follow, depending on the way in which the wall has been designed and reinforced. The concrete will also usually be chosen to have a slower rate of strength development in order to ease the problem of cutting one pile into another.

The intermediate piles are often formed through a heavy temporary casing which can be rotated by an oscillator device, the cutting edge of the casing being toothed in a way that enables the casing to cut into the concrete of the initial piles on either side. However, in recent years it has become common to install such walls using heavy continuous flight auger equipment (augercast) and although this may offer less control over tolerances of construction, the method is now widely accepted.

Obviously these piles can be used to form a continuous watertight wall or nearly watertight wall, but this depends on the constant control of tolerances for plan position and boring direction. A failure to ensure intersection will quickly make the wall non-watertight.

The increasing use of continuous-flight augers over the last two decades has provided yet another variant in the area of bored pile walls. Using this type of equipment it is possible to construct 'interlocking' bored pile walls in much the same way as for secant piles. The difference is that rotary bored 'soft' concrete piles, frequently made from concrete with the addition of a few per cent of hydrated bentonite mixed into it, are first installed. The strength of the soft material is generally in the region of a few MPa at 7 to 14 days old, at which time the structural piles are installed to intersect with them and form a continuous wall, which is capable of retaining water and soil pressures. The permanence of such soft concrete is frequently not accepted for long-term walls or those that will remain unlined by structural concrete, mainly because under conditions of freezing and thawing or wetting and drying some disintegration takes place. However, because of the ability to use continuously flighted augers, the process is rapid and economical, but the maximum available pile diameters may be more limited than for traditional secant bored piles.

Recently there has been a growing use of concrete made using ground granulated blast furnace slag for the 'soft' piles of secant or interlocking walls. These require careful design and good job planning but they offer better durability and may be used with success.

Guide walls are generally constructed to ensure the correct plan position of the tops of contiguous and secant bored pile walls, and in the construction of diaphragm walls they are essential.

In this latter form of diaphragm construction, a grab or cutting-mill is used to prepare a series of slots between a pair of parallel guide walls, set at just over the wall thickness apart. The guide wall depth is usually not less than 1 m and during excavation of each slot the stability of the ground is ensured by means of a bentonite or polymer suspension. When the trench reaches the required depth a 'stop end' is

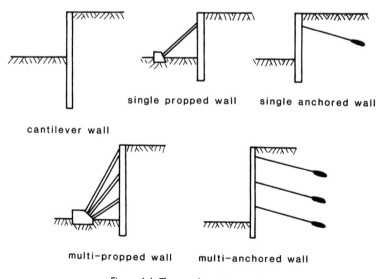

single propped wall   single anchored wall

cantilever wall

multi-propped wall   multi-anchored wall

Figure 6.1 Types of retaining wall.

placed at one end of it, or occasionally at both ends depending on the sequence of working. The function of this long element is that it can serve as a shutter and be removed after concreting to leave a close contact face for the concrete of the next adjacent panel (Sliwinski and Fleming, 1975).

When a panel has been completed and the concrete has set, the construction of the next panel in the line may be commenced in exactly the same way.

The maximum depth to which contiguous bored piles may be constructed is in practice about 18 to 20 m though in certain circumstances and particularly with wide pile spacings, it can be more. Secant piles can be bored to depths of the order of 30 m but the difficulties of construction greatly increase as pile depth goes below about 20 m and this is especially so in soils where the piles cannot be made 'dry' during construction. Diaphragm walls can be constructed using heavy rope-operated grabs to depths of the order of 50 m or more, though the difficulties of extracting very deep stop ends, where these are necessary, are considerable. This problem can to a large extent be overcome by the installation of permanent precast concrete stop ends. Alternatively 'peel off' stop ends are sometimes used, the stop end being peeled from the set concrete of a completed panel by a lateral pull.

Clearly the practical section of a wall depends very much on whether it is required to retain water or not, and it is expensive to choose a watertight type of wall, including perhaps the insertion of water barriers, when it is not required.

The reinforcement of all these types of retaining wall is conventionally by round steel reinforcing bars, but there are occasions, especially in secant piling, when the use of structural steel I-sections is preferred.

The way in which walls are used has a substantial influence on their dimensions and cost. In practice the majority of cantilever walls are constructed using bored piles, but propping or tying the wall back by means of ground anchors may reduce penetration length, deflection and bending moments. However, props tend to obstruct the working area of a site, and when other operations such as excavation, piling and slab construction have to be carried out with the prop in place, it is clear that many contractors would prefer to spend more on the wall and less on the following works.

This problem can be alleviated by the use of ground anchors, in which case a clear unobstructed working space can be given on the side of the wall where excavation has to be carried out, but because of adjacent structures or services or because other complications arise, it is frequently not acceptable to use ground anchors.

As the required retained height of a wall increases various solutions present themselves for strutting it, for example, by partially constructing each floor in a basement as the excavation is carried out, and using that section of floor as a stiff waling beam around the site. Alternatively, flying shores across a site are also used quite frequently.

The complications of strutting or tying back walls are such that the costs of deep retaining walls increase rapidly with the depth of wall required, and of course if the wall is of a diaphragm or secant type and is being used additionally as a water cut-off; the simple act of tying walls back or propping them may save nothing on the length of pile which has to be constructed in the ground.

Where the ground-embedded length of wall is in dense or medium-dense sands and gravels, in stiff clay, or in soft rocks, walls of up to about 10 m in retained height have been constructed as cantilevers, but this requires concrete walls of perhaps 1.2 m or more in thickness. Commonly 500 mm or 600 mm diameter contiguous cantilevered pile walls can be used for retained heights of approximately 5 m.

Tied or propped walls may be essential where hard rock occurs at such a depth below the dredge or excavation level that the penetration of a cantilevered pile wall is either impracticable or excessively expensive.

A special case arises with deep circular excavations, often used for pump houses for example, and sometimes for car parks. In this case a diaphragm wall can be used to surround the excavation and this can be constructed and excavated to substantial depths without the need for internal props or ties, simply relying on the self-propping action of the continuous ring so formed for support and probably a capping beam to even out non-uniformity of upper soils.

In terms of the performance of bored pile or diaphragm retaining walls it is evident that the requirements are related to some extent to the purpose for which the walls are used. It is not possible to construct any retaining wall without some deflection occurring in the whole soil system as a result. The consequences of failure of a wall depend on its height, the dangers to life that would result, and the damage to adjacent property. Clearly the attention and care in design and construction has to be greater in some cases than in others, and a sound and adequate design depends on a fundamental understanding of the soil mechanics of the problem and on experience of performance in similar circumstances.

The whole question of the design of such retaining walls is a matter which has received insufficient attention until recent years, and the complexities are such that at the present time a variety of different design methods are in use, each with its shortcomings. For major excavations, detailed analyses of the resultant soil movements have been made using finite element computer programs, but these only yield information of a quality corresponding to the input information. However, there seems little to be gained in the majority of cases by using over-complex solutions, and the basic procedures outlined in this chapter are not intended to apply to those special cases where a number of solutions need to be compared and evaluated in order to minimize damage in specially sensitive surroundings.

6.2 Soils and the data needed for wall design

The point has been made that a first priority in every piling problem is a proper understanding of the soil conditions, and in this respect the design of retaining walls makes its own special demands.

It is first of all necessary to appreciate fully the conditions of the site, its general topography and layout. If the site is one, that is basically unstable in the first place, then the installation of a retaining wall may be a waste of time and effort, or this factor may influence the design so that a much stronger or different type of wall is used. Sloping hillsides, for example, at slopes of perhaps 10 degrees or more in stiff clay, may be subject to surface soil drift in the top 2 m or so, and a result can be that passive rather than active soil pressures apply near the wall head. The site may be close to a river and the water levels in the ground may be dependent on the fluctuations in its level. Even the cutting down of trees in overconsolidated clays immediately beside a wall on the retained side may have significant effects on wall deflection and soil pressures in that vicinity.

The geological conditions need to be appreciated – the wall may be required to retain overburden over a rock stratum, in which case the direction of dip is important. If in such a case the rock is so hard that bored piles cannot penetrate it by a sufficient amount at reasonable cost, then a propped or tied wall must be used. If for example the piles or diaphragm walls are to be constructed using bentonite or polymer suspensions, the fluid could be lost, to great embarrassment, in a cavernous rock formation. These are examples of situations, which sometimes arise, where circumstances may control or dictate the type of solution possible.

The history of the site is also of interest – clearly it is very relevant to know if a site was excavated, part-excavated or filled in the past, or whether its ground has been severely contaminated by chemicals; perhaps it is underlain by old sewers or tunnels, or indeed it may have had old basements with heavy floors which, although no longer seen, would lead to extra expense in piling for new works.

In practice the fill that occurs on many sites is not examined in sufficient detail in site investigations, and yet for a retaining wall it may be one of the most important factors in deciding the basic dimensions of a wall and its cost. Soil testing is often not easy in such conditions, but good descriptions can enable reasonable assumptions to be made about the strength parameters that have to be used.

Most site investigations are concerned primarily with the bearing capacity of the ground which that is to support a new structure, and are therefore concerned with the immediate strength of the soils rather than the long-term strengths, which because of the compressive stresses are bound to be increased. The differences between short- and long-term strengths are really a function mainly of clay behaviour, and it is here that the most important problems lie.

The strengths of clay soils depend mainly on their geological history, the amount of compression they have undergone in the past and any subsequent release or partial release of that compression. Normally consolidated clays are basically as laid down, whereas overconsolidated clays have been subject to a cycle or cycles of overburden change in the past.

In the type of problem for which bored pile or diaphragm walls can provide an appropriate solution, the piles are usually installed in the ground and the earth is then taken away on one side. It is unusual, though not impossible, for the soil retained by such a wall to be placed after the construction of the wall, and setting this case aside, the effect of excavation in a clay is to release some of the applied vertical stress. In consequence, softening takes place on the passive side, with its most immediate effects being near the excavated level. Likewise, the deflection of the wall, however small, has a softening effect on the active side of the wall.

It is very unusual in the design of walls to choose to rely on passive pressure from soft normally consolidated clays in order to maintain the stability of a wall. In the first place the soil strengths are low and would require deep penetration, but in addition the deflection of a wall in the long term in such conditions is likely to be larger than can be tolerated.

A site investigation carried out before any excavation has been done and giving quick undrained strengths for a firm to stiff clay is not very relevant to the long-term stability of a wall after excavation. It is for this reason that most engineers now agree that the design of retaining walls in overconsolidated clay soils should be carried out using effective stress parameters derived either from undrained triaxial tests with pore water pressure measurements or from drained triaxial tests.

In carrying out such tests the strain rate should be slow enough to allow virtually all excess pore water pressures to equalize or dissipate in the sample as the test proceeds. The strength parameters as defined by ϕ', the frictional component, and c', the cohesive component, will be obtained, and since c' is a powerful parameter in the calculation of earth pressures with significant influence on the results, exaggerated values for this would lead to unsafe retaining wall designs. Such exaggerated values are obtained if the tests are made at too fast a rate of strain.

The choice of an appropriate stress level for testing in these clays is of some importance. In the top of the passive zone in front of a wall after excavation the conditions are such that the samples should be tested at low lateral stress levels, but in practice it is not very satisfactory to test at confining pressures below about 50 kPa using standard laboratory equipment. The range of confining pressures in which interest lies is generally between say 50 and 150 kPa, and a great deal of valuable reference data can be obtained from specialist papers on soils such as Chalk, Barton Clay, Gault Clay, Glacial Tills, Mercia Mudstone, London Clay, Oxford Clay and the Upper Lias Clay (see Table 6.1).

A good deal of debate centres around the effective stress parameter c', as stated above, since it appears to be a parameter of transitory importance. Its value depends on the degree of over-consolidation, the stress level of the test, the degree of chemical weathering to which the soil has been subjected, and the amount of sample swelling that occurred before the test. A small part of the inferred value of c' is due to real inter-particle bonds and these are destroyed by the dilation of the sample at strains beyond the peak value in the triaxial test.

This parameter is therefore one to be used with a degree of caution, but to eliminate it entirely from wall designs where the strains are small does not seem to be justified. In selecting parameters for wall design, c' should be reduced below any measured laboratory values in those strata where the largest strains are to be expected, and this

Table 6.1 Reference sources for specialist papers in various geological formations

Chalk	Hobbs and Healey (1979)
Gault Clay	Samuels (1975)
Glacial Tills	Gens and Hight (1979)
	Skempton and Brown (1961)
	Vaughan et al. (1975)
	Weltman and Healy (1978)
Mercia Mudstone	Chandler (1969)
London Clay	Bishop (1971)
London Clay	Bishop et al. (1965)
	Skempton (1970)
	Skempton (1974)
Lower Oxford Clay	Parry (1972)
Lias Clay	Chandler (1972)

means appropriate reduction in the upper part of the passive zone and probably in the active zone down to the dredge level. This practice has been successfully followed in many designs in combination with appropriate selection of factors against wall failure by rotation.

It is suggested that on the passive side of a wall the value of c' should be taken as zero at the dredge level where, in the case of both cantilever and propped walls, the strain will be at its maximum value due to both swelling and wall movement, and that it may gradually be increased to the peak laboratory value at the toe of the wall or over a depth of 1 to 2 m below the dredge level whichever is the lesser depth.

On the active side of the wall because of the wall deflection and consequent softening, it is suggested that above the dredge level c' should be reduced to about half its laboratory value.

It is sometimes argued that for temporary works in stiff clay, the undrained triaxial test results may be used in design, perhaps modified or reduced by some arbitrary factors to take into account the partial reduction of strength with time. The rate of change of intergranular stress in the soil is dependent on the soil fabric: fine sandy or silty partings, fissures, laminations and joints. Failed cuttings and trenches in stiff clay are often seen to have a blocky appearance. The material beneath the surface of a block may have a strength near that which would be measured in an undrained triaxial test, so that comparatively small volumes of absorbed water only may be involved in softening along joints, and yet a face has failed that would not have done so at normal undrained strengths. For this reason the time taken for the transition between short- and long-term strength is very unpredictable, and in designing walls for a temporary state it seems wiser to use realistic long-term effective stress parameters together with a low factor on stability, rather than undrained strengths and a much higher but uncertain factor.

As in the case of soft clays, it is usually undesirable to rely on loose sands for wall support in the passive zone because of the large strains that may occur. In order to arrive at an appropriate angle of internal friction for a sand, the current practice is usually to carry out Standard Penetration Tests, bearing in mind the corrections that may be necessary in the N value for tests at shallow depth and for fine soils in submerged conditions. The general quality and consistency of Standard Penetration Tests should also be considered bearing in mind possibilities of 'blowing' conditions in boreholes or the presence of larger soil particles obstructing the test tool.

Various proposals have been made for shallow depth correction but it is suggested that the corrections given on Figure 2.8 are appropriate.

The correction of N for submerged conditions applies only to very fine sand or silt below the water table, and is due to the excess pore water pressures set up during driving of the sampler. In this case

$$N_{corrected} = 15 + \frac{1}{2}(N_{measured} - 15) \tag{6.1}$$

The approximate relationship given by Peck *et al.* (1974) may then be used to determine the angle of internal friction ϕ', but as this does not fully take into account various features of cohesionless soils it is probably better to use the system proposed by Stroud (1989) (Chapter 2, section 2.5.4).

Having established the basic soils parameters ϕ' and c', the other parameters that have to be considered apart from the physical geometry of the system are

- active side wall friction angle δ_a;
- passive side wall friction angle δ_p; and
- wall adhesion c_w.

6.2.1 Wall friction and adhesion

Piles are frequently required to carry load by shaft friction and the existence of such friction is unquestionable in all manner of soils. In the case of a load-bearing pile it is the lateral stress that is locked in between the soil and the pile that gives rise to a force acting through friction on the pile surface in the axial direction. Bored pile and diaphragm-retaining walls have surfaces of similar roughness and are acted upon by lateral earth forces in the active and passive senses. Hence, provided there is differential vertical motion between the wall and the soil, frictional forces must be mobilized which affect the stability of the wall, and these forces should be taken into account in wall design.

Corresponding to the shear strength parameter ϕ', the forces acting on a wall are related to the parameters δ_a for the active side, δ_p for the passive side, and corresponding to c', c_w as wall adhesion. To neglect these factors is certain to lead to uneconomic designs.

In order to justify the normal reduction of active, and the increase of passive, earth pressure due to these effects, it is first necessary to ensure that the movements of the soil around the wall in any potential failure mechanism are directionally such as to give benefit.

The soil on the active or retained side of the wall has to move down relative to the wall as shown in Figure 6.2 in order to mobilize friction in the beneficial direction and, on the passive side, the displaced soil has to move upward. These conditions arise naturally with most walls, but if the wall itself is going to suffer settlement under sufficient vertically applied load, then δ_a will diminish; similarly if a heavy strip footing is placed in front of the wall on top of the passive zone it may prevent upward

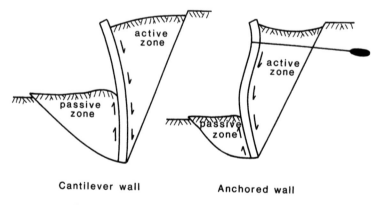

Cantilever wall   Anchored wall

Figure 6.2 Directions of soil movement at failure.

movement of this wedge and reduce δ_p. In this latter case, however, the surcharge of the footing would probably compensate for this effect on earth pressure, increasing the resistance to failure.

The question of evaluating the parameters δ_a, δ_p and c_w raises the problem that these forces have not been measured in practice but are derived rather by inference. Experience with load-bearing bored piles in granular soils suggests, from a good many instances of load testing, that the ratio δ/ϕ' is unlikely to be less in practice less than 0.7. Current practice in virtually all soils is to take $\delta_a = 2/3\phi'$ and this leads to satisfactory analysis.

The most vital part of the soil mass from the point of view of both wall stability and deflection is the passive zone in front of the wall, and because errors on this side are of much greater consequence than on the active side, it is current practice here to take $\delta_p = 1/2\phi'$.

There is no ready means of inferring values of c_w in relation to c' in clays, but by analogy with adhesion factors α, which are used in pile design in stiff clay, where α has values of the order of 0.5, so also the assumption is made that $c_w = \frac{1}{2}c'$. Because c' must be influenced by remoulding of the clay in the area immediately adjacent to the wall, it is felt that c_w should be limited in stiff clay soils to a maximum of about 15 kPa.

6.2.2 Sensitivity analysis

Having determined the basic values of all the soil parameters to be used in a design, it is often useful to consider potential errors and mis-selection in these values in the specific design in question, in order to ensure that in a worst credible situation the factor against rotational instability of a wall exceeds unity. If, however, this practice is taken to extremes and combinations of several concurrent worst credible situations are applied simultaneously, then the result will be uneconomic walls.

6.3 The basic design principles for bored pile or diaphragm walls

Before the installation of any retaining wall commences, there exists a state of lateral stress in the ground at that location. In normally consolidated soils, expressing the ratio of lateral effective earth pressure to vertically applied effective earth pressure at any point as K_0, then $K_0 = 1 - \sin\phi'$ in cohesionless soils and is perhaps a little lower in clay soils.

In overconsolidated clays, which have been subject to high overburden pressures in the past, the value of K_0 is related to the normally consolidated K_0, and the over-consolidation ratio (OCR), as indicated in Figure 6.3. The OCR is expressed as the maximum vertical stress that the soil has experienced in the past divided by the present vertical effective stress.

In practice the value of K_0 in stiff clays may be as high as 3, but lower in the near surface layers because these may have failed in a passive sense in the course of overburden removal, (i.e. K_0 cannot exceed K_p, where K_p is the coefficient of passive earth pressure).

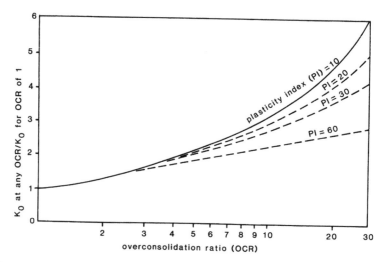

Figure 6.3 Values of K_0 related to over-consolidation ratio for soils of varying plasticity. (After Brooker and Ireland, 1965.)

If, in the course of carrying out excavation in front of a retaining wall, steps are deliberately and carefully taken to prevent movement of the wall, then the forces that would apply to that wall would correspond to the earth pressures at rest. Such a case might arise, for example, in a deep circular excavation supported by 'rigid' diaphragm walling panels, but in practice no system is so rigid as to fully maintain this pressure state and estimates have to be made of wall stresses based on some intermediate assumption – ensuring that if the worst happens a collapse will not occur. Even in the circular diaphragm wall case, the very act of digging the trench will have had significant reducing effects on the locked-in soil stresses.

In the majority of cases it is either impractical or uneconomical to resist movements in the ground, and walls can happily be permitted to undergo small movements so long as these produce no unfortunate consequences in other structures adjacent to or connected with the wall. The economic design is one that permits movement but limits it to the values necessary to develop active pressure conditions on the retained earth side.

Considering an element of soil in the ground subject to an effective principal stress σ'_v, this element may be failed either by reducing the horizontal stress σ'_h acting on it to a value where the Mohr Circle intersects the failure envelope as shown in Figure 6.4, or by increasing the effective stress until failure similarly occurs.

If this concept is applied to the earth in front of and behind a retaining wall then the soil fails in front of the wall in the passive mode and behind the wall in the *active* mode as shown in Figure 6.5 and where the cohesion intercept $c' = 0$, then

$$\frac{\textit{active} \text{ effective pressure}}{\text{vertical effective pressure}} = K_a = \frac{1 - \sin\phi'}{1 + \sin\phi'} = \tan^2\left(45 - \frac{\phi'}{2}\right)$$

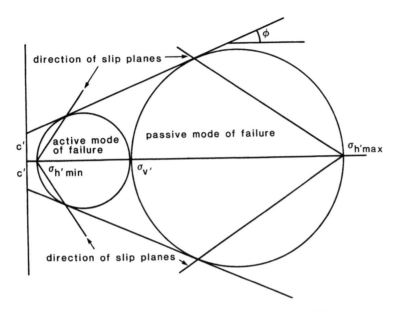

Figure 6.4 Mohr diagram corresponding to active and passive failure states.

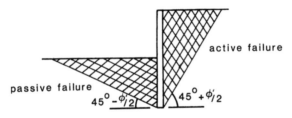

Figure 6.5 Smooth retaining wall mobilizing full active and passive pressures.

Likewise,

$$\frac{passive \text{ effective pressure}}{\text{vertical effective pressure}} = K_p = \frac{1 + \sin\phi'}{1 - \sin\phi'} = \tan^2\left(45 + \frac{\phi'}{2}\right)$$

It will be observed that the influence of $c' > 0$ is to increase the size of the Mohr Circles, to decrease active pressure, and to increase passive pressure. The decrease of active pressure due to change in c' from zero is

$$2c'\sqrt{K_a} = c'K_{ac}$$

The increase of passive pressure is correspondingly

$$2c'\sqrt{K_p} = c'K_{pc}$$

and the general expressions for earth pressure, as developed by Rankine, may be expressed as

active earth pressure $\quad P_a' = K_a(\gamma z - u) - K_{ac} \cdot c'$

passive earth pressure $\quad P_p' = K_p(\gamma z - u) + K_{pc} \cdot c'$

where γ is the soil bulk density, z is the depth below surface in a uniform soil and u is the pore water pressure. While the Rankine method displays the principles involved, the method cannot readily be extended into particular cases where wall friction exists on the faces contacting the plastic zones. Coulomb adopted a different and more versatile approach to the same problem; analyzing the stability of a non-plastic active soil wedge at the point of failure, as shown in Figure 6.6 for a smooth wall and for the level retained ground case, he obtained a solution identical to that of Rankine. Happily the form of the active and passive pressure equations may be retained for cases where inclined surfaces and wall friction are both taken into account.

The equations developed from the Coulomb approach and due to Mueller–Breslau show that the active earth pressure coefficient for horizontal pressures, K_a, is

$$\frac{\cos\delta_a \cdot \sin^2(\alpha - \phi')}{\sin^2\alpha \cdot \sin(\alpha + \delta_a)\left\{1 + \sqrt{\dfrac{\sin(\delta_a + \phi') \cdot \sin(\phi' - \beta)}{\sin(\alpha + \delta_a)\sin(\alpha - \beta)}}\right\}^2}$$

and the passive earth coefficient for horizontal pressure and level ground on the passive side, K_p, is

$$\frac{\cos\delta_p \cdot \sin^2(\alpha - \phi')')}{\sin^2\alpha \cdot \sin(\alpha + \delta_p)\left\{1 + \sqrt{\dfrac{\sin(\delta_p + \phi') \cdot \sin(\phi')}{\sin(\alpha + \delta_p) \cdot \sin\alpha}}\right\}^2}$$

where ϕ' = angle of the internal friction.
$\quad \alpha$ = inclination of the wall to horizontal (frequently 90°).
$\quad \delta_a$ = angle of wall friction on the active side of the wall (soil down drag).
$\quad \delta_p$ = angle of wall friction on the passive side of the wall (soil rising).
$\quad \beta$ = angle of retained soil surface to horizontal.

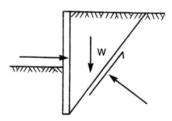

Figure 6.6 Forces acting on Coulomb active wedge at failure.

While the values of K_a calculated by the Coulomb method are widely accepted and, for $\alpha = 90°$ and $\beta = 0°$, are very similar to values for K_a from other sources, the values of K_p have been shown to be significantly too high except in cases where $\delta_p = 0$, and work by Sokolovski has led to values of K_p which are as shown in Table 6.2.

The angle of the surface of soil in the passive zone, β, is expressed in Table 6.2 in relation to the horizontal, being positive for soil and rising with increasing distance from the wall face. The coefficients, K_p, represent the coefficient of horizontal pressure on a vertical wall.

The reason for the need to calculate K_p by means other than the Coulomb wedge method is that as δ_p increases above $0°$, the lower boundary of the passive wedge becomes curved and goes deeper into the soil – rather like a bearing capacity type of failure mechanism. This leads to a reduced value for K_p, the Mueller–Breslau value for K_p without any modification is substantially too optimistic where the value of ϕ' is above about $20°$ to $25°$.

The relationship between K_a and K_{ac} and similarly between K_p and K_{pc} may be expressed as in the Rankine method of analysis, but when wall adhesion is also present a further modification has to be made.

Thus the values for the coefficients which refer to soil cohesion become

$$K_{ac} = 2\sqrt{K_a\left(1 + \frac{c_w}{c'}\right)}$$
$$K_{pc} = 2\sqrt{K_p\left(1 + \frac{c_w}{c'}\right)}$$

With the respective values for K_a, K_{ac}, K_p and K_{pc} known it is then possible to use the vertical effective stress at any point adjacent to the wall to calculate the horizontal limit of active or passive earth pressure on that element.

However, in addition to considering earth and water pressures, pressures from other imposed effects such as surcharges must be added to the problem before wall stability can be examined. The methods of dealing with some of these features are outlined as follows.

Table 6.2 Passive earth pressure coefficients (after Sokolovski, 1965)

β	ϕ'	10°			20°			30°			40°		
	δ	0°	5°	10°	0°	10°	20°	0°	15°	30°	0°	20°	40°
−30°	K_p	1.04	1.10	1.14	1.26	1.47	1.62	1.49	2.01	2.42	1.86	2.98	4.15
−20°	K_p	1.18	1.28	1.33	1.51	1.80	2.00	1.90	2.69	3.29	2.50	4.42	6.30
−10°	K_p	1.31	1.42	1.50	1.77	2.16	2.41	2.39	3.50	4.35	3.37	6.36	9.42
0°	K_p	1.42	1.55	1.63	2.04	2.51	2.86	3.00	4.46	5.67	4.60	9.10	13.9
+10°	K_p	1.49	1.64	1.73	2.30	2.88	3.32	3.65	5.62	7.29	6.16	13.06	20.4
+20°	K_p	1.53	1.69	1.80	2.53	3.26	3.79	4.42	7.13	9.27	8.34	18.3	29.9
+30°	K_p	1.52	1.70	1.82	2.76	3.61	4.24	5.28	8.76	11.69	11.3	26.7	43.4

6.4 The treatment of groundwater conditions

Although superficially water might appear to be easier to deal with than earth pressure, there are a number of problems associated with it that are difficult to quantify.

On many sites the best way of determining true groundwater levels is to use piezometers, rather than to rely on short-term borehole observations. This is especially so in the case of clay soils.

The stability of a retaining wall depends not only on the earth pressures applied to it, but also on the groundwater regime that exists around it, and the installation of a wall can itself influence and bring about changes in this. In the case where a wall is installed in a soil with a groundwater gradient, the wall, if it is impermeable, may act as a dam to produce a higher water level than existed prior to its installation and this is a difficult condition to assess. In such cases it is usual to install drains behind the wall in order to collect and divert water that would otherwise overtop it, and this can define the design condition on the 'active' pressure side. On the contrary, in the case of a permeable contiguous bored pile wall, vertical drains are sometimes installed in the gaps between piles, leading to a toe drain, and the effective water level on the 'active' and 'passive' sides will to all intents and purposes equalize.

To fully analyze the water pressures on a wall and their effects on the potential failure mechanism, it is proper to draw a flow net and, using the pore water pressures derived from this, to analyze a series of active wedges, relating the shear stresses along the planes to the net effective normal pressures. This rigorous procedure is rarely followed in practice because it would lead to complex and tedious analysis, particularly in cases where the soil is multi-layered, with strata of differing permeability. Two conditions are therefore worth considering for computational convenience in the case of impermeable walls:

1   It is assumed that the final water levels on the active and passive sides of the wall can be defined, and the water pressure diagram is drawn as in Figure 6.7 This condition would apply if at the toe of the wall as constructed there existed an impermeable stratum, but elsewhere above this the ground is considered to be permeable. This reduces the groundwater to a simple model that is widely used.
2   It is assumed that the wall is surrounded by a soil of uniform permeability, in which case, with the water levels on the active and passive sides defined, a flow net can be drawn to indicate the long-term seepage around the wall (Figure 6.8).

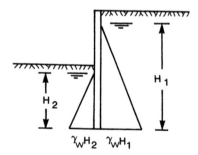

Figure 6.7 Water pressure diagram on active and passive sides of the wall.

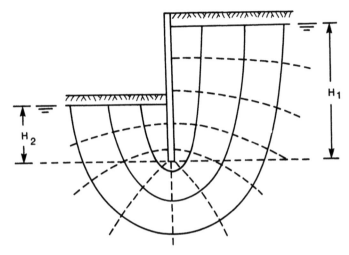

Figure 6.8 Seepage around an impermeable wall in soil of uniform permeability.

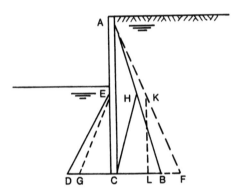

Figure 6.9 Derivation of water pressure diagram.

Rather than become involved with analysis of this kind it is convenient to make an approximation and to assume that the water head $(H_1 - H_2)$ is dissipated uniformly along the surface length of the wall $H_1 + H_2$, so that a water pressure diagram as shown in Figure 6.9 is derived.

It will be observed that this now eliminates the water pressure jump that occurs at the toe of the wall C, and the pressure at the wall toe on the active side CB equals that on the passive side CD. The lines AF and EG represent the condition of the approach described above. The net water pressure diagram now is represented by AHC whereas the system outlined in (1) above would produce a net diagram $AKLC$. The total water pressure applied to the active side of the wall is therefore significantly reduced.

However, it will also be observed that this is equivalent to reducing the bulk density of water γ_w on the active side of the wall and increasing it on the passive side, and this in turn increases the soil bulk density on the active side of the wall and reduces it on the passive side. Thus because of upward seepage pressure the passive soil pressure

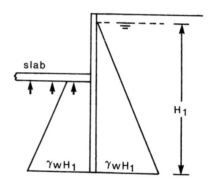

Figure 6.10 Impermeable slab subject to uplift water pressure.

is reduced and because of downward seepage the active soil pressure is increased. The question that arises therefore is one of how much difference these assumptions make in any particular design case.

It has also to be remembered, in the case of deep basements for example, that where an impermeable slab is cast in the bottom, this slab may be subject to uplift pressures in the long term unless provision is made for drainage. Thus seepage may cease and full water pressure on both sides of the wall may return to its original active side level as indicated in Figure 6.10.

Examination of cases involving cantilever walls in clay soils reveals that, where the difference in water level is say 3 or 4 m between active and passive sides, the effective difference in the finally determined wall dimensions and bending moments between methods (1) and (2) is negligible. Where the difference is say 6 m, the case (2) analysis leads to slightly deeper walls and slightly greater bending moments.

The difference in approach is probably of most concern in relation to walls that are propped or tied back by ground anchors, which are generally of greater retained height than normal cantilever walls, and which are often subject to greater water differential pressures than cantilever walls. In this case toe penetration below the dredge or excavation level may be relatively small depending on the selected position of the tie or anchor, and although the uplift seepage effect on the passive side may be small, it is prudent to recognize the sensitivity of penetration to this effect and to ensure that toe penetration is adequate to deal with any uplift seepage pressure that may exist.

In general terms therefore method (1) above is widely used and for most walls proves to be a satisfactory way of dealing with water pressures, though the assumptions made are far from rigorous.

In the case of contiguous bored pile walls, where the wall does not provide an impermeable barrier, and where high water pressures cannot normally exist on the active side of the wall, it is usual only to take into account a small nominal difference of water head between active and passive sides in the temporary state. It should be recognized that in such cases, if the excavation is subsequently lined and water rises to a higher level in the permanent state, then bending in the wall may be more severe. In all cases both the temporary and permanent states of the wall need to be considered.

6.4.1 Tension cracks

On many sites, there is overburden of fill or granular material on top of clay on the active side of a retaining wall, or alternatively the area behind the wall has been built on or paved over. These features should prevent the formation of shrinkage or tension cracks in the clay behind the wall in periods of dry weather. However, in some instances the surface of clay is exposed, and where a value of cohesion c' is used in the design it is implied that a negative active earth pressure can exist in the vicinity of the wall head. This feature in these circumstances could lead to surface cracking in the clay and the cracks could fill with water, which would in turn cause local pressure corresponding to the head of water retained in the crack. The possibility of such cracks would seem mainly to be confined to cantilever walls rather than to propped or anchored walls where the position of the anchoring force is usually located at some distance below the wall head. In this case the deflected form of the wall is such as to produce a closure of any cracks so formed.

For reasons associated with the requirements of the old British Code of Practice for Retaining Walls (CP2), it has been common to stipulate in designs a minimum equivalent fluid pressure on the active side of retaining walls of 4.7 kPa/m depth, measured below the wall head. While this feature is not of much other effect in wall design when using effective stress parameters, it does serve the useful purpose of providing a reasonable positive-pressure condition where tension cracks might otherwise occur in the immediate vicinity of the wall head.

A new British Code of Practice is now available (BS 8002) which does not specify a minimum equivalent pressure but on the other hand specifies over excavation and minimum surcharge on the retained soil. The effects may therefore be better directed to lateral increase of active soil pressure though in general they do not differ very materially.

6.5 Earth pressures due to ground surcharges

6.5.1 Uniform surcharges

A uniform surcharge acting on the active side of a wall produces a uniform vertical stress throughout the retained soil, and this is assumed to give rise to a lateral earth pressure equal to the surcharge multiplied by the active earth pressure coefficient or coefficients on the whole of the active face of the wall.

Occasions arise in which, for example, a road extends across the width of the active soil wedge, and in this case it is normally assumed that standard road loadings apply and these are taken as a uniform surcharge. In clay soils where the design parameters are chosen on a long-term basis, this approach may be rather too conservative and some modified lesser surcharge may on occasions be justified.

6.5.2 Point loads and line loads

The presence of a point, line or strip load resting on or in the soil being retained by a wall gives rise to additional lateral pressures on the wall. The usual methods for assessing the lateral stresses due to these conditions are those of Boussinesq

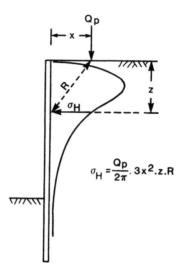

$$\sigma_H = \frac{Q_p}{2\pi} \cdot 3x^2 . z . R$$

Figure 6.11 Lateral pressure distribution produced by point load Q_p. Rigid unyielding wall leads to double the value of σ_H.

and Terzaghi. The figures given by Terzaghi (1954) are in fact a modification, based on some experimental work, of the Boussinesq method.

The Boussinesq theory is formulated to give the stresses in a semi-infinite elastic medium due to the application of a load at the surface, and if the Boussinesq formulae are used with reference to loads applied near a rigid boundary the effect is to constrain the deformation of the elastic medium and to double the horizontal stresses acting on the boundary. Thus for a point load, and taking a Poisson's ratio of 0.5, the Boussinesq relationship is as shown in Figure 6.11.

However, it is clear that the deformability of the wall and how it relates to elastic theory is dependent on the wall type. Thus a cantilever wall may deform much more than a multi-strutted wall, and the section of an anchored wall above the top anchor may act very much as if it were rigid. It is therefore important, if using the Boussinesq method, to consider carefully how it should be applied.

The lateral soils pressures acting on a wall due to a point load and as calculated by Terzaghi are shown in Figures 6.12 and 6.13; similarly the lateral pressures due to line loads are shown in Figure 6.14. Lateral pressures due to a strip load, based on the Boussinesq method as modified by experiment and given by Teng, are shown in Figure 6.15.

6.6 The use of berms

The passive pressure acting on the embedded part of a wall is a function of the vertical soil pressure acting in the potential failure zone near the wall, and this can be increased significantly by leaving a berm against the foot of the wall at the excavation level. Since in retaining-wall design the temporary condition during construction is often worse than the permanent condition, and since berms can often be accommodated

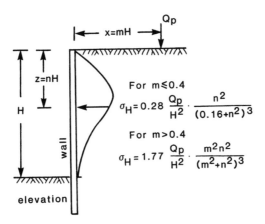

Figure 6.12 Lateral pressure due to point load (Boussinesq method modified by experiment). (After Terzaghi, 1954.)

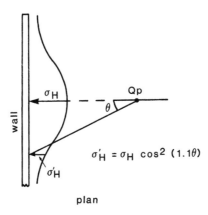

Figure 6.13 Lateral pressure due to point load (Boussinesq equation modified by experiment). (After Terzaghi, 1954.)

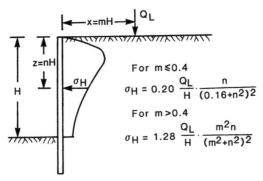

Figure 6.14 Lateral pressure due to line load (Boussinesq method modified by experiment). (After Terzaghi, 1954.)

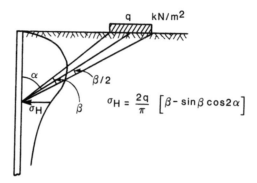

Figure 6.15 Lateral pressure due to strip load (Boussinesq equation modified by experiment). (After Teng, 1962.)

at reasonable cost in that condition and can be removed either in short sections as permanent work proceeds or at a stage when propping from the permanent works is possible, they offer a valuable means of minimizing costs in certain instances.

There are various ways in which the effects of berms can be assessed, and while little detailed research has been carried out into their behaviour, experience confirms that the approximate methods used in design are adequate. It would in theory be possible to analyze the passive failure mechanism for a wall including a berm by an iterative method involving a series of trials with an appropriate plastic failure mechanism, but this would not remove all uncertainties regarding the state of stress in the berm, effects of disturbance, softening, etc. and would lead to a protracted method that would be inconvenient in everyday design. For this reason, several approximate methods are used in practice as indicated in the following paragraphs.

1 An empirical method is to treat the berm as causing an increase in the effective ground level on the passive side of the wall. In this method the height of the berm is treated as not more than $1/3x$ the berm width as shown in Figure 6.16, and the effective ground level is then taken as half the maximum height of the berm at the point where it contacts the wall. Any stable soil existing above the 1:3 slope may then be treated as a surcharge acting on the effective ground surface over the approximate width of the potential passive failure mechanism. While this method is open to criticism on the grounds that it is highly empirical, it appears to yield relatively satisfactory and conservative results by comparison with other methods, both in terms of wall depth and of bending moment. Unless such a berm is wide, it is unwise to terminate any tied wall above or near to the final excavation level on the passive side.

2 An alternative method of treating a berm is to convert it to an effective surcharge acting on the potential passive failure zone. In this method the effective self-weight of the berm is calculated and is then distributed over the approximate width of the passive failure mechanism at the general final site excavation level (see Figure 6.17), thus increasing the passive pressure available accordingly. In addition

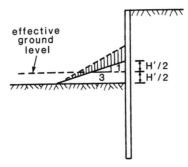

Figure 6.16 Relation of effective ground level to berm dimensions.

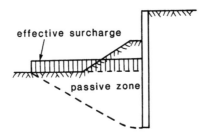

Figure 6.17 Treatment of berm as surcharge.

to the enhanced passive pressure, some shear forces must exist in the berm above the dredge level and these will contribute to the stability and to moment reduction in the wall. If the shear forces acting within the berm are not taken into account, and it is perhaps difficult to do so reliably because of softening of the material involved in the berm, then this second method must be regarded as conservative. Berms are usually and sensibly blinded with concrete in order to diminish the risk of softening in clay soils and to inhibit erosion and damage in granular soils.

6.7 Wall stability analysis

It is important to recognize that a wall can only be as stable as is the soil mass in which it exists. If there is a possibility that, as a consequence of the excavation, the whole body of soil in which a wall is installed can become unstable, then the solution to this problem becomes an overriding requirement and must be analyzed using normal slope stability methods. This potential failure mechanism is most likely to occur in cases of shoreline or waterfront structures and with walls that are placed in sloping hillsides. For waterfront structures the possibilities of scour at the dredge level must also be borne in mind, since the construction of a wall may in itself give rise to water currents, which will erode the zone of passive resistance. In the case of large swift-flowing rivers, the effects may be dramatic.

Having established the general stability of a site, it is then necessary to consider the need to satisfy the equilibrium of the vertical, horizontal and rotational forces on the wall.

Two basic concepts have been used in calculating the stability of walls, namely the so-called 'free earth' and 'fixed earth' methods. In the 'free earth' method the stability of a wall is considered without making the assumption that the deflected form of the wall is controlled by fixity in the ground. In the 'fixed earth' method it is assumed that the rotation of the toe of a wall is constrained by fixity in the soil mass and a consequence of this can be a reduction of moment in propped walls, but an increase in the embedment length. However, for such a wall there is no relevant mechanism that may be used to analyze the condition of rotation about the prop, and the method relies on a pure empirical approach.

In a cantilever wall, equilibrium is not possible without involving a reaction force at the wall toe and in practice the wall is lengthened after the basic analysis, in order to provide this. Thus the analysis in this case involves 'fixed earth', but here a mechanism of failure can be postulated to accord with the rotation of the wall at some point below the excavated level.

The idealized and simplified soil pressure diagrams associated with the analysis of cantilever and propped walls are as shown in Figures 6.18 and 6.19. In the initial stage of calculation this is simplified further (Figure 6.20).

The procedure adopted in the calculation of retaining wall stability is in general to progressively increase the embedment of the wall, analyzing the disturbing and restoring moments, until a state of stability is reached, at which point the equilibrium of horizontal forces determines the value of the reaction R (in the case of a

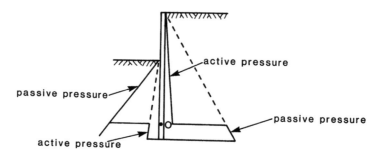

Figure 6.18 Cantilever wall: idealized pressure diagram for rotation about point O.

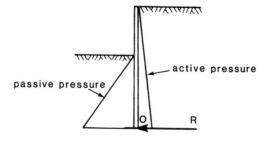

Figure 6.19 Simplified pressure diagram at rotational failure.

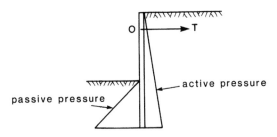

Figure 6.20 Anchored or propped wall: idealized pressure diagram for rotation about point O.

cantilever wall) or the reaction T (in the case of an anchored or propped wall). This procedure establishes the penetration of the wall below excavation that represents an incipient failure condition. However, this is of passing interest in terms of the required wall penetration for stability and is used, for reasons to be described later, only in the establishment of bending moments in the wall.

The procedure is then continued until some desired ratio is achieved between the restoring moment acting on the wall and the disturbing moment. This ratio is frequently described as a factor of safety, but for the reason that at this stage in design the wall length has not finally been determined in the case of a cantilever, and because there are other ways in which this factor can be expressed in other methods of analysis, the term 'factor of safety' is not entirely satisfactory and hence it should be regarded rather as a method related 'rotational stability factor'.

In practice this factor, F_p, can conveniently be introduced by simply dividing the passive earth pressure at all levels by F_p, but of course the water pressures acting on the passive side of the wall are not so treated.

Once the notional rotation point has been found in the case of a cantilever under this condition, the force R is determined, and reversing the roles of active and passive pressure, so that passive pressure appears on the active side below this point, an additional length of wall is added to provide the balancing force R.

In the case of the anchored wall, the rotation point in the design is constrained to be at the anchor level, and hence the length of the wall can be determined simply using a similar factor F_p applied to the passive earth pressure. Once this has been done, the force in the prop or anchor is determined as before from the horizontal equilibrium conditions.

When using an effective stress method of design such as is outlined above, the values F_p that are applicable to both cantilever and anchored walls are necessarily related to the soil properties; appropriate values for F_p are in the ranges shown in Table 6.3.

There are a number of limit state equilibrium programs that can deal with cantilever and single propped or tied retaining wall cases. Many of these have been developed by companies 'in house' but one which is generally available with a variety of options has been developed by British Steel and Geocentrix Limited under the name REWARD. In addition to the normal limit state equilibrium method it has features that allow approximate wall deflections to be calculated.

Table 6.3 Values of rotational stability factor, F_p, for various angles of friction

Effective stress parameter	Range of F_p values	
ϕ'	Temporary works	Permanent works
40	2.0	2.25
35	1.75	2
30	1.5	1.8
25	1.35	1.65
20	1.2	1.5

where F_p is rotational stability factor.

6.7.1 Walls with two or more props or ties

While the sections above deal with the cases of cantilever walls, or walls that are in part supported by a single prop or tie, it is frequently found that to reduce and make realistic bending moment in walls it is necessary to insert more than one prop or tie as the excavation proceeds. This introduces a degree of redundancy into the problem and means that a simple statical analysis is no longer adequate.

Obviously, if a wall were allowed to rotate or translate to such a degree that full active and passive pressures were mobilized, the problem would resolve itself to that of a beam on elastic foundations in conventional structural terms and the stiffness of the props would govern the bending moments in the wall and the prop forces.

However, some forms of construction such as that termed 'top-down', where floors are inserted to prop the wall as downward excavation proceeds, preclude this degree of movement, and indeed such movements would rarely be tolerable. Retained soil pressures, when sufficient movement to lead to active pressures is restrained, will lie somewhere between earth pressure at rest and active pressure, depending on the stiffness of the wall and the stiffness of the propping system.

It is for this reason that a number of computer programs have been developed, aiming to characterize the wall and props by their elastic properties and the soil as an elasto-plastic medium, with the limits of elasticity prescribed by the full active and passive states.

Various computer programs have been developed specifically for the analysis of retaining walls, and in addition there are a number of general purpose Finite Element or Finite Difference codes, which may be used to analyze retaining structures. It should be pointed out, however, that there are significant difficulties in defining accurate numbers for all the stiffnesses, especially that of the soil. Some props themselves rely on soil reactions, and concrete materials also have time-dependent stiffnesses. The best way to arrive at soil stiffnesses is to use back analysis of cases which closely resemble the problem in hand and in which the bending and prop forces have been monitored. A selection of software is given below:

CRISP-90: This is a Finite Element program developed at Cambridge University with the support of the British Transport and Road Research Laboratory. CRISP stands for 'Critical State Program' and is approved by the Department of Transport for the analysis of geotechnical problems. It is written in American National Standards Institute Standard Fortran and requires any modern desk-top computer. It uses an incremental tangent stiffness approach and has a variety of other features.

FREW: This is a program in the OASYS suite, developed by Arup, London. A line of nodal points and 3 stiffness matrices, relating nodal forces to displacements, represents the wall. The soil may be described by any one of three methods:

1   a subgrade reaction method;
2   stiffness matrices calculated by a Finite Element program; and
3   the Mindlin method, representing the soil as an elastic solid – this method is restricted to the case of constant stiffness with depth.

The program presents results for earth pressure, shear forces, wall-bending moments, strut forces and displacements. Full details are given by Pappin *et al.* (1986).

FLAC: This is a program for the solution of general geomechanical problems based on a Finite Difference method. The letters of the title stand for Fast Lagrangian Analysis of Continua and the program has been developed by ITASCA Consulting Group, Inc., PO Box 14806, Minneapolis, Minnesota 55414. Amongst other uses, it is capable of solving a range of earth retention problems, and any type of non-linear soil stress/strain relationship can be followed. It is suitable for the solution of retaining wall, tunnel-lining and rock bolting problems. It is claimed that the Finite Difference method is as flexible as the Finite Element method and requires less computational effort. The published works of Wilkins (1963), Jaeger and Starfield (1974), Cundall (1976) and Marti and Cundall (1982), are relevant to this program.

WALLAP: This program is available from Geosolve, London and offers a limit equilibrium analysis for calculating 'factors of safety', and a Finite Element analysis for calculating bending moments, displacements, and strut forces. The stability analysis is carried out only for cantilever or singly propped walls. The finite element analysis is elasto-plastic, with limits defined by active and passive states. Minimum effective earth pressure can be prescribed as an equivalent fluid density in accordance with certain codes of practice and fluid pressure within tension cracks can be specified.

PLAXIS: This program is available from Plaxis of Delft. It is a general geotechnical engineering finite element package which is suitable for modelling retaining walls and other structures. It allows graphical input of soil layers, structures, construction stages, loads and boundary conditions. Special beam elements are used to model the bending in retaining walls, tunnel linings and other slender structures and elasto-plastic spring elements are used to model ground anchorages and struts. Pre-stressing of anchorages and struts is permitted and various soil models and pore-water pressure distributions may be applied.

   The output has enhanced graphical features. All displacement and strain graphs can be visualized. Graphs and tables are available for axial and shear forces and the moments in structural elements.

It should be emphasized that the results obtained from such programs are very dependent on the relative stiffness of the chosen components of the model and that simple arbitrary rules for determining soil stiffnesses can lead to large variations. Depth, soil grading, and over-consolidation ratio can have considerable influence on the stress/strain properties of soils and simple rules for arriving at these parameters should be shunned in favour of proven experience. Programs, apart from needing to be mathematically correct, also need to be calibrated against known case evidence.

Considerable experience has been gained through studies carried out at Imperial College, London where a close association has been maintained with a series of major excavations. Their experience can be utilized through the Imperial College Finite Element Program (ICFEP).

Each stage of any multi-propped excavation must be studied, mimicking the stages of excavation and propping as closely as possible. In this way envelopes of bending moment, prop force and required penetration can be derived for the purpose of calculating, for example, reinforcement in reinforced concrete walls. Care should be taken in the matter of wall penetration below the excavation level, since upward seepage can lead to significant reduction in passive pressure over the permanently embedded length of a wall, even in stiff clays. In several major cities the effective groundwater table is rising due to the decline of industries that previously extracted water from boreholes. This warrants consideration and a report has been produced on the subject by the Construction Research and Information Association (CIRIA) mainly in relation to London.

It may be noted that, because there was a certain apparent divergence between measured prop loads and those derived from design programs such as described above, Construction Industry Research and Information Association (CIRIA) members initiated a study to consider whether in many situations prop loads were being overstated. The study involved a considerable number of recorded cases and produced a Funders Report (Twine and Roscoe, 1996). Basically it applies a method similar to that of Peck (1943) based on some 81 cases and redefines methods for drawing the characteristic distributed prop load diagrams.

Attention may be drawn to a general principle in multi-propped wall design. A wall could be designed by dividing it into sections and placing a pin joint at each prop except the top one. In this case the bending moment diagrams would be all simple positive moments except for any cantilever part at the top. Prop loads would take up the values depending on the load applied due to the soil pressures acting on the wall and generally active reinforcement would be close to the exposed wall face. Likewise the wall might be designed with a pin joint at the centre of load application between each two props, again with the top prop being a special case. In this case the wall is now a double cantilever developing maximum negative moment at each prop position. The prop loads will take up values according to the earth pressures applied and generally the moments are negative and reinforcement would be required in the earth retained side of the wall. These are two different scenarios, but the earth pressure is effectively the same and the prop loads are not very much different. The moral is that prop loads are very important, but bending moments can vary considerably with little consequence provided they account for the full anticipated pressure on the retained earth side of the wall. The solution from a given detailed analysis of moments in a wall based on stiffnesses of wall, prop and soil corresponding to prescribed assumptions, is but one of a number of possible solutions; there is certainly room for readjustment of designs to balance moments within boundaries as indicated and particularly to accommodate sensibly adjusted reinforcement in the concrete.

6.8 Structural wall design

6.8.1 General considerations

When a retaining wall has been analyzed for stability, it remains to design the structural form of the wall, and at this stage appropriate bending moments and shear forces must be used. This aspect of design has an important influence on the economy of the design as a whole. Temporary and permanent states must be considered. For example, in the case of many cantilever walls, a ground-bearing slab is cast to abut the wall and has the effect of propping the wall in the post-excavation stage. In many circumstances this effectively stops wall deflection and further moment development. In most instances, therefore, the temporary state proves to be the critical one. It is unusual to design such walls for pressures on the retained side other than the water pressure and the (effective stress) active pressure. Even in overconsolidated clays it would appear that the horizontal stresses in the soil mass, which are relieved at the excavation stage, do not become restored. The force lines in the soil mass almost certainly reorientate around and beneath the excavation (Figure 6.21).

Experience with many walls in such circumstances in stiff clay appears to indicate that wall bending moments in the slab vicinity do not in the long-term significantly exceed those calculated at the excavation stage, when only active and passive soil states are normally considered, although the position of maximum moment may change a little.

6.8.2 Determination of bending moments and shear forces in a cantilever wall

The conventional method of determining maximum bending moment in a cantilever wall is to consider the active and passive pressure diagrams as used in the

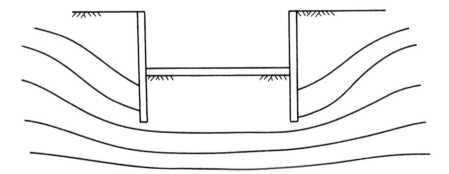

Figure 6.21 Force lines indicating redistribution of pressures on wall with time.

stability calculations. The full ultimate passive pressure is used on the excavated wall side, rather than any reduced value. This is because maximum passive pressure must be mobilized near the excavation level through wall rotation, and because excavation will, by reduction of the vertical soil stresses at that level, have produced an ultimate or near-ultimate passive state. Hence in practice maximum bending moment appears in most circumstances not far below the excavation level.

This means that from the head of the wall down to the point of maximum bending moment the design moments correspond most nearly to ultimate active and passive conditions.

However below the point of maximum bending moment, because of the extension of the wall to provide a rotational stability factor, the full ultimate pressures on the passive side will no longer apply and, in consequence, if the ultimate passive state were assumed here, the moments would be curtailed too rapidly and excessive shear forces would be evident.

The method of determining moments below the maximum moment point is therefore empirical and as shown in Figure 6.22. It should however be pointed out that where a significant change in the stiffness of the soil occurs below the excavated surface level, for example, if a wall toe is embedded significantly into rock, then this simple method may have to be modified in the light of judgement and of the fixity that the rapidly increased stiffness may provide near the wall toe.

6.8.3 Determination of bending moments and shear forces in tied or propped walls

For walls with a single prop or tie, the earth pressures and prop or tie force can be used simply to determine the bending moment and shear forces in the wall. It is commonly assumed for the purpose of bending moment determination that full passive pressure acts on the wall below the excavation level, rather than some reduced pressure that might be derived because of the extension of the wall to provide a factor of safety or rotational stability factor. This is because of the usually significant reduction of

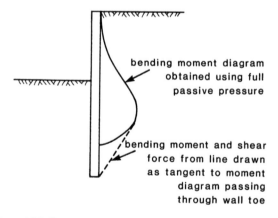

bending moment diagram
obtained using full
passive pressure

bending moment and shear
force from line drawn
as tangent to moment
diagram passing
through wall toe

Figure 6.22 Determining moments below the maximum point.

effective overburden pressure at and below the final excavation level, which initiates the passive condition.

For multi-propped or tied walls, where the moments have been derived from one of the available Finite Element or similar computer programs, the significant stiffness difference between anchor ties and, for example, floor props will already have been taken into account and bending moments corresponding to the assumptions will be available. While steel reinforcement may be varied and curtailed to some extent in reinforced concrete walls, the temptation to believe in bending moments being exactly certain in value at given levels should be resisted as discussed above, since very often the variation of, say, the stiffness of one prop relative to another can alter the diagrams in a significant way. It is often the case, however, that the spacing between floors that prop a wall in the final condition is less than the spacing between the props or ties in the temporary state, and consequently bending moments and shear forces can be at their most severe in the temporary state.

It will be noted that, for relatively flexible walls, pressures are increased at the propping points and diminished midway between propping points, the soil effectively arching between the props. This feature will be apparent in the analysis if carried out by one of the appropriate computer programs.

Most reinforced concrete diaphragm walls are not designed in the structural sense in accordance with conventional codes applicable to above ground conditions. Historically this is because concrete for use in piles and walls has to be very free flowing because, if stiff, it would not flow properly around steel reinforcement. Shear does not appear to be a problem in such cases. It is notable that to the authors' knowledge, no case of shear failure in such walls has ever been reported. This may be because shear in a wall would pre-suppose shear in the soil behind it, and it would appear that, in this respect, the soil and the wall may therefore act in a composite manner. Alternatively, and particularly in sandy soils, arching between props may mean that soil loads concentrate locally to prop positions with the result that shear elsewhere is diminished.

6.9 Retaining wall deflection and associated soil movements

Although the act of excavation to install a bored pile or diaphragm wall panel must in itself lead to some small ground deformation, it is the excavation that subsequently takes place adjacent to the wall, and the following changes in the general soil profile, which account for most of the movement that occurs.

When a deep excavation is carried out, this affects the stresses in the whole immediate area, both within and without the plan area of the works involved. Such stress changes lead to general movements in the soil mass, and for large excavations, particularly in clay soils, this is a time-dependent and significant feature.

Secondly, relief of lateral pressure due to excavation adjacent to a retaining wall leads to lateral wall movements, and the magnitude and timing of these are very dependent on soil properties, the way in which excavations are staged, the stiffness, position and timing of installation of props or ties, the pre-stress if any that is provided in the lateral support system, the presence of berms and how and when they are removed, and the stiffness, general dimensions and plan arrangement of the wall.

It is clear therefore that there is no single simple answer to the questions that may be asked regarding predicted wall movements but at the same time there are available methods of analysis which range from an advanced finite element type to simple reliance on extensive experience. However good and thorough the analysis, any failure to comply with the design concepts in final construction can completely invalidate conclusions that might have been drawn from calculation.

A simple model for assessing the magnitude of general ground movements due to excavation may be based on the whole mass acting elastically. Movements may be estimated by treating the excavation as negative load as for an embankment with vertical sides (Poulos and Davis, 1974). The analysis will then be based on Young's modulus E and Poisson's ratio v_u, the undrained parameters if short term conditions are being considered, or E' and v' where long-term analysis is sought.

It is useful to remember that where G is the shear modulus,

$$\frac{E'}{2(1+v')} = G' = G_u = \frac{E_u}{2(1+v_u)}$$

For undrained conditions $v_u = 0.5$, whereas v' is in the range 0.1 to 0.33 and may be taken as about 0.2 for stiff clays.

In so far as the local movements in the vicinity of the wall are concerned the general pattern of behaviour is as indicated in Figure 6.23.

The interaction between the soil and a retaining wall is complex. Rigorous analysis is not feasible at the present time. The conventional calculations for stability are based on lumped 'factors of safety' and provide no explicit information regarding deflection. It has been shown by Bica and Clayton (1989) that there are some 25 variants on the normal stability calculation method, so it is hardly surprising that deflection is difficult to determine even on an empirical basis.

Provided the deflected form of a wall can reasonably be established, either by calculation or experience of similar cases, then reasonable deductions may be made concerning the deformations that occur in the retained soil, using the method proposed

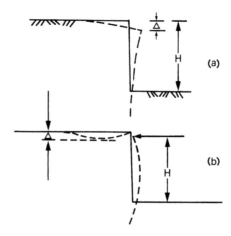

Figure 6.23 Local movements in the vicinity of (a) a cantilever wall, and (b) a propped wall.

by Bransby and Milligan (1975) and Milligan (1983). This method predicts that significant movements will occur within a zone bounded by a line at 45° from the base of the wall to the ground surface, and that any point within that zone will have equal displacement components in the vertical and horizontal directions. Hence the soil surface deformation is a direct reflection of the deflected shape of the wall.

If the soil is not going to behave in an undrained manner, then the zones of deformation are reduced in extent, with the ratio of horizontal to vertical movement near the surface in the retained soil increasing to as much as 3 in the case of a cantilever wall and falling to as little as 1/3 behind a propped wall.

6.10 Lateral movements of embedded walls

Whereas the various finite element and finite difference programs now available allow deflections of walls of all types to be calculated, there is always difficulty in having to estimate parameters and in particular soil stiffnesses. The necessary parameters may be relatively uncertain unless back-figured from similar works in the same conditions.

An extensive survey of wall deflections has been carried out by Long (2001), containing some 300 records of ground movements associated with deep excavations worldwide. It may be observed from his work, and that of other workers such as Fernie and Suckling (1996), that even with the stiffest wall systems, movement is inevitable. Controlling basal heave has a marked effect for all wall types. Prop positioning rather than fundamental prop stiffness appears to be more important in movement control and there is no discernible difference in the performance of propped or anchored systems. In stiff soils, changing the wall thickness (stiffness) has little effect, unless there is a significant structural change in section, say by providing diaphragm wall T-panels.

These findings appear to hold subject to workmanship and installation procedure being satisfactory and water flow not being a cause of ground movement. Deformations in general appear to depend on the fact that the major earth movements around an excavation are directly related to the ground unloading effect in front of the walls which in turn influences the lateral wall movements (as referenced by Poulos and Davis (1974) earlier).

The factor of safety against basal heave may be defined after Bjerrum and Eide (1956). This assumes that the unloading caused by excavation is analogous to the performance of a single large footing at the final excavated level. In practice for a clay soil c_u is taken as the undrained shear strength at this level and γH is the total stress removal. The factor of safety against basal heave would commonly be about $7c_u/\gamma H$ where H is the depth of typical excavation (the multiplier possibly varying from 6 to 9). Base heave in cohesionless soils is in general not a problem unless there is water inflow, but since deformation is primarily a matter of soil stiffness, it may be considered that a dense sand is equivalent to a stiff clay as a first approximation for the purpose of deflexion assessment.

Clough and O'Rourke (1990) prepared a chart that has become widely used. It relates the maximum wall movement, as a percentage of excavation depth, to a system stiffness number defined as $(EI)/(\gamma_w h_{avg}^4)$, where E = Young's modulus, I = moment of inertia per unit length of wall, γ_w = unit weight of water, and h = the prop spacing. Alternatively it has been suggested by Addenbrooke (1994) from finite element analyses

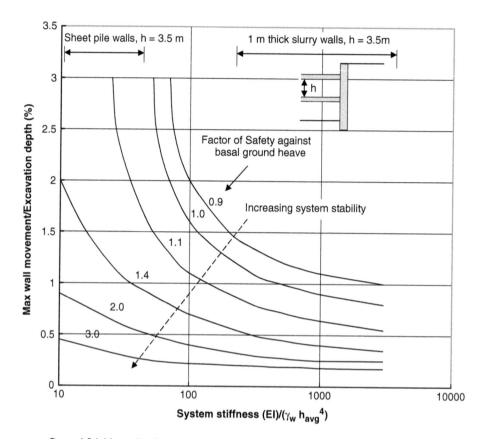

Figure 6.24 Normalized maximum wall movement as a function of system stiffness.

that walls with the same flexibility number, $\gamma h^4/(EI)$ will give rise to the same $\delta_{h\,max}/h$, $\delta_{v\,max}/h$ and sum of the normalized prop forces, $F/\gamma d^2h$, where d is the excavation depth and h is the average prop spacing (or maximum spacing if significantly different) or in the case of a cantilever is the depth to an effective point of fixity.

It is concluded that in relation to the Clough and O'Rourke chart and *for stiff soils* with a large factor of safety against excavation base heave:

1   Normalized maximum lateral movement values $\delta_{h\,max}$ are frequently between 0.05% H and 0.25% H where H is the excavation depth.
2   Normalized maximum vertical settlement values, $\delta_{v\,max}$ are usually lower with values frequently between 0 and 0.2% H.
3   There is no discernible difference in the performance of propped, anchored or top-down systems.
4   The values recorded are somewhat less than would be expected from the Clough and O'Rourke (1990) chart, possibly because the soils are on average stiffer than originally assumed by the authors.

5   Deflection values seem relatively independent of system stiffness and are apparently more controlled by excavation base heave.

In the case of walls that retain significant thicknesses of *soft soil* ($> 0.6H$), where there is stiff soil at the excavated level and below and there is a large factor of safety against base heave.

1   The $\delta_{h\,max}$ and $\delta_{v\,max}$ values increase significantly from the stiff soil cases.
2   The values are close on average to those predicted by Clough and O'Rourke (1990).
3   Addenbrooke's (1994) flexibility number may be helpful for comparisons.

For retaining walls embedded in a stiff stratum but retaining a significant thickness of soft soil ($> 0.6H$) and having soft material at the excavation level and also a large factor of safety against base heave.

1   The $\delta_{h\,max}$ and $\delta_{v\,max}$ values increase significantly from those where stiff soil exists at the excavated level.
2   The Clough and O'Rourke (1990) charts considerably underestimate movements.

In cases where there is a low factor of safety against base heave, large movements ($\delta_{h\,max}$ up to 3.2% H) have been recorded. It is not desirable to have soft soils in the upper passive zone of soil in front of a wall.

For cantilever walls:

1   For cantilever walls the normalized maximum lateral movements average about 0.36% H.
2   They are surprisingly independent of excavation depth and system stiffness possibly because the basis of design usually adjusts wall depth and structural requirements together, although it is self-evident that reliance on embedment in soft soils is not desirable.
3   It appears that the inclusion of a cantilever stage at the beginning of a construction sequence, can often lead to excessive movements.

The use of capping beams on cantilever walls, particularly where the plan shape of the wall involves changes of direction, is effective in restraining lateral and vertical deflections and unifying wall behaviour, provided such beams are adequately reinforced for the shears and tensions which arise.

It should also be noted that the use of berms and their removal just before casting a slab up to the wall can significantly reduce deflections of the wall in both the short and the long term, and that increase of the embedment of walls below dredge level in the case of tied walls helps to reduce the deflections near the bottom of the wall.

6.11 Measurement of the deflection of retaining walls

In recent years the availability of good-quality inclinometer instrumentation has allowed many researchers to undertake detailed measurements of wall movements during following stages of construction. This has provided much worthwhile information relating to the back analysis of wall behaviour, and has improved the general knowledge of how walls behave in practice.

In using modern inclinometers it is usual to install the necessary ducts with the reinforcement cages during construction of the retaining wall, and it is essential to avoid damage to these ducts thereafter. It is not always easy to ensure that duct heads will remain un-mutilated in following construction work, so the purpose of the duct must be explained clearly to those who may be concerned with its subsequent preservation. Inclinometer ducts may also be placed in boreholes in the ground to measure movements other than at the wall position.

It is then necessary to establish the true plan position of each duct with accuracy, bearing in mind that total deflections to be recorded may be no more than say 20 mm. Thus it is necessary to seek to achieve positional measurement accuracy within 1 or 2 mm, which may necessitate a sturdy reference point close to the duct position that will remain accessible throughout the works. The use of a good-quality precise surveying practice is advisable.

The inclinometer consists of a pendulum arrangement within a watertight torpedo-shaped housing. The pendulum is strain gauged so that the output from the torpedo presents a record of the angle at which the pendulum is held. The instrument is so designed that the component of inclination is recorded in the plane defined by four wheels on the torpedo, and the wheels are guided in vertical slots formed in the specially extruded duct tube. The wheels on the probe are lightly spring-loaded.

The gauge length of instruments may vary but is usually between 0.5 m and 1.0 m and the probe is generally lowered one gauge length at a time so as to obtain a continuous and accurate measure of any duct deviations. Resolution is often of the order of 0.1 mm on the gauge length and it has been found that, with careful work, accuracies of about 25 mm can be achieved over lengths of 300 m run. Errors can arise due to spiralling of the duct, lack of repeatability of inclinometer positions, temperature changes and operator skill.

In the field, a check can be made on the accuracy by taking two sets of readings in the same vertical plane, but separated by 180°; the sum of these readings should be constant if an error does not exist. When used in walls it is normally assumed that the movement of the wall is in a plane at right angles to the wall, so that the duct must be correctly orientated.

Vertical movement of walls and of the soil around them is also of interest during the processes of excavation and thereafter. In order to measure vertical movements at or near the surface, no more is required than well established reference points in the soil, preferably founded below the immediate surface, and good-quality surveying methods. If, however, it is required to measure vertical ground movements at various depths below the surface, then magnet extensometers may be most suitable. These instruments consist of a number of small ring magnets, secured at required positions in the ground by a combination of springs and grout, with an internal PVC tube passing through them. The position of each magnet is recorded by the operation of a reed switch that is lowered down through the tube with a steel tape. The system and movements within it can be related either to the surface survey or a deep magnet reference point as required.

It is fairly obvious that to fully instrument and record data for a large retaining wall is expensive, but quite apart from the improvement of knowledge that it may bring, it can be used to establish whether significant movements may have occurred in adjacent properties and this can sometimes help resolve or avoid contentious claims.

Chapter 7

Problems in pile construction

With any civil engineering work, care is necessary at all stages of design and construction. In particular, workmanship and supervision is of prime importance. Construction of piled foundations is a specialist activity, calling for considerable expertise and reliable workmanship, the more so as the completed element can rarely be inspected for defects. Remedial work to piling found to be wanting at a later date can be extremely time consuming and expensive, if not well-nigh impossible.

Two main methods of installation of piles are in common use: driving using a pile hammer or boring. Each system has particular advantages and disadvantages, and selection of the correct pile type calls for considerable experience.

The various methods of constructing and installing piles are discussed in Chapter 3: each system gives rise to particular constructional problems.

7.1 Driven piles

Preformed driven piles take the form of prefabricated elements driven into the ground. Materials commonly used for the manufacture of piles include steel, concrete and timber. This type of pile has the advantage that the pile element can be thoroughly inspected before driving and that quality control is easy to organize and enforce.

7.1.1 Design and manufacture

The piles should be designed to withstand handling and driving stresses. Stresses arising from these temporary conditions are usually more severe than the stresses in service.

During the manufacture of concrete piles adequate curing of the piles is necessary, and a continuous water spray is recommended. The formation of shrinkage cracks can lead to corrosion of the reinforcement. Similarly, premature lifting of the pile can cause cracking. Such defects are usually easy to detect. However it is rather subjective to decide when a crack is too wide. Piles with major cracks due to excessive bending moments should be condemned. Experience and personal judgement are necessary when assessing the severity of cracking; the aggressiveness of the environment should also be considered. It is not necessary to reject all cracked piles, and a series of transverse hair cracks of a width 0.15 to 0.3 mm (DIN 4026, BS 8110) may be considered acceptable.

This chapter was written by the late Ken Fleming. The authors felt that it would be a fitting tribute to leave it largely unchanged for this new edition.

With segmental pile systems special precautions are necessary to ensure that the ends of the segments are square. Out-of-squareness will cause difficulty in making the joints and will cause eccentric transfer of load across the joints. A lack of fit of the joints will rapidly become unacceptable on driving, owing to the slender geometry of pile segments. It is suggested the limit on out-of-squareness should be less than 1 in 300, in good practice. The first section driven determines the direction of pile motion and following sections must follow the same path. Unless the surrounding soil is very weak the straightening of the pile caused by soil forces will cause cracks in the section adjacent to the joints.

The design and manufacture of steel H-piles is quite straightforward. If heavy driving conditions are expected, the pile may be fitted with a shoe or a rock point. Preparation of tubular or hexagonal piles may involve a considerable amount of welding. As far as possible, the weld should be carried out in the fabrication shop under the supervision of suitably qualified personnel. Pile driving will subject the welds to high repetitive stresses and any defects can easily lead to the propagation of brittle fractures. It is important that the pile sections are correctly aligned, as lack of straightness will cause problems on driving and result in high bending stresses being locked into the pile.

Buckling of slender piles is not usually a critical design consideration, as even soft alluvium provides sufficient lateral restraint to prevent buckling of the pile. However, the selection of unduly light sections can cause problems in handling and pitching of the piles.

Similarly there are few problems with the manufacture of timber piles, provided the correct species of wood is selected for the job in hand. The piles should be fitted with a steel or cast-steel shoe to protect the toe of the pile. A steel band is normally fitted to the head of the pile to prevent 'brooming' during driving.

7.1.2 Installation of driven piles

7.1.2.1 Pile driving

The most obvious cause of damage during driving of concrete piles is distortion or spalling of the head of the pile (Figure 7.1). Such damage is frequently the result of insistence on driving to a predetermined length. The driving of piles may be controlled by a specified final set, and driving to a length may entail the adoption of expedients such as jetting or pre-boring.

Depending on the soil profile, heavy driving may be necessary to penetrate thin dense layers. The piles selected should be capable of accepting this heavy driving, indeed it may be considered necessary to drive test piles to check the performance of the pile in this respect. Steel piles will accept heavy driving as they rarely break. However, severe distortion of the pile head may occur. Damage to the pile toe may also occur and a pile may appear to be driving normally although the set may not build up as expected. If this is suspected the piles should be load tested and/or extracted for examination (see Figures 7.2 and 7.3).

It is not usually practicable to inspect the toe of a driven pile, except in the case of tubular piles. If there is any doubt as to the ability of the pile toe to penetrate the various

Figure 7.1 Damage to head of precast concrete pile. (Photograph courtesy of W. A. Dawson Ltd.)

strata, a rock point should be fitted to assist penetration. Problems can be expected when piles are driven through strata containing boulders or other obstructions. In such ground concrete piles may be broken and steel H-section piles deformed or split (Figure 7.3).

It is important that the designed geometry of the pile group is maintained. Positive control is possible only at the ground surface, and piling frames should be set on stable foundations. If guide trestles are used they should be of substantial construction, and positive guides for the piles should be fitted. It is not possible to correct deviations once pile driving has started. Piles, particularly flexible piles, may wander off line considerably below ground level. An extreme case of an H-pile turning through 180° has been reported. However, a greater danger is that piles will collide in a closely spaced group.

Misaligned piles will not generally reduce the bearing capacity of the group significantly. However, O'Neill *et al.* (1977) reported that the stiffness of experimental pile groups was reduced by 30% over that of the designed geometry. Very little can be done to prevent piles wandering, except to choose a stiff of pile section and possibly

Figure 7.2 Buckling of sheet steel piling. (Photograph courtesy of CIRIA, London.)

experiment with different types of shoe. If obstructions in fill are causing the problem, pre-boring may be resorted to.

7.1.2.2 Control of final set

In some soil types difficulty may be experienced in achieving the desired set, and both increases and decreases of the set some time after piling have been noted. The pile driving process frequently results in an increase in pore-water pressures within a few diameters of the pile. Dissipation of these pressures leads to an increase in adhesion, and hence set, with time. If the set-up is rapid, cessation of driving to splice the pile, etc., may cause difficulty in re-starting the pile. In extreme cases refusal can result (Figure 7.4).

Piling into chalk or other soft carbonate rocks can present special problems, particularly with low-displacement piles, such as H-piles. Breakdown and lique-faction of a thin layer of rock around the pile may lead to very low driving resistances. Table 7.1 shows the driving records for steel H-piles in Upper Chalk at Newhaven. Driving resistances were very low compared to the values expected from the Standard Penetration Test results. Pile tests showed that the piles had an

Figure 7.3 Buckling of steel H-pile driven into boulder clay. (Photograph courtesy of Sir Robert McAlpine & Sons Ltd.)

Table 7.1 Pile driving records for typical steel H-pile: Denton Island Bridge, Newhaven, Sussex (Courtesy East Sussex County Council)

Date	Depth (m)	Driving resistance (blows/0.25 m)	Comments
29.3.83	12.5–27.5	3–5	Hammer: 4 tonne drop weight, fall about 1 m
11.4.83	27.5–32.5	8–10	Soil profile: 0–30 m silty clay
11.4.83	32.5–36.0	12–15	alluvium
11.4.83	36.0–41.25	9–12	30 m—upper chalk SPTSs in top 6 m
11.4.83	41.25–47.0	15–18	of chalk gave $N = 35$ to 50
	47.25	68	blows/300 mm, i.e.
	47.5	59	Grade I material
	47.75	63	Below 36 m no data
	48.00	47	
15.4.83	48.25	81	Design length of pile
	48.5	78	pile 36 m
	48.75	68	
	49.00	43	Pile completed

inadequate bearing capacity at the designed length, when tested 14 days after driving. The piles were lengthened by about 10 m, and re-testing proved them to be satisfactory.

In dense silts and some weathered rocks, the reverse effect may be observed. It is postulated that negative pore-water pressure may be induced by pile driving, leading

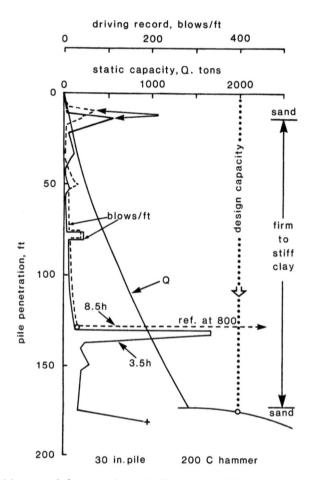

Figure 7.4 Driving records for two piles in the German North Sea (McClelland et al., 1969).

to a temporary increase in strength and driving resistance. Subsequent dissipation of the negative pore pressure is then reflected in a loss of set. Such an effect was recorded on Teesside, where tubular steel piles were driven into highly fractured hard Mercia Mudstone. The problem was overcome by re-driving the piles after a few days. In some cases several re-drives were necessary and the loss of set was eventually reduced. Unfortunately no pile load tests were undertaken in this case and the effect of the reduction in set on the bearing capacity could not be assessed.

Very occasionally, piles that have been driven into rock derive a suitable set but on load testing settle excessively. Where the founding strata is a blocky rock with the potential presence of clay infill between the blocks, it has been suggested that the blocky rock acts as a pile extension during driving but on static loading the clay consolidates or is squeezed out from between the blocks leading to settlement (see Figure 7.5).

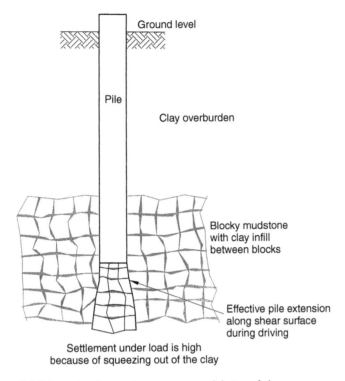

Ground level

Pile

Clay overburden

Blocky mudstone
with clay infill
between blocks

Effective pile extension
along shear surface
during driving

Settlement under load is high
because of squeezing out of the clay

Figure 7.5 Schematic of squeezing-out or consolidation of clay.

7.1.3 Associated ground movements

When large groups of displacement piles are to be driven, heave of the ground may be a significant factor. Careful levelling of the pile heads and the ground surface should be instigated if significant heave is expected. Cases have been reported of heave reducing the bearing capacity of piles, or in extreme cases inducing tensile failure in the shaft. Provided the pile shaft is not damaged, bearing capacity can usually be recovered by re-driving the piles to an acceptable set. If tensile failure of the pile shaft occurs, repairs can be expensive and may involve drilling down the axis of the pile to grout the fractures. The problem of heave can be alleviated by driving the centre piles of a group first and working out to the perimeter piles, selecting a low displacement type of pile, or pre-boring holes to remove a volume of soil equal to 50 to 100% of the volume of the piles. It should be noted that pre-boring can lead to a loss of skin friction, although this may not be significant in large groups of closely spaced piles.

The effects of heave may extend several metres beyond the perimeter of a pile group and damage to adjacent structures has been recorded.

When piling close to sensitive structures it is usual to adopt low-displacement piles to minimize heave.

In sensitive clays heave is localized and the remoulded soil around the pile is extruded at the ground surface. Reconsolidation of the soil in this situation may result in down-drag on the pile. Similarly, in loose sands and gravels little heave is recorded as the pile driving operation usually causes compaction of the deposit. Insensitive clays and silts are the soil types most prone to heave, and the volume of the soil displaced is approx-imately equal to the volume of the piles. Based on a study of case histories, Hagerty and Peck (1971) suggested that the surface heave for such soils would be equivalent to one-half the displaced volume of soil, the remainder of the displaced volume causing ground heave in the soil surrounding the pile group.

In particular circumstances lateral soil displacement may occur, leading to the dam-age of adjacent structures or the displacement of previously driven piles. The walls to the Clyde Dry Dock at Greenock were overloaded and cracked when cast-in-place piles were driven adjacent to the wall (Geddes *et al.*, 1966). In this instance high pore pressures were also thought to have been induced in the silty alluvium. Significant lateral soil displacement may also occur when piles are driven through river banks or other sloping sites.

In addition to soil displacement, heavy ground vibrations are transferred to the soil close to the pile. In silty alluvium, substantial pore pressures may be developed as a result of pile driving. On one site, piezometers adjacent to the piling were seen to overflow for a short period. Dissipation of the excess pore pressures may lead to the development of downdrag forces. In a recent case, pore pressures set up by pile driving re-activated an old landslip, causing 600 mm displacement of the pile heads.

7.1.4 Noise and ground vibration

7.1.4.1 Noise from piling operations

Pile driving is an inherently noisy operation and severe environmental restrictions may be included in the conditions of contract. Noise levels of over 85 decibels within 10 m of the piling plant are quite common. Typical data on noise levels produced by piling operations have been published by CIRIA (Report No. 64). These data are reproduced in Table 7.2.

Investigations into the sources of the noise have shown that a large proportion arises from secondary effects, rebound of the hammer, rope slap, engine noise, etc. Improved design of the components of the piling rig can considerably reduce the high-frequency components of noise. To dampen the noise sufficiently to be accept-able in urban situations, it may be necessary to enclose the hammer or the leaders in an acoustic chamber. Several piling companies have developed such chambers to suit their particular piling machines. The use of such devices usually results in some reduction in the efficiency of the pile hammer and creates difficulties in handling and pitching piles.

Driving steel sheet piling is exceedingly noisy, as the driving cap usually involves steel to steel contact. In areas where severe restrictions are placed on noise levels, noise outputs may be reduced by using an enclosure around the hammer, the use of pile driving vibrators or employing the Giken hydraulic pile driver. However alternative pile driving methods such as these involve the use of an auxiliary power unit which may, in itself, emit a high level of sound.

Table 7.2 Noise data for different pile types (after CIRIA Report 64)

Pile type	Dia. (or with) (m)	Method	Size	Dolly type and condition	L_w (A) (dB(A))	Soil	Leq (dB(A)) at 10 m (1 cycle)
Sheet steel	0.48	Diesel hammer	310 MN m energy	None	133	Clay	104
H-section steel	0.36 square	Diesel hammer	620 MN m energy	None	125	Fill/clay/ sandstone	96
Precast shell	0.535 dia.	Drop hammer	6 tonne, 0.5 m drop	Wood (good)	124	Fill	83
Precast concrete	0.275 square	Drop hammer	4 tonne, 0.5 m drop	Wood (average)	116	Chalk/clay	87
Cast-in-place driven casing	0.4 dia.	Drop hammer	4 tonne, 0.5 m drop	Wood (good)	119	Fill/clay sand-marl	89
Impact bored cast-in-place	0.5 dia.	Tripod winch	18 kW	None	104	Rough fill/clay/ limestone	76
Bored cast-in-place	0.75 dia.	Crane-mounted auger	Crane engine 99 kW Donkey engine 125 kW	None	117	Clay/silt	79

7.1.4.2 Ground vibration caused by piling

Related to noise is the potentially more serious problem of ground vibrations, and the consequent hazard posed to adjoining structures. Many investigations have been made into the problem, firstly to predict the likely peak ground accelerations, velocity or displacements, and secondly to establish safe limits of ground vibrations for various types of structure. To date, none of these investigations has proved very conclusive with respect to piling operations. Prediction of peak-to-peak acceleration or velocity in real situations is most complex. Firstly, the energy transfer to the soil is poorly understood and attenuation of high-frequency components is rapid. Secondly, the response of various forms of construction to the ground vibrations is difficult to predict, and some structural details, e.g. floor spans which resonate, may lead to a considerable magnification of the effect.

Several indices have been proposed as a measure of the intensity of ground vibrations (Geiger, 1959). The most widely accepted of these criteria are based on the peak particle velocity or the energy intensity of the vibrations induced in the soil adjacent to the foundations of a building. Empirical guidelines have been drawn up using these criteria to define various levels of damage. One of the most popular of these is included in the German Codes of Practice DIN 4150 (1970) (Table 7.3).

The recommendations in Table 7.3 have not been drawn up specifically for ground vibrations induced by piling, and it is considered that they are overly stringent. For structures which are not of great prestige or historical value it is considered that the limits in this table could well be doubled. When considering reasonable limits on ground vibrations, the background level of vibration should be assessed. In built-up areas heavy traffic can cause surprising intensities of vibration, and peak-to-peak velocities exceeding 3 mm/s have been recorded 10 m from a road.

Table 7.3 Maximum allowable velocity, from DIN 4150

Class	Description	Max. velocity* (mm/s)
I	Ruins and buildings of great historic value	2
2	Buildings with existing defects (visible cracks in brickwork etc.)	5
3	Undamaged buildings in technically good condition except for minor cracks	10
4	Strong buildings, industrial buildings of R.C. concrete or steel	10–40

* Particle velocity in soil adjacent to foundations of structure.

The use of empirical limits on velocity or acceleration necessitates the use of field instrumentation to observe the induced vibrations. Suitable equipment is expensive to purchase, and several levels of recording are possible. The simplest scheme is manual recording of peak-to-peak phenomena, and the most complex is a full record of the ground vibrations to enable a frequency analysis to be carried out. Siting of the accelerometers is important, as amplitudes differing by a factor of 10 can easily occur due to structural magnification. It is also highly desirable to use triaxial accelerometers, as it is not usually clear in which direction the most damaging vibrations will propagate. The human body is extremely sensitive to vibrations and very low intensity shock waves may be detected. For this reason, people frequently conclude that ground vibrations are more severe than measurements show them to be. To limit unreasonable claims for damage to adjoining structures, it is highly desirable to carry out a detailed structural survey of the buildings. Good-quality photographs of sensitive areas or existing defects form useful records to counter or confirm particular assertions.

A common situation is for piles to be driven through a soft soil into a bearing stratum, with adjacent buildings founded on the soft material. Practical experience in such situations suggests that the soft layer effectively damps the ground vibrations. As a crude rule little or no structural damage is likely to be experienced if the pile and rig is at least 5 m from an adjacent building. The occupants of such a building may, however, suffer severe nuisance. Various expedients may be adopted to reduce the intensity of ground vibrations caused by pile driving. These include pre-boring, the use of low-displacement piles, or in particularly sensitive areas, the adoption of alternative pile types (see also section 10.4.5).

7.2 Bored piles

7.2.1 Excavation of the pile bore

In favourable soil conditions, i.e. stiff clays, bored piles can be constructed without casing, except for a short guide length. Under less satisfactory conditions temporary or permanent casings may be used to support the pile, or bentonite drilling fluid may be resorted to. Excavation of the soil is by augers, buckets, or a variety of grabs (Chapter 3). Continuous-flight auger equipment, which can place piles in many soil conditions without the use of any casing, is in common use.

Figure 7.6 Bulbous projection on pile shaft caused by overbreak. (Photograph courtesy of CIRIA, London.)

7.2.1.1 Overbreak

The formation of cavities or overbreak outside the nominal diameter of the pile is a significant problem in cohesionless soils (Figure 7.6), particularly below the water table. Under such conditions it is essential that a temporary steel casing is used and that the cutting edge of the casing is driven below the base of the advancing bore. If the casing is not at the base of the hole, slumping or washing of soil into the bore can occur, with subsequent problems on concreting the pile. The cavities may or may not be filled with water. Even with well-controlled drilling it is possible for the inflow of water through the base of the advancing bore to seriously loosen the surrounding soil, with a consequent loss of skin friction.

In very clean sands or rounded gravels, a soil arch may not form around the borehole, even in dry conditions. Under these circumstances collapse of soil into the bore will form large irregular cavities. However, the problem is easily overcome by the use of a temporary casing.

The problem may be alleviated by drilling under water and maintaining an excess water head in the casing or by resorting to the use of a bentonite drilling mud or in appropriate circumstances, a polymer fluid. Alternatively the boreholes may be permanently cased.

Severe boring problems may occur, with resultant overbreak in pile shafts, in boulder clay. In such heterogeneous materials it is very difficult to assess the maximum size of any rock fragments from site investigation data. It is frequently advantageous to adopt large-diameter piles, drilled with heavy equipment, to enable the largest boulder to be removed from the bore. The alternatives of chiselling to break up boulders or to redrill the pile at another position, can be expensive and serious delays to the piling programme may result.

Figure 7.7 Blocks of clay fallen from underream of pile. (Photographs courtesy of CIRIA, London.)

7.2.1.2 Base of boreholes

To develop the available end-bearing capacity it is necessary to ensure the base of the bore is clean and undisturbed. In clay soil the boreholes can be cleaned satisfactorily with special tools, but, in stiff fissured clays with sand partings and perhaps with water seepage, blocks of material may fall into the borehole. Underreamed piles are particularly liable to this problem (Figure 7.7).

The base of boreholes should be inspected before concreting and any debris removed. High-intensity spot lamps now available may be used for this purpose. The importance of cleaning the base of boreholes cannot be overstressed: cases of pile failure have been attributed to neglecting this step.

In granular soils, loosening of the base is almost impossible to avoid, even when drilling under water or bentonite. For piles founded in such deposits it may be wise to reduce the contribution of end bearing when computing the allowable capacity of the pile. Special measures may be adopted to consolidate the loosened soil. These usually involve placing a stone fill pack at the base of the borehole, followed by high pressure grouting after the pile is concreted. The main effect of the grouted packing is to preload the pile and reverse the direction of the shaft friction. The overall load/settlement performance of the pile is enhanced, but the technique has little effect on the ultimate bearing capacity of the pile. Sand or silt may accumulate at the base of a pile before concreting, leading to the effect shown in Figure 7.8. The fines may arise from local seepages into the borehole, or from sediment falling from the drilling water or mud.

For piles deriving bearing capacity from a rock socket, cleaning the base in a water-filled hole is obviously crucial to the satisfactory behaviour of the pile. In such circumstances final cleaning may be effected by the use of an air-lift or by adopting the reversed-circulation drilling technique.

Figure 7.8 Toe of small-diameter pile filled with silt prior to concreting. (Photographs courtesy of CIRIA, London.)

Debris such as small tools, boots, bags and timber can easily fall down a borehole and care should be taken to avoid this. A temporary steel casing projecting above the ground should be used at the top of all holes. Empty bores should be covered at all times with a substantial cover to prevent personnel accidentally falling into the shaft.

7.2.1.3 Effects of water in boreholes

The ingress of small quantities of water into pile bores is not a serious problem. However, a build-up of water at the base of the hole can cause problems on concreting. A surprisingly small amount of water can result in segregation of the concrete at the toe of a pile. Figure 7.9 illustrates the defects found in practice arising from this cause.

If water is seen to accumulate at the base of the borehole, the depth should be measured. For accumulations more than a few centimetres deep the water should be removed before concreting, or alternatively a tremie pipe may be used to place the concrete. Expedients sometimes adopted, such as placing a dry batch of concrete, are not usually successful and should be discouraged. Water will cause softening of the surface in clay soils, and for this reason piles should be concreted as soon after boring as possible. If concreting is delayed, slumping of softened soil from the shaft may occur in extreme cases, and low values of adhesion will certainly result.

7.2.2 Concreting the pile

Concreting cast-in-place piles is a special operation calling for considerable skill and the correct design of the concrete mix. More problems are caused by using a mix which is too stiff than by using one with too high a slump.

7.2.2.1 Quality of the concrete

The desirable properties of the concrete for forming cast-in-place piles are discussed in Chapter 3. It is important to appreciate that the method of placing concrete in bored

Figure 7.9 Toe of pile after concreting with water in bore. (Photograph courtesy of CIRIA, London.)

piles places severe restraints on the type of concrete that may be used. In general it is preferable to ensure the pile shaft is free from defects by using a highly workable mix, than to specify one with a low water/cement ratio in the expectation of producing a concrete of high durability. Using concrete with too low a workability is a major cause of defects in bored pile construction (Figure 7.10).

It is not practicable to vibrate concrete in a pile shaft, hence reliance is placed on the energy of the falling stream of concrete to do this. The free-fall method of placing concrete will result in the segregation of the concrete if the mix is unsuitable.

In general terms concrete suitable for constructing cast-in-place piles should be free flowing yet cohesive; the mixes are designed to be 'self-compacting'. The slump test is quite adequate to assess the former property; however, cohesiveness or resistance to segregation is difficult to define. If available, rounded aggregates should be chosen and a reasonably high sand content adopted. Plasticizers may be used to improve the workability of the mix with advantage. Experiments with super-plasticizers have

Figure 7.10 The effect of low slump concrete. (Photographs courtesy of CIRIA, London.)

shown that these additives are not worthwhile owing to their high cost and the limited time for which the effect occurs. Similarly, air-entraining agents are not effective, except for concrete to be placed by tremie pipe, as the entrained air is lost on impact of the falling concrete.

The durability of the concrete is difficult to reconcile with the high workability required. Very few cases of deterioration of cast-in-place piles have been reported in the literature, and no cases of failure are known. The site investigation should reveal the presence of aggressive ground conditions. In aggressive conditions it is recommended that the cover to the steel is increased, and the outer 50 mm or so of the shaft may be considered as a protective coating. Sulphate attack, the most common problem, may be overcome by using sulphate-resisting or suitably blended cements. Extremely aggressive conditions may be encountered in sabka deposits of the Middle East, with highly saline groundwaters. In these situations an expedient, albeit expensive, is to case the pile with a permanent plastic sleeve, mainly to prevent chloride attack and steel corrosion in that part of a pile which is near to or above the groundwater level. The combined presence of water, chlorides and oxygen can lead to rapid steel corrosion in such an environment.

7.2.2.2 Placing concrete

Concrete is normally placed in the bore by allowing it to fall freely from the ground surface. It is important to make arrangements to prevent the concrete impinging on either the reinforcement or on the sides of the hole. Segregation and inclusion of soil debris can be the result. The concreting process should be continuous and he completed without a break.

If prolonged delays between batches occur, water and laitance bleeding from the previously placed concrete may cause a weakness in the pile shaft. This emphasizes the need for good mix design.

Raking bored piles present special problems and it is suggested such piles should be concreted using a tremie pipe or chute.

For boreholes filled with water or drilling mud, the tremie method of concreting is a necessity. Underwater bottom-discharge concreting skips are not normally accepted because of the difficulty of ensuring complete discharge at each withdrawal of the skip.

7.2.2.3 Extracting temporary casing

After concreting, extraction of the temporary casing can create problems particularly if delays occur and partial separation of the pile shaft may result (Figure 7.11). Friction between the concrete and the casing is obviously aggravated by the use of dirty or dented casings. With unsuitable mix designs, it is not unknown for the casing to be withdrawn still filled with concrete. Difficulties arise in pulling the casing if a low-slump concrete is used, or if the aggregate is particularly angular. Insistence on vibrating the concrete can lead to arching within the casing. However the use of external casing vibrators has been found to be beneficial.

Bleeding of mixing water from the joints of screwed casings may cause local consolidation of the concrete, leading to disturbance of the reinforcement cage and difficulty in extracting the casing.

Cavities outside the casing can lead to serious defects in the pile, particularly if they are water filled. Figure 7.12 shows the mechanism of the formation of the defect and Figures 7.13 and 7.14 show examples of defective piles. If air-filled voids exist outside the casing, defects are not structurally as serious (Figure 7.15). Normally the concrete in the shaft would slump to fill the cavity resulting in bulbous projections. There is a

Figure 7.11 Separation of pile shaft caused by extraction of casing. (Photograph courtesy of CIRIA, London.)

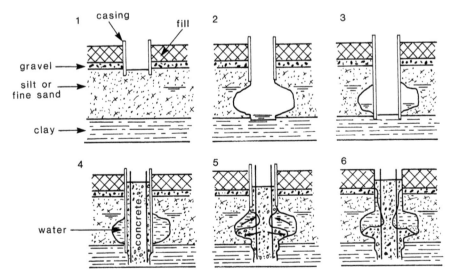

Figure 7.12 Schematic presentation of formation of water-filled cavities. (Illustration courtesy of Cementation Piling and Foundations Ltd., Rickmansworth.)

Figure 7.13 Defect in pile shaft caused by water. (Photograph courtesy of CIRIA, London.)

Figure 7.14 The effect of a large water-filled cavity. (Photograph courtesy of CIRIA, London.)

danger that debris can be dragged into the pile shaft, particularly if large cavities exist. Dry voids may form where thin gravel layers exist or where piles are formed in fill. The formation of such voids can be reduced by the use of a vibrator to insert the pile casing. Obviously, if there is any doubt that satisfactory pile shafts will be constructed, thin permanent casings should be used to support the soil.

Where slumping of concrete into voids has occurred, topping up the pile shaft with fresh material frequently leads to problems (Figure 7.16).

Debris may be dislodged by removal of the casing or by the placement of additional concrete and incorporated in the pile shaft.

Figure 7.15 Defect caused by concrete slumping into a dry cavity. (Photograph courtesy of CIRIA, London.)

Figure 7.16 The consequence of 'topping-up' after extraction of casing. (Photograph courtesy of CIRIA, London.)

7.2.2.4 Problems in soft ground

A variety of defects may arise when forming cast-in-place piles in very soft alluvium exhibiting undrained shear strengths less than 15 or 20 kN/m^2. These defects relate to the lateral pressure exerted by the fresh concrete, which in turn is dependent on the workability of the mix. The lateral pressure of the concrete can easily exceed the passive resistance of soft soils, and bulges on the pile shaft will almost certainly occur. Such defects may be detected by a close check on the volume of concrete used or by sonic integrity testing.

Near the head of the pile the lateral pressure of the concrete may be low, and further reductions in pressure can be caused by friction as the casing is removed. In such situations it is possible for soft soil to squeeze the pile section, leading to local waisting of the concrete (Figure 7.17). Again, the likelihood of this type of defect may be overcome by casting concrete to ground level or by the use of permanent light steel casings.

These risks can be analyzed if reliable soils information is available and steps can be taken to avoid them.

7.2.2.5 Effects of groundwater

The main effect of groundwater seepage, particularly in soils containing thin bands of silt, is to cause local slumping of the bore. The resulting debris may not be completely removed before concreting, or the material may become included in the body of the concrete.

Leaching of cement from fresh concrete is most unusual as high water velocities are required. Occasional cases of damage have been reported in made ground (Figure 7.18). A more common effect is illustrated in Figure 7.19, where the pile

Figure 7.17 Waisting of a cast-in-place pile. (Photographs courtesy of CIRIA, London.)

Figure 7.18 Erosion of wet concrete by groundwater flow: pile in fill. (Photograph courtesy of CIRIA, London.)

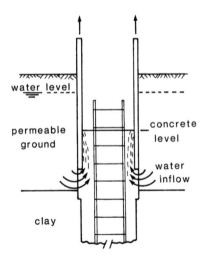

Figure 7.19 Effect of water inflow at head at pile.

cut-off level is below the groundwater level. For this situation to develop, the water pressure would have to exceed twice the head of the concrete at the base of the casing. The problem may easily be overcome by overfilling the pile or by filling the casing above the pile with a little sand and topping it up with water.

If groundwater causes instability of the pile bore, drilling using a permanent casing or a supporting fluid such as bentonite slurry should be resorted to.

Figure 7.20 Concrete retained by reinforcement cage in underreamed pile. (Photograph courtesy of CIRIA, London.)

7.2.3 Design of pile reinforcement

A relatively common problem with the construction of cast-in-place piles is unsuitable detailing of the reinforcing cage. In some instances the spacing of the bars may be so close as to completely retain the concrete (Figure 7.20). Even if this is not so, free flow of the concrete may be sufficiently impeded to prevent complete embedment of the steel. This problem is related to the workability of the concrete and is more likely to occur with low-slump concrete. A few large-diameter bars are to be preferred, with a minimum spacing of at least 100 mm.

The lateral outward pressure on the main steel caused by the flowing concrete may be considerable, and hoop steel should be adequate to restrain the main steel. Displacement of the reinforcement due to this cause has been observed; the defect is usually evidenced by the downward movement of the projecting bars at the pile head. Attempts to compensate for a slumped cage by the insertion of individual free bars which are pushed into the wet concrete can result in the defect shown in Figure 7.21.

Where the pile shaft contains a large amount of steel, e.g. in tension piles, widely spaced welded steel bands may be used instead of the normal helical binding. Such welded bands will also prevent twisting of a heavy cage which can occur when setting the steel or concreting of the shaft. The use of steel bands is quite common practice in some European countries.

Experience has shown that reinforcement cages should extend at least 1 m below the level of any temporary casing. This will prevent movement of the cage caused by drag as the casing is extracted.

7.2.4 Piles constructed with the aid of drilling mud

Excavation of bored piles with the aid of a bentonite-based drilling mud may result in particular defects associated with the method of construction. The conditions under which such defects can occur are discussed in the following sections.

Figure 7.21 Defects associated with displacement of reinforcement. (Photograph courtesy of CIRIA, London.)

Slurries with a concentration of bentonite clay in the range 4 to 6% are normally used to support the borehole. The fresh slurry can become contaminated with soil arising from the excavation, resulting in an increase in density and viscosity of the drilling fluid. The borehole is basically stabilized by the pressure exerted by the drilling mud. If this pressure is less than the active soil pressure, failure of the sides of the bore will result. Such failures are most likely to occur near the head of the pile, especially if the water table is high. Short guide casings should be used and the level of the drilling mud should be maintained at least 1 m, and normally 1.5 m above the groundwater level to avoid problems of instability.

Excessive viscosity of the drilling fluid may occur with heavy contamination of soil. In this situation rapid movement of the drilling bucket may lead to a 'piston effect'. On withdrawal of the bucket, significant reductions in pressure can occur and local collapse of the borehole may result. Clearly control of the viscosity of the drilling fluid should be enforced to avoid this problem and digging buckets should be provided with a fluid by-pass.

Cleaning the pile base should be carried out with extra care as it is not possible to inspect the base. Cleaning may be effected by means of a special bucket or with an air lift, and loosening of the base should be avoided. Fine sand and silt suspended in the bentonite slurry will gradually settle to the base of the pile, hence final cleaning of the hole should be carried out shortly before concreting.

An interesting example of bored piles constructed under bentonite slurry in Italy was quoted by Fleming and Sliwinski (1977). In this case several methods of construction were adopted and the piles were load tested. Details of the tests and the soil strata penetrated are shown in Figure 7.22. It is clear from the test results that both bucket excavation and drilling of the bores developed a loosened layer of sand at the base. The maximum available performance was only achieved after grouting the base of pile No. 6 to consolidate the loosened base soil and partially reverse the direction of shaft friction. The groundwater level was very high at this site and problems with the stability of the excavation are the most likely cause of observed behaviour.

For placing concrete under a bentonite drilling mud, the tremie method is used and it is important that a suitably designed mix is adopted. In other respects, concreting bentonite-filled holes gives rise to problems similar to those of concreting open bores (section 7.2.2).

It has been found that the process of forming a pile under bentonite suspension does not materially reduce the shaft friction. Normally a very thin and weak filter-cake layer will form on the wall of a bore in any reasonably fine permeable soil, and the rising column of concrete will largely remove this and any loose debris from the hole. However, if the bore is left filled with slurry for long periods, or if the bentonite suspension is heavily contaminated with soil particles, an appreciable thickness of filter-cake may accumulate. The filter-cake may be strong enough to resist abrasion during the placing of concrete, giving rise to a significant reduction in shaft friction on the completed pile. In clay soils the concrete becomes contaminated with debris

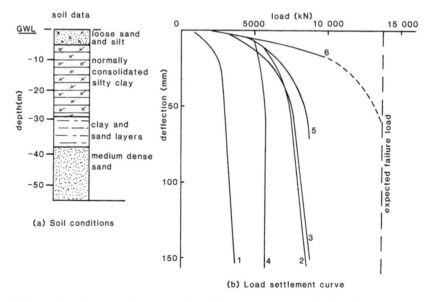

(a) Soil conditions

(b) Load settlement curve

Figure 7.22 Results of tests on piles bored with drilling mud at Porto Tolle, Italy. Maintained load test:
(1) Pile dia. 800 mm, depth 42 m, bucket excavation
(2) Pile dia. 1000 mm, depth 42 m, bucket excavation
(3) Pile dia. 1000 mm, depth 42 m, bucket excavation
(4) Pile dia. 1000 mm, depth 43.5 m, bentonite circulation
(5) Pile dia. 1000 mm, depth 43 m, bentonite circulation
(6) Pile dia. 1000 mm, depth 46 m, bucket excavation with grouted toe. (After Fleming and Sliwinski, 1977.)

and intermixed with slurry. It is therefore necessary to overfill the piles sufficiently to ensure sound concrete is present at pile cut-off level. In practice, overfilling by about 1 m has been found satisfactory.

These comments also apply in a general way to the use of polymer fluids but the experience of use of these support fluids is not so extensive as with bentonite. Given proper pH control, selection and mixing of the polymer the most likely problem relates to inadequate cleaning of the pile toe. As has been mentioned in the section dealing with the use of polymer fluids, aggressive cleaning of any solids collected at the pile toe is required, and air-lifting is unlikely to be sufficient.

7.3 Driven cast-in-place piles

There are many proprietary systems of constructing driven cast-in-place piles. In those systems which incorporate a permanent lining of steel or concrete, defects in placing concrete are likely to be few, providing the lining system is watertight. Such piles are very similar in behaviour to driven precast piles. Piling systems which use a driven temporary casing, basically to form a hole, may also suffer from some of the defects associated with bored cast-in-place piles. One of the major advantages of driven cast-in-place piles is that the driving records provide a valuable, if crude, check on the stiffness of the bearing stratum.

As discussed in section 7.1.3, driving steel casings for cast-in-place piles may result in soil heave or lateral displacement of adjacent piles. Once the piles have been cast it is not possible to correct the effects of heave by re-driving the piles, and pre-boring or other measures may have to be adopted to alleviate the problem.

In some instances casings are driven by an internal drop hammer acting on a gravel or dry concrete plug at the base of the casing. If the dry mix is of the wrong consistency, driving may not cause the plug to lock in the casing. In this situation any end plate will be broken off, and ingress of soil and water into the casing cannot be prevented and the tube may have to be withdrawn and re-driven using a correct mix. Heavy driving with an internal hammer may result in bulging of the casing, or in extreme cases, splitting of the steel if it is of inadequate thickness.

With some proprietary piling systems, the internal hammer is used to compact low-slump concrete as the casing is withdrawn. With this system the hammer may catch on the reinforcement and damage it, or over-vigorous compaction may cause the cage to spread. It is highly desirable to use a robust steel cage and to check the level of the reinforcement of such piles before and after concreting.

Concrete placed in driven cast-in-place piles is liable to suffer many of the problems associated with bored piles and the reader is referred to sections 7.2.2 and 7.2.3. With this type of pile it is recommended that concrete should be cast to ground level in order to maintain a full pressure head and prevent contamination of the pile.

7.4 Continuous-flight auger piles

The main problem with continuous-flight auger piles is related to the matching of concrete supply with the withdrawal rate of the auger and it is desirable to over-supply concrete at all stages of pile construction to a limited degree. It is for this reason that major specialists in the United Kingdom now fit instrumentation to all the machines used for this type of work. The items normally measured are the rotations of the auger, the concrete pressure at the auger head, the concrete volume being supplied and the depth of auger penetration below ground. All of these items are measured against a real-time base and, on completion of any pile, a record of the main parameters can be printed in the machine cab. The data are stored and can be analyzed in greater detail using an office-based computer.

Pressure alone is not an adequate measure, since it is frequently lost when the auger is withdrawn to a shallow depth (say 7 or 8 m below ground), due to the considerable head of concrete effective at the auger base and the relatively loose state of the soil within the auger flight at this stage.

Most of the problems which have occurred in the past with this type of pile are believed to be due to over-rapid extraction of the auger in the latter stages of construction. When concrete is seen to issue from the ground, wrongfully giving the impression that all soil has been displaced, there is a tendency to extract the auger too quickly. If this is done, soil may be carried back into the body of the pile by the reverse motion of concrete on the auger and the upper part of the pile can become contaminated.

Other problems which can occur are related to excessive rotation of the auger when it encounters a hard stratum ('over-flighting') and particularly when the ground above consists of water-bearing sands or silts. Slow excavation relative to the number of rotations will cause the auger to start loading from the sides rather than from the base, thus loosening the surrounding ground and, in some cases, creating water-filled cavities (shown diagrammatically in Figure 7.23). A great deal can be done to avoid this situation by the selection of a drilling machine with sufficient power to drill the harder strata at a steady and sufficient rate. A well-designed boring head is also necessary.

Care must be exercised when concrete placing commences so that the auger is not lifted too rapidly in the initial stages. Failure to do this may result in significant loss of end-bearing or possible segregation of the concrete in water-bearing ground.

The use of integrity testing, coupled with good on-board instrumentation and an understanding of the process, provides a high level of assurance in this type of work. The monitoring of the concrete supply to ensure that concrete volume supplied always exceeds the volume vacated in the ground by the auger, remains the most important item in the process. Where upper soils are very soft, it is possible to use short permanent casings with this method, though this is rarely necessary.

This type of pile is usually constructed using a pumpable concrete with a slump in the region of 150 mm. It is necessary to ensure that the supply pipe does not permit grout leakage, if line blockages are to be avoided. Reinforcement cages should be robust and preferably welded or otherwise firmly fixed at bar intersections, main steel bars should be heavier than in conventional bored piles, depending on the depth to which they have to be inserted, and lateral reinforcement or hoops should be reduced to a reasonable minimum except in the top 3 or 4 diameters of pile where shear due to lateral loading may be an important consideration. A generous number of spacer wheels should be provided to ensure that the cage remains concentric in the pile. For long cages, say more than 10 m, it may be necessary to use a vibrator to assist insertion.

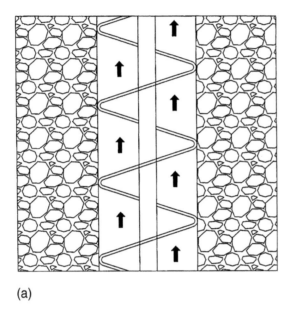

(a)

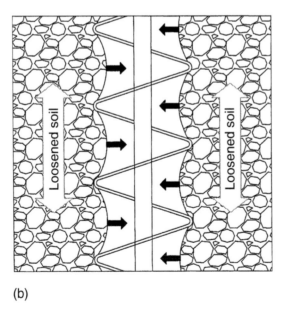

(b)

Figure 7.23 Diagrammatical representation of normal CFA auger-filling (a) and the effect of 'over-flighting' (b).

7.5 Cast-in-place screw piles

These piles share some of the potential problems that can occur with CFA pile in as much that the extraction of the auger is required at a controlled rate in order that concrete fills the void created by the screw auger completely and that the screw-form is created correctly. Unless the screw-form is properly formed it will not contribute to the pile capacity. It is possible for the threads to be badly or partially formed and these will shear off from the pile shaft at low stresses. This is illustrated in Figure 7.24 which shows severe pile necking likely to have been the result of poor control of extraction rate of the auger. It is interesting to note that ultra-sonic integrity testing did not reveal this fault.

Rig instrumentation is therefore essential to control the auger extraction and concreting process. In those screw piles in which there is no attempt to form a thread by back-screwing in the original downward path cut by the auger, the soil is remoulded during extraction and concreting and whilst the dimension of the pile is greater by the outer screw diameter, cohesive soils are at a remoulded strength. Pile capacities

Figure 7.24 Necking of a screw pile shaft likely to have been the result of poor control of extraction rate of the auger.

that are less than expected values based on measured of non-remoulded strengths are therefore possible.

7.6 General comments and conclusions

Many piling systems are available and some of these systems have been developed to deal with particular ground conditions. It goes without saying that an adequate site investigation should be carried out and that an experienced engineer's advice be sought as to the most suitable piling systems to be adopted. In a few marginal ground conditions often associated with organic soils, there is always a risk of defective construction with cast-in-place piles unless permanent casings are adopted. Similarly, problems may be experienced with driven piles in boulder clays or other soils containing obstructions.

Piles are normally lightly reinforced, and damage to the head of the pile may occur due to impact with or soil movement caused by heavy plant. Such damage can lead to protracted arguments between the main contractor and the piling subcontractor. Similarly, incautious excavation around the piles may lead to a slope failure with consequent damage to the piles. Such a case was recorded during excavation for a conical sedimentation tank in soft alluvium. Piles previously cast to support the edge beam were sheared when failure of the excavation occurred.

In practice, defects in piles are not very common, and it is hoped that as integrity methods of testing are more widely used, defects may be detected at an early stage. The wider use of such methods should also encourage higher standards of construction and supervision of piling contracts. Piled foundations usually comprise a large array of piles, and defects in a few piles, although highly undesirable, will not have a significant effect on the overall behaviour of the foundation. Failure of piled foundations as a consequence of failure of the piles is extremely rare.

In the early 1980s, Cementation Piling and Foundation Ltd carried out a survey of bored piles, using TNO sonic equipment. The results of the 1981 and 1982 surveys are given in Table 7.4. From this table it is evident that the major cause of defects was damage to the piles after construction as a consequence of incorrect breaking down or overloading by site traffic. Damage due to construction defects was about 0.5%.

Table 7.4 Results at bored pile surveys*

	1981	1982
Number of piles tested	5000	4550
Number of piles to show faults	73	88
Fault		
Soil contamination 0–2 m	24%	5%
2–7 m	9%	9%
Poor-quality concrete	6%	3%
Voids adjacent to pile shaft	3%	2%
Damage subsequent to construction	58%	80%
Total percentage of piles with defects	1.5%	1.9%
Percent failure due to construction defects	0.6%	0.4%

* Courtesy of Cementation Piling & Foundations Ltd.

Obviously the experience of one contractor may not be representative of the industry as a whole. Nevertheless, such surveys contribute to confidence that quality of construction is high, and the accumulation of such data over a period should encourage a higher standard of workmanship. No such comprehensive data on the quality of driven piles is available, although defects in the pile body before installation are more readily detected.

Construction of piled foundations is a skilled operation, and calls for a high level of training and experience on the part of the contractor. Inexperience on the part of the operatives, incorrect specifications or wrong selection of pile type are the sources of many problems. Specialist contractors can avoid many of the pitfalls that lie in wait for the unwary, and in difficult ground conditions advice from such firms should be sought at the design stage.

Chapter 8

Integrity testing

8.1 Introduction

Over the past two decades, there has been a steady improvement in the quality of both bored cast-in-place and displacement types of piling. This has been brought about by a gradually increasing awareness of the ways in which defects in piles may occur, better specifications, a greater degree of supervision, and an interest by the leading piling contractors in improving methods and efficiency. The introduction of various integrity tests may also have promoted self-imposed improvements in installation techniques by piling contractors, although it should not be forgotten that certain leading piling contractors have been actively involved in the introduction and use of integrity test methods.

The cost of rectifying defective piles discovered at a late stage in building construction can be enormously high, and for this reason alone, an early indication of problems from the use of a suitable integrity test is highly cost effective, and a strong incentive for the development of straightforward and reliable tests.

Whilst established piling practice has improved, a number of newer methods of pile installation have been developed in recent years. Concurrently there have been advances in rig automation and pile installation monitoring such that there has been a general improvement in quality and the likelihood of pile failure through poor installation methods has to some extent diminished. However, pile integrity tests, provided they are appropriate to the pile type and soil conditions, still have a role in providing a means of quality control.

Before the introduction of integrity testing, the only check on pile quality was to load-test a small proportion of the piles in a group. Exceptionally, a small group of piles would be tested in total, when, for example, the load/settlement performance was critical. The cost in the latter case was high, and in the former case, the check was limited. If the proportion of defective piles in a group was high, the selection of a small proportion for load test would be more likely to detect a defective pile, than if only a very small proportion were defective, as would perhaps be the normal situation. Therefore the primary purpose of a load test is to establish the settlement characteristics of a pile. It is not a good method of quality control, except as an incentive to good installation when it is known that piles are to be selected at random for load testing. Integrity testing on the other hand is primarily a quality check on a product that is for the most part not accessible for visible inspection. The choice is between high-cost intrusive methods of inspection (such as coring a pile), and cheaper indirect methods (e.g. seismic methods) when all the piles in a group can be inspected at a modest cost. The intrusive methods can be more specific in their diagnosis of an integrity problem than an indirect method, but many piles in a group would be left untested. It should also be appreciated that despite the high cost of, for example, core drilling a pile shaft, this method of inspection is unlikely to detect defects such as lack of cover to reinforcement on the perimeter of a pile although it is appropriate for inspecting the conditions at a pile toe.

The statistical probability of detecting a faulty pile in a group with a given number of defective units for various testing levels is shown in Table 9.1 (Chapter 9).

Today, low-cost methods of integrity testing applied to the entire pile group have become accepted as routine procedure. It is however important to be wholly satisfied with the skill of the testing operative and the equipment employed as there is an element of interpretative skill involved in understanding the output from such tests that is generally outside the experience of engineers engaged in civil engineering construction.

In recent years a number of experiments have been carried out with artificial 'faults' within piles. The results of the application of various methods of integrity test have been published. Whilst such experiments have a certain limited validity, decisions about the outcome of integrity tests are frequently facilitated by the study of a family of results. One or more anomalies lead to a search for the potential cause and its implications. This may involve pile head excavation, coring, and very particularly, a study of the ground conditions and construction equipment. Under these circumstances, well-instrumented construction equipment and integrity testing can prove to be a powerful and informative combination.

To apply these tests to the best advantage, the merits and limitations of the methods of integrity test available need to be known in order to select a test appropriate to the pile type and ground conditions. Where a particular type of defect is considered more likely than another, a test can be specific to this particular problem, be it transverse cracking in driven concrete piles, or lack of end bearing in bored cast-in-place piles.

8.1.1 Occurrence of faults in piles

The occurrence of faults in piles is infrequent, but when they are discovered, perhaps in one pile, doubt is then likely to be cast on the other piles on a site. The result may be delays to the contract, dispute as to causes and remedies, and probably an increase of cost. As shown by Table 7.4, the largest single cause of damage is that inflicted on pile heads after completion of the pile.

It is wrong to imagine that modern integrity testing can discover each and every fault that may exist in a pile. For example, cover to reinforcement, small zones of segregated concrete or small voids may be difficult to detect, but most of the successful systems can give strong indications of the presence of serious problems. The consequences of a major defect in an important pile are such that costs of remedial measures may far exceed the cost of an integrity survey, whereas early detection may lead to relatively inexpensive repairs or indeed modification of the method used so as to obviate further difficulties.

8.2 Types of integrity test

In some kinds of integrity test, pre-selection of the pile(s) to be tested is necessary. For example, access tubes for sondes have to be cast-in. The cost of this operation is generally significant, and equipping all the piles in a foundation in this way is usually prohibitively expensive. A proportion of the piles is therefore selected, and while this is usually a higher proportion than that usually selected for load testing, the selection process carries with it the same disadvantages, i.e. a risk of missing a defective pile, and the possibility of greater care being taken with the installation of the piles to be subjected to testing. Therefore, the confidence that can be placed on the results of a particular type of test does not depend on the technical merit of the system alone.

The tests which are generally recognized as 'integrity tests' are as follows:

(i)   acoustic tests
(ii) radiometric tests
(iii) seismic (sonic echo tests)
(iv) stress wave tests
(v)   dynamic response tests
(vi) electrical tests.

Other, more straightforward, examination of the integrity of a pile can be made by excavation around the pile head (or adjacent to a pile by excavation of a shaft), or exploratory drilling of the shaft, supplemented by a television inspection of the drill-hole.

Whilst not an integrity test as such, the results of a pile load test do provide information that relates to the modulus of the pile shaft relative to the modulus of the soil in which the pile is installed. A back-analysis of the apparent modulus of a pile shaft, given some knowledge of the modulus of the supporting soil at an appropriate strain should match the modulus of the pile shaft material. If a pile is cast-in-place or jointed, a lower than expected modulus can be indicative of cracking or joint opening. Such an analysis would probably be computer aided, but might be considered where pile load test results are anomalous.

8.2.1 Acoustic tests

8.2.1.1 Single-hole test

In order to carry out this test, a sonde has to be traversed the length of the pile. A tube has either to be cast in the pile for this purpose, or a hole drilled in the pile. The sonde contains a transmitter of acoustic energy at one end, and a receiver at the other, with an acoustic isolator in-between. The dimension of a typical sonde is 55 mm diameter by 2 m long (see Figure 8.1). The sonde is raised and lowered on an electrical cable which supplies power and also carries electrical signals. In order to obtain good acoustic coupling to the pile shaft, the hole in the pile (which should be around 70 mm diameter for the sonde as described), is fluid-filled. Water is suitable for this purpose, but may be subject to a high rate of leakage, and a drilling mud may then be more suitable.

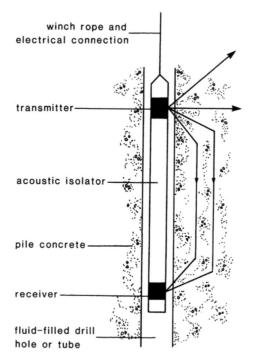

Figure 8.1 Schematic diagram of a single-tube sonic logging apparatus.

The sonde is generally lowered to the base of a pile, and raised at a steady rate. Ultrasonic signals from the transmitter radiate in all directions, some arriving at the receiver via the pile medium. If this is sound concrete, the delay time from transmitted pulse to the received signal is small. The compressive wave is received first, followed by the shear wave. If there is a void, porous concrete or an inclusion of soil, the delay time is increased. At the receiver, a transducer converts the pulses to electrical energy which is received at the surface as an amplitude modulated (a.m.) voltage. For interpretation, the a.m. signal is used to produce a variation in intensity of an oscilloscope scan. A peak in amplitude of the signal corresponds to a brightening of the trace, and a trough to a darkening. A voltage which is proportional to the depth of the sonde is derived from the winch unit. The final display may appear as shown in Figure 8.2. In this example, defects are indicated at depths of 4 and 7.5 m, as signified by the increased time of transmission. The profile from a sound pile shaft is also shown.

The cost of drilling access holes is high, and is approximately equivalent to the cost of forming the original pile, metre for metre. The prior installation of an access tube in a cast-in-place pile is cheaper, but then a pre-selection has to be made. Steel tubes should not be used, as steel is a better transmitter of acoustic energy than concrete. For the same reason the tube or drill hole should not be adjacent to a reinforcing bar. Plastic tubes can fail under the hydrostatic head of wet concrete combined with the elevation of temperature that occurs in a massive concrete section as the concrete hydrates. High quality ABS tubes may be necessary, rather than uPVC.

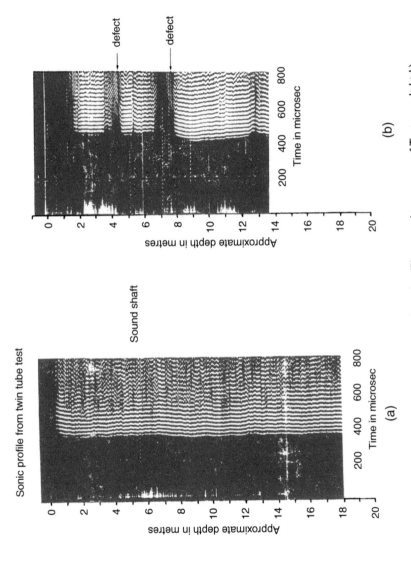

Figure 8.2 Acoustic test – display of results. (Photographs courtesy of Testconsult Ltd.)

In the single-tube test, the radial extent of the zone of concrete examined increases as the separation between the transmitter and receiver is increased. The resolution decreases however, and the limiting size of a detectable defect increases. With a spacing between the transmitter and receiver of 30 mm, Kennet (1971) quotes a vertical resolution of 50 mm. A number of traverses can be made at different separations to increase the volume of concrete included in the inspection, enabling finer detail to be detected in the zone immediately around the access hole. There is, however, some doubt as to the actual volume of material effectively scanned in the single sonde test, and the method is not in common use.

8.2.1.2 Double-hole test

This test is essentially similar to the single-hole test, except that the transmitter and receiver are traversed in individual holes, whilst being held level with each other, or obliquely. Defects which are present between the two tubes are well defined, but there is no radial scanning as with the single-tube method. If the transducers are held obliquely, additional information may be obtained, which can assist in the delineation of the extent of a zone of anomalous response.

For this test, holes are usually preformed with steel or plastic tubes, as it is unlikely that a pair of holes could be drilled sufficiently parallel to each other over the length of a pile. The test arrangement is shown in Figure 8.3. Some errors can occur at the base of the pile if the holes are not of the same depth, if the cables are not precisely

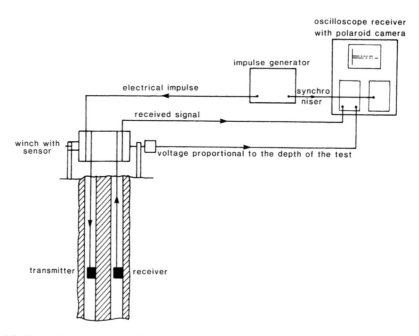

Figure 8.3 General arrangement of equipment for twin-tube acoustic test. Redrawn from material provided by Testconsult Ltd.

paired in length, or if the tubes are of significantly larger diameter than the transmitter or receiver units.

8.2.1.3 3D testing

Representing the extent and form of a defect in 3D is a significant improvement on 2D outputs from other forms of sonic logging and can be achieved by employing four tubes within a pile shaft and producing six 2D profiles for the planes that exist between the tubes. By combining the data a 3D image can be achieved with a good idea of its extent and location across a pile shaft. The system has been developed in the United Kingdom by Tesconsult Ltd.

8.2.2 Radiometric tests

These test methods employ nuclear radiation techniques for the detection of variations in density of the pile material, and hence the location of apparent defects.

In many respects, radiometric test methods are similar to those employed in acoustic tests. Both generally require pre-selection, as access tubes require to be cast in the piles for test. As a probe traverses the pile, radiation is received and measured, either from a source mounted on the probe in the same tube, or from a source and detector in separate tubes. The method can however be employed in fresh concrete. At the present time, radiometric methods are not commonly used in the United Kingdom.

8.2.2.1 Backscatter test (gamma-ray)

When a single tube is used (or a number of individual tubes in the same pile), the probe is constructed as shown in Figure 8.4. A probe of this type is described by Preiss and Caiserman (1975); overall dimensions of 48 mm diameter by 600 mm long are quoted. A 54 mm-diameter access hole is required. Such a hole could be drilled in a pile by rotary percussion; or more usually preformed with a steel tube.

The radioactive source produces gamma rays which travel into the concrete of the pile. A proportion of the rays emitted are scattered in the pile material and reflected back towards a detector which is separated from the source by a lead shield within the probe. The proportion of backscattered particles counted by the detector varies with the density of the pile material. Low-density zones, such as may occur at an inclusion of soil, increase the count rate, as less radiation is adsorbed. A density profile with depth can be produced in this way. The probe is lowered on a cable, and count rate observed and recorded at the surface. With a probe of the dimensions described, a zone around the access tube of approximately 100 mm radial extent is examined. In order to satisfactorily examine the majority of the concrete in a large-diameter pile, a number of parallel access holes would be required.

8.2.2.2 Neutron test method

A neutron source is employed for checking between pairs of tubes in a pile, as a gamma-ray source capable of being used in this way could pose a radiation hazard. Neutron transmission varies with the moisture content of the material through which it passes.

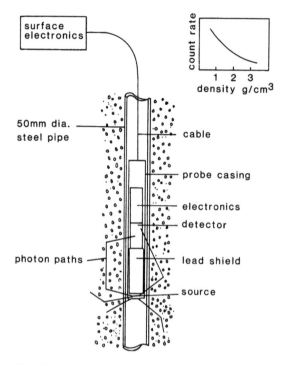

Figure 8.4 Gamma ray backscatter equipment for integrity testing.

This method is also suitable for detecting soil or water-filled/empty voids in a pile shaft. In order to carry out the test, the source and detector are raised simultaneously from the base of the pile, and a moisture content profile obtained from the traverse.

8.2.2.3 Precautions

The radiocative materials employed in these methods will require that the equipment is used in accordance with the Ionising Radiations (Sealed Sources) Regulations, 1969. Further assistance may be obtained from the National Radiological Protection Board, Rowstock, Oxon.

8.2.3 Seismic test methods (sonic echo method)

A hammer blow struck at the top of a pile produces a low-strain stress wave which can be detected by a surface-mounted accelerometer at the pile head. At the moment the pile is struck, the adjacent accelerometer first picks up the surface Rayleigh waves, and later responds to the wave which has traveled to the base of the pile (or a discontinuity) and has been reflected back. A constant of proportionality (C) may be assumed between the velocity of the surface waves, V_r, and the velocity of the longitudinal waves, V_l. Hence surface wave measurements at the pile head may be used to deduce the longitudinal wave velocity, assuming that the concrete in the pile is consistent. This may not be the case as the concrete at the pile head in a cast-in-place pile may be less dense. The relationship is given by $V_l = C \times V_r$, where C is approximately equal to 1.90. In some circumstances, the surface waves are not well defined and V_l may be assumed to be of the order of 4000 m/s. The distance to the base of the pile or a discontinuity (l) is given by $2l = t \times V_l$, where t denotes the delay time.

The recording system comprises a triggering device to start the time trace, with the accelerometer connected to an oscilloscope by way of an amplifying filtering system. Signals from the accelerometer are digitally processed such that background 'noise' is filtered out and a clean trace, suitably amplified, may be obtained by striking the pile head a number of times with a small nylon-headed hammer. It is necessary only to remove loose or weak embedded material from the surface of the pile in order to obtain an acceptable signal and therefore under ideal conditions, a single operator can test 200 to 300 piles per day. The analyzing equipment is compact, comprising a small robust portable data-logger and computer with an integral display (Figure 8.5). For each pile tested, a fully-annotated record is produced which records the estimated length of sound pile. Data recorded in the field can be downloaded onto a PC operating Windows for further analysis. Equipment commonly in use is that marketed by IFCO of the Netherlands, who are specialists in this form of integrity testing.

Whilst the travel time of the stress wave gives a measure of the length of the pile under test, or the distance from the pile head to a distinct break, the form of the trace provides information on the geometry of the pile, such as reductions or increases in cross-section. Results of the test are either presented in the time domain or the frequency domain. Both use the same pile response data and the difference is only in the method of presentation. Additional data is recorded in the frequency domain method of presentation only because it is necessary in this method to eliminate the characteristic components of the hammer/pile blow response from the plot. Typical outputs are shown in Figure 8.6. The test may be carried out after the concrete in the pile has cured for at least 4 days.

Figure 8.5 Seismic integrity test (sonic echo method) in operation.

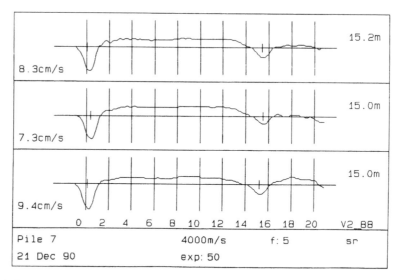

Figure 8.6 Result of integrity test on cast-in-situ pile carried out using TNO-FPDS equipment. (Illustration courtesy of Cementation Piling and Foundations Ltd.)

The method is not generally suitable for pre-cast jointed piles or piles with jointed permanent casings. In dense soils the stress wave signal is attenuated more rapidly with depth than in weaker soils and there is a limitation on depth of the order of 20 to 30 m. As piles are more likely to be damaged at the pile head than anywhere else, any defects below this level would not be revealed as the pile response is obscured by the upper defect. Whilst the interpretation of the pile 'signature' requires special expertise there is now a considerable fund of experience to call upon both in the United Kingdom and abroad and it is feasible to distinguish between cracks, necking or bulging of the pile shaft.

8.2.4 Stress wave tests

In the stress wave (or dynamic) load test, the static behaviour of the pile is predicted from the results of the response of the pile to a dynamic impulse. This can be considered as a form of integrity test, but is expensive as such. It is usually seen as a replacement for static load testing but given the difficulties in interpretation, extreme caution is advised in this respect, even if correlation with an actual static load test at the site in question is carried out.

The impulse applied to the pile has to be large enough to induce a significant deflection of the pile, yet not so large the pile is damaged. A 'stress wave' is produced at the head of the pile by applying a force to the head of a pile (e.g. by the drop of a pile hammer) or a purpose-made sliding mass as shown in Figure 8.7. The force at the pile head is monitored by strain gauges and accelerometers on the pile, and the displacement is simultaneously measured using an electronic theodolite. Various theoretical approaches have been developed to relate the dynamic response of the pile to its static capacity. There are however major problems to be overcome in this transition.

Figure 8.7 Arrangement of sliding mass (2000 kg) for stress wave integrity test. (Photograph courtesy of PMC Limited, Durham.)

The analysis of the progress of the compressive stress wave as it travels down the pile and its subsequent return as a tensile reflected wave, modified as it is by the restraint of the soil is complex. The difference between the upward force in a pile in free space and the actual upward force represents the soil restraint.

To obtain a realistic measure of this force, all the soil around the shaft has to simultaneously attain the condition of ultimate resistance. Although this may be the situation at the top of a pile when the blow is struck, it is unlikely to be so towards the base of the pile and at the toe. By creating a computer pile/soil model the actual deflections and forces measured can be compared with the computer predictions, but because of time-dependent soil properties this is not straightforward. In the early methods (e.g. by the Case Organization in the United States and the Institute TNO) an empirical damping factor (J) is employed. Later workers (e.g. SIMBAT – Testconsult

CEBTP Ltd.) have used the response of the pile to different strain rates employed in the impact. The end result is a calculation of the pile capacity in terms of shaft adhesion and end bearing.

8.2.5 Dynamic response tests

In the vibration test, an electrodynamic vibrator is placed on the prepared pile head, and the pile is subjected to a continuous vibration over a broad frequency band. Although the method of pile excitation is different, the results of the dynamic response test are considered generally similar to those from the seismic (sonic echo) test. Both tests provide similar information regarding the basic pile geometry. The differentiation between the tests lies largely in the electronics involved in the signal processing. The response of the pile is monitored via a velocity transducer. Resonance peaks are produced, which may be used to deduce the effective length of the pile and to reveal apparent defects. The method was developed by CEBTP, and has been widely used.

A block diagram of the test arrangement is shown in Figure 8.8. The vibrator supplies a constant-amplitude sinusoidal input over a frequency band of, say, 20 to 1000 Hz. The voltage from the velocity transducer is fed to the y axis of an x-y plotter which is provided with a signal on the x axis proportional to the frequency, f. By dividing the magnitude of the pile head velocity, v, by F_0, the applied force, the mechanical admittance of the pile head is obtained. In the idealized response curve for a pile shown in Figure 8.9, the slope of the early part of the curve, m, is indicative of the stiffness of the pile–soil system as presented at the pile head. This may be related to the behaviour of the pile under load, but an apparent relationship can occur for various reasons unrelated to static conditions and absolute correlation is not to be expected.

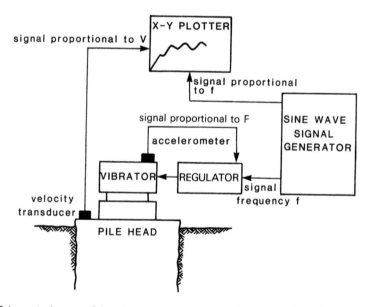

Figure 8.8 Schematic diagram of the test arrangement for the vibration method of pile integrity testing.

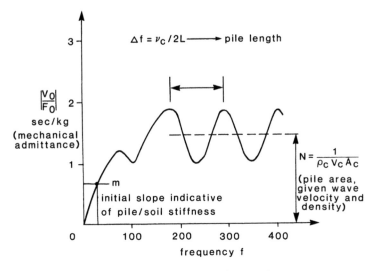

Figure 8.9 Idealized response curve from a vibration test.

The distance between resonance peaks at higher frequencies is related to the effective pile length, l, by the expression:

$$f_2 - f_1 = V_c/2l$$

where V_c is the velocity of longitudinal wave propagation in concrete. Hence l, may be estimated. A value for the cross-sectional area of the pile, A_c, may be obtained from the mean mechanical admittance of the pile head, N, over the higher frequency range, from the relationship

$$N = 1/p_c \times V_c \times A_c$$

where p_c is the average density of the pile concrete.

A development of this test is to record the transient response of the pile to a single blow, and to carry out a Fast Fourier Transform analysis of the waveform. A test of this type is now offered commercially by Testconsult Ltd.

An advantage of this type of test is that it can be carried out only 4 days after casting. The pile head is prepared 24 hours in advance of a test by rough levelling using a mortar screed where necessary, and final levelling with a thermosetting resin just prior to the test. It is usual to carry out the test with the pile head freely accessible, but in some soil conditions it may be possible to carry out the test from the pile cap. The advice of the specialist testing house should be sought if the cap has been cast and the pile is to be tested in this condition. Expert interpretation of the results of the tests is required, but as with the seismic test, there is by now a great deal of experience in the use of the test, to guide the interpretive process.

There are some soil type/pile length-diameter limitations with the test. In stiff clays, a minimum length to diameter ratio of 20:1 may be found, with a maximum of the order of 30:1. A general figure of 30:1 can be used, possibly extending to 50:1 where soft materials overlie a stronger founding strata. Both cast-in-place and driven piles are suitable for testing, but jointed piles will give erratic results, depending on the final condition of the pile joints, i.e. whether they are in tension or compression. Major changes in the cross-section of a cast-in-place pile, whether by design (at an underream) or accident (because of a defect), will produce resonance peaks and hence given an indication of their distance from the pile head.

It is however pointed out that because of the short duration of the test, little or no information is contained in the low-frequency part of the spectrum. The meaning of the initial slope of the frequency against v/F_0 plot is speculative in this instance when using a single blow.

8.2.6 Electrical testing methods

Various methods of using the response to electrical inputs of a pile and/or the pile–soil system, or detection of electrical currents inherent in the pile–soil system are available. The following techniques are generally recognized:

(i) *self-potential* – measurement is made of the small potential present between the pile reinforcement and the soil;

(ii) *resistivity testing* – uses methods similar to those in geophysical exploration, with the pile reinforcement as one of the electrodes;

(iii) *resistance to earth* – a known current is passed though the pile–soil system and an electrode set some distance away. The potential difference between the pile and an intermediate electrode is then measured; and

(iv) *induced polarization* – a square-wave electromagnetic signal is supplied to a pile and an electrode in the soil. The decay of the signal with time measured via a second electrode.

Of these, the methods based on measurement of resistance to earth have been most promising. A virtue of electrical methods is the lack of any special preparation of the pile, and the tests can be carried out on fresh concrete, although the results may vary with the age of the concrete. In addition, electrical methods tend to be sensitive to exposure of the pile reinforcement rather than to inclusions in the pile shaft for instance. When pile reinforcement is exposed (as may occur in a voided pile), a pile is liable to fail in the long term, even if its immediate load-carrying capacity is adequate, and therefore electrical tests are potentially advantageous in this respect. Consideration may be given to combining this form of testing with another test that is insensitive to exposure of the reinforcement, but gives good information on, say, pile length and thereby give a more complete coverage.

8.3 Summary of methods

The wide diversity of testing methods may make selection difficult. The criteria which are applied may well include cost, whether or not pre-election can be tolerated, delay in testing before some methods may be used, and, in some cases, the possible type or location of a probable defect, which may render some tests more suitable than others.

In common use are the double hole sonic test, the seismic (sonic echo) test and the dynamic reponse test. The first of these requires tubes to be installed in the pile, and is considerably more costly than the other two, which are carried out from the pile head and require little preparation. Although tests carried out from the pile head do not quantify features within a pile which repeat with depth, such features may only be of academic interest if, for example, discontinuity is present at a high level.

A summary of the methods and the considerations discussed is given in Table 8.1, which also gives an indication of relative costs.

Table 8.1 Summary of methods of carrying out integrity tests on piles

General method	Type	Relative cost	Advantages	Disadvantages
Acoustic (section 8.2.1)	Single hole ultrasonic logging	High cost if inspection hole has to be drilled	The system scans the pile concrete radially for a distance of approximately 100 mm around the drilled hole or installed tube. The method is relatively rapid, some tens of tests being possible daily	Pre-selection is generally required as it is usual to install purpose made tubes in the pile for the equipment. It is necessary to employ plastic tubes for this purpose, with the attendant risk of collapse, steel tubes being unsuitable for single hole logging. The installation of the tube adds to the reinforcement steel in restricting free concrete flow and may in itself lead to defects. Poor bonding of the tube to the concrete may yield an anomalous response. Unless a hole is drilled, it is not possible to inspect right to the base of a pile by this method. Specialist interpretation is generally necessary
	Twin tube ultrasonic logging	Medium cost, depending on expense of casting-in inspection tubes	By using a suitable number of inspection holes, the major portion of a pile core may be inspected. The inspection may be carried out to the base of the pile. Speed of testing similar to single-tube method (per traverse) at about 500 to 1000 m/day	As for single-tube method except that inspection to the base of the pile is routine. The number of inspection tubes required increases with pile diameter, and multiple tubes are usually necessary for large-diameter bored piles
Radiometric (section 8.2.2)	Single-tube radiometric logging; backscatter technique (gamma ray method)	Medium to high cost if hole requires to be formed	The system may be used on freshly cast piles and is fairly sensitive to voids or inclusions. Almost the full depth of the pile may be inspected. A zone approximately 100 mm radially from the inspection hole is examined (not uniformly). For the backscatter method, the radioactive source is weak, and no special safety precautions are required. The method is rapid and permits a large number of piles to be checked in one day	It is usual to use cast-in tubes with the disadvantage of pre-selection and added congestion in the pile in the pile shaft. The number of inspection holes required increases with pile diameter. Specialist interpretation is required

Method	Cost	Description	Limitations/Remarks
Twin-tube radiometric method; transmission technique (Neutron method)	Medium to high cost	By using a suitable number of inspection holes, the major portion of a pile core may be inspected. The inspection may be carried out to the base of the pile. The method is suitable for use on freshly cast piles. The method is rapid and a number of piles may be checked in one day	As for single-tube method with the added disadvantage that a higher intensity gamma ray source is required than with the backscatter technique requiring special safety precautions
Seismic (section 8.2.3) Sonic echo	Low cost	Rapid method which does not require pre-selection. Cracks in long slender piles detected	Only suitable for continuous piles i.e. not jointed types. Concrete cast-in-place piles should be at least 1 week old
Stress wave (section 8.2.4) Stress wave test	High cost	No pre-selection is required. An approximate estimate of the bearing capacity of the pile may be obtained	The method may require special preparation of the pile head and may be rather costly and time consuming. The method is not widely used in the UK. Expert interpretation is required
Dynamic response (section 8.2.5) Vibration test	Medium to low cost	Method does not require pre-selection. The pile may possibly be tested 3 to 4 days after casting. From the results it is possible to compute pile length or distance to a major discontinuity. In certain circumstances, pile/soil stiffness at low stresses may be estimated	Some preparation of the pile head is necessary 24 h before the test. The pile should be trimmed level with the surrounding ground and be clear of blinding concrete or other external structures
Transient response test	Medium to low cost	Rapid method. Other advantages as for vibration test	There are limitations on the geometry of the pile to be tested which vary with the ground conditions. Specialist interpretation is necessary
Electrical (section 8.2.6) Self-potential resistivity testing	Not applicable	Rapid methods using standard geophysical equipment may be used on low cut-off piles and possibly on freshly cast piles	Experimental methods, not in routine use. The methods are unsuitable for piles reinforced in the top few metres only. Specialist interpretation required

Chapter 9

Pile testing

9.1 Introduction

Since the load testing of piles is expensive, the cost clearly needs to be carefully weighed against the reduction in risk and assurance of satisfactory behaviour that the pile test provides. Piles cannot be readily inspected once they have been constructed, and variations in the bearing stratum are difficult to detect. Hence it is not easy for the engineer to be assured that the piled foundations comply with the specifications and drawings. It goes without saying that a comprehensive site investigation is essential.

The pile test programme should be considered as part of the design and construction process, and not carried out hurriedly in response to an immediate construction problem. Pile tests may be performed at various stages of construction; these are conveniently separated into tests carried out prior to the contract and tests carried out during construction. Pre-contract piles may be installed and tested to prove the suitability of the piling system and to confirm the design parameters inferred from the site investigation. Normally the installation of piles, particularly of the driven pre-formed types, will involve disturbance or re-moulding of soils, and high pore water pressures may be induced in the immediate vicinity of the pile. The bearing capacity of a pile may therefore increase or, occasionally, decrease with time after installation. It is common practice not to test piles immediately after placing but to specify a delay of perhaps 1 to 3 weeks or even more, depending on soil type and experience.

Testing of contract piles may involve integrity testing to check the construction technique and workmanship and load testing to confirm the performance of the pile as a foundation element.

The extent of the test programme depends on the availability of experience of piling in a particular geologic environment and the capital cost of the works. It is thus difficult to be specific as to the number of piles to be tested and the extent to which the piles are instrumented. Pile testing may range from 'spot checks' on pile quality to comprehensively instrumented piles bordering on a research exercise. For small contracts it may not be possible to justify the cost of any pile tests and the engineer must rely heavily on his own or the piling contractor's experience.

There are two main aspects of the pile quality to be considered:

1   the integrity of the pile and its ability to carry the applied loading as a structural unit; and
2   the load bearing and deformation characteristics of the pile-soil system.

Table 9.1 Probability of selecting at least one defective pile from a group containing 100 piles

No. of piles not meeting specification	No. of piles tested	Probability of selecting at least one low grade pile
2	2	0.04 (1 in 25)
2	5	0.1 (1 in 10)
2	10	0.18 (1 in 5.5)
2	20	0.33 (1 in 3.0)
10	2	0.18 (1 in 5.5)
10	5	0.41 (1 in 2.5)
10	10	0.65 (1 in 1.5)

It is noted that the errors in the position and alignment of the piles within a pile group may significantly influence the behaviour of the completed foundation.

The selection of, and the number of, piles to be tested obviously greatly influences the reliability of any conclusions drawn from the test results. Table 9.1 illustrates the effect of sample numbers in the likelihood of detecting a defective pile. From this table it is evident that to achieve a reliable assurance of the satisfactory quality of the group of piles, a large number of piles have to be tested. In general, load testing is unsatisfactory for quality control on the grounds of cost. It is usual to select at least one 'typical' pile for testing, together with any pile that, from its construction records, is thought to be unsatisfactory. Quality control of pile construction may be checked by adopting one of the non-destructive testing techniques that are now available. A great deal of effort has been put into the development of dynamic loading and integrity testing of piles; the former to check, as reliably and simply as possible, the bearing capacity and the latter to confirm soundness and structural continuity. These methods are now sufficiently well-developed to be considered useful tools in any quality control procedure.

The behaviour of a single test pile may well be very different from that of the pile working in a group. The results of observations of load distribution down the shaft of a pile at a bridge in Newhaven, England, are shown in Figure 9.1 (Reddaway and Elson, 1982). These results clearly demonstrate that under short-term test loading the soil resisted a large proportion of the load as friction in the alluvium, in spite of a bitumen slip coat which was applied to the upper 10 m of the pile. On completion of the bridge, the same pile (now working in a group of 32 piles), transmitted the entire loading to the underlying chalk. In such circumstances one would seriously question the usefulness of the pile test. Subsequent analysis of the situation, using an elastic continuum model, demonstrated that the observed behaviour was a function of the very large difference in stiffness between the alluvium and the chalk. It should be noted that the designers had, not unreasonably, assumed that friction would be developed in the gravel, which was moderately dense.

It is emphasized that any pile test programme should be carefully designed, and that the purpose of each test should be clearly stated. A large amount of information can be obtained from carefully conducted loading tests on test piles. The data can be further augmented by instrumentation of the pile. The information may well lead to refinements of the foundation design with a consequent possible cost saving, or certainly greater assurance of the satisfactory performance of the foundation. On the

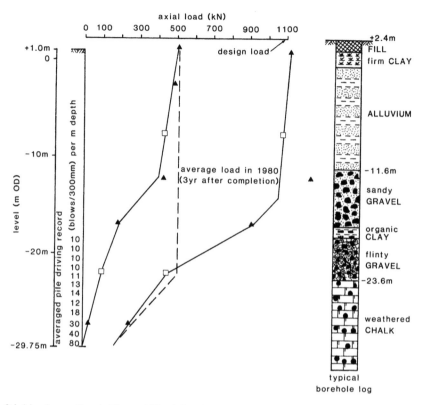

Figure 9.1 Newhaven Overbridge: axial load distribution in pile No. 122. Upper 10 m of pile slip coated. □, average of two active gauges; ▲, single gauge. Date of test 22 Nov 1976.

other hand, poor-quality test pile data may lead to an uneconomic design and are obviously wasteful of resources and money.

9.2 Static load testing of piles

9.2.1 General comments

Pile load testing, using the maintained load or the constant rate of penetration test procedures, is the most commonly adopted method of checking the performance of a pile, although dynamic testing of driven piles, using 'wave equation' analysis, is also common. For offshore installations it is usually not possible to carry out static load testing and reliance is placed on dynamic analysis to predict the static ultimate load capacity from the measured driving resistance. Dynamic analysis is discussed in more detail later in this chapter.

Any test pile and loading system should be properly designed, bearing in mind the large forces and high strain energy in the system under full load. The test pile should be typical in all respects of the piles in the foundation. This is particularly important for pre-contract piles, as small variations in construction procedure and workmanship can cause disproportionate effects on the behaviour of the pile. An obvious disadvantage of pre-contract piles is that the main piling contract should perforce be awarded to the pre-contract piling firm. Test piles that are not part of the permanent works, constructed and tested prior to the commencement of the main piling programme, avoid double mobilization charges. However, the purpose of a preliminary test pile is often defeated if contract piling continues immediately after test pile installation and before the loading test can be carried out. Pile testing requires some considerable time and engineering input to carry out and to interpret the results; this process should not be rushed at the expense of quality of the data and reliability of the results.

It is important that the test piles are loaded in the way envisaged by the designer; in particular it maybe necessary to sleeve the pile in any overburden. Such a free-standing length should then be designed as a long column. Any instrumentation built into the pile should be designed to have a minimal effect on the pile stiffness and ultimate load carrying capacity. This is not always easy to arrange. Inclinometer ducts or rod extensometers may affect the axial and lateral stiffness of the pile shaft, and built-in load cells may form physical discontinuities which appreciably alter the behaviour of the pile-soil system.

The objectives of a preliminary pile test programme are usually to determine:

1   the ultimate bearing capacity of the piles, relating this to the design parameters;
2   the relative magnitudes of shaft and end-bearing capacities;
3   the stiffness of the pile-soil system at design load. A back analysis of this data will enable the soil modulus to be evaluated, and hence the deformation of pile groups may be predicted with greatly increased confidence.

These various objectives will necessitate a carefully chosen test procedure and instrumentation programme.

The preparation of a test pile will usually involve cutting down the pile to the required level, exposing the reinforcement, and casting a small pile cap or bedding a 25-mm thick steel plate directly on to the pile. The surface of the pile cap or plate should be flat and should be perpendicular to the axis of the pile, to minimize eccentricity of loading. In particular circumstances it may be necessary to extend the pile by casting or welding on an additional length. Obviously the pile should be cut free of any blinding concrete. If the pile is to be loaded to its ultimate capacity, it may be necessary to provide additional reinforcement at the head of the pile to absorb eccentricities of loading and prevent bursting of the head of the pile. If additional reinforcement is required it is obviously impracticable to select test piles at random.

A pile test arrangement is shown in Figure 9.2, and the components of the equipment are discussed below.

9.2.2 Test equipment

9.2.2.1 Provision of a reaction

The ultimate load of a pile may range from a few tens of tonnes to as much 2500 tonnes, and provision of a reaction against which to generate the load requires careful consideration. The geometry of the pile-reaction arrangement should be such as to minimize

Figure 9.2 A 2500-tonne pile load test using tension piles to provide reaction. (Photograph courtesy of Cementation Piling and Foundations Ltd., Rickmansworth.)

interaction between the pile and the reaction, and to avoid movement of datum beams used as references to measure settlement.

Kentledge is commonly used to provide reaction, and more or less any material available in sufficient quantity can be used (Figure 9.3). Specially cast concrete blocks or pigs of cast iron may be hired and transported to site. The cost of transport to and from the site is a significant factor. Regular-shaped blocks have the advantage that they may be stacked securely, and are unlikely to topple unexpectedly. Sheet steel piling, steel rail, bricks or tanks full of sand or water have to be adopted as kentledge from time to time. The important criterion is that the mass of material is stable at all times during and after the test.

Clearance is required above the pile cap for the jack, load cell and reaction beam, and it is necessary to support the whole weight of the kentledge on a timber or other appropriate cribbage. The area of the cribbage platform should be sufficient to avoid bearing failure, a severe problem on soft ground. The kentledge is usually supported on a deck of steel beams or a timber mattress. The cribbage should also be spaced at least 3 or 4 diameters from the pile to avoid an unacceptably large interaction between the soil and the pile as load is transferred to the pile. This spacing determines the minimum length of the reaction beam required.

The reaction beam is subjected to high bending and buckling stresses and needs to be designed to carry the maximum load safely. The maximum safe load should be clearly marked on the beam so that it is not inadvertently exceeded during a test. A thick

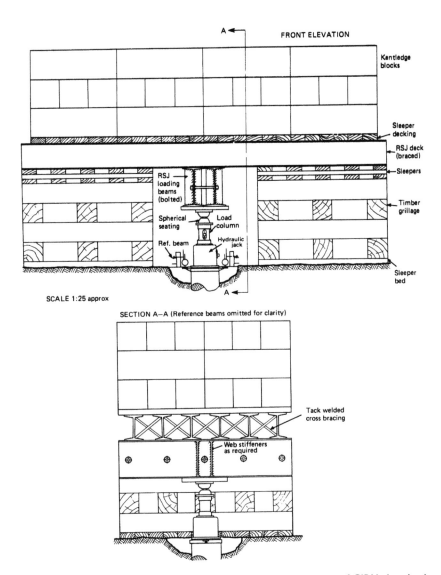

Figure 9.3 Test arrangement using kentledge. (Illustration courtesy of CIRIA, London.)

spreader plate is normally welded to the centre of the beam to disperse the point load from the jack.

The whole system should be firmly wedged, cleated or bolted together to prevent slipping of any member. Frequently the piling contractor or specialist testing concern will have a ready-made set of equipment which is designed to suit their system of working. During the test it is important that none of the timber work or decking is unloaded completely, to avoid it becoming unstable. It is therefore usual to provide extra kentledge, and 20% above the ultimate load of the pile would be considered a satisfactory margin. Not so much overcapacity need be provided if the stack is built carefully from regular units. Common sense dictates that the kentledge is stacked uniformly, and interlocked if possible, and that the centre of gravity of the mass is as near as possible to the pile axis.

Raking piles are sometimes tested using kentledge as a reaction, provided the rake is not too great. A steel wedge of the correct angle is bolted to the reaction beam to provide a face parallel to the pile head. The horizontal reaction is provided by friction and the mass of the kentledge should be increased accordingly. This system of testing raking piles is not very satisfactory and it is difficult to ensure that eccentric loads are not applied to the pile head, leading to premature failure of the pile head.

Tension piles may be utilized to provide a satisfactory reaction, particularly for raking piles (Figure 9.4). It is obviously most convenient if adjacent permanent piles can be used. If only two adjacent piles are suitable they would often have to be lengthened to provide adequate pull-out resistance, and severe problems of lateral instability can occur under load. Some specifications do not permit this, while others insist that if it be done, uplift on the tension piles is measured and needs to be of very limited magnitude.

It is important that the spacing of the piles is as large as practicable, as significant interaction between them can occur at spacings of less than 5 pile diameters. In practice, spacings of 3 to 4 diameters between the centres of the test pile and the reaction piles are commonly adopted. The effect of the pile interaction is to reduce the observed settlement of the pile, and it may be necessary to make corrections for this effect where close spacing is unavoidable. It is suggested that the pile layout is arranged to limit the effect of pile interaction to an absolute maximum of 20% of the settlement of the test pile.

Various systems are in common use to connect the tension pile to crossheads. The reinforcement of the anchor piles may be left projecting sufficiently to allow a welded washer connection to be used (Figure 9.5). The fillet weld should be designed to carry the anticipated load and each weld should be carefully inspected before the system is loaded. High tensile steel reinforcement should not be extended by welding, and bent or distorted bars should be avoided. Accidental bending and subsequent straightening of such bars can lead to brittle failures. Alternatively, straight, high tensile, McCall or Dywidag bars may be cast into the top of the piles, using patent anchor systems supplied by these firms. Locking collars or nuts should be provided with such systems to guard against slippage under high load. The top of the tension pile should be adequately reinforced against the stresses imposed by the anchors. The tension piles should normally be reinforced for the full depth of the pile, although the reinforcement in the pile may be reduced with depth taking into account the reduction in tensile stress along the pile shaft towards the pile toe.

Connection to precast concrete anchor piles is not as easy as in the case of cast-in-place piles. It is necessary to break down the head of the pile to expose the reinforcement, and either cast on a special anchor block or extend the main steel using sleeved connectors or welding. Connection to steel H-section piles is straightforward, and suitable cleats may be designed and fixed to the piles by welding or bolting.

The use of *ground* or *rock* anchors to provide a reaction is sometimes convenient, depending on the particular geology of the site. The installation of the anchors usually requires the mobilization of a special drilling rig and its associated equipment. The main disadvantages of ground anchors as a reaction system are the axial flexibility of the anchor tendons and the lack of lateral stability of the system.

Multiple ground anchor systems are to be preferred and each anchor should be proof tested to 130% of the maximum load before use. To reduce the extension of the cables during the test it is usual to pre-stress the anchors to as high a proportion of the maximum load as possible. Hence a suitable reaction frame and foundation must be provided to carry this load safely.

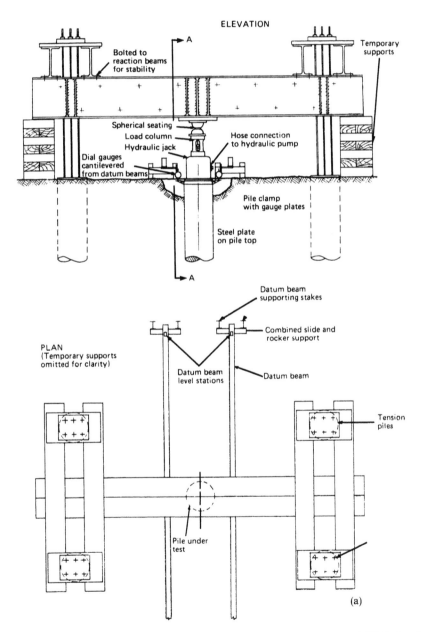

Figure 9.4 Test arrangement using tension piles. (Illustrations (a, b) courtesy of CIRIA, London.)

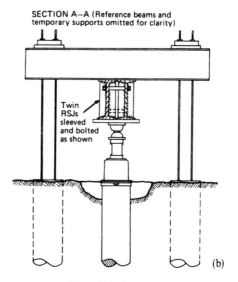

SECTION A–A (Reference beams and
temporary supports omitted for clarity)

Twin
RSJs
sleeved
and bolted
as shown

(b)

Figure 9.4 Continued

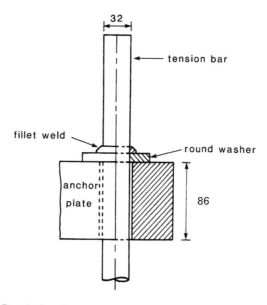

32

tension bar

fillet weld

round washer

anchor
plate

86

Figure 9.5 Detail of washer connection to reinforcement (dimensions in mm).

The tensile forces may be transmitted to the cross-head using a looped cable and steel saddle, or individual wires in the cable may be locked separately in an anchor block.

Owing to the high extensibility of the tendons, very large amounts of energy are stored in the system, creating a potentially hazardous situation. Thus each component of the system should be designed and installed to avoid any risk of sudden load reduction or over stress.

9.2.2.2 Load application and measurement

The application of the load to the test pile needs to be closely controlled, and a range of hydraulic jacks is readily available. Short, large-diameter low-pressure jacks are to be preferred to the high-pressure type as they are more stable. The capacity of the jack is not particularly important provided it exceeds the maximum applied load by a generous margin, and has sufficient travel of the ram (at least 100 mm). For very high loads it may be necessary to use a cluster of jacks. These should be set in a cradle to prevent relative movement of the jacks or uneven loading.

The jack may be actuated by a hand pump for loads up to approximately 300 tonnes. However, for very high loadings or constant rate of penetration tests, it is preferable to use a motorized pumping unit. Pressure gauges should be fitted to the system to allow the operator to check the load in the system at a glance, as he may be some way from the test pile.

The line of action of the load should be co-axial with the pile and time is well spent setting up the jack, load cell, etc., carefully (Figure 9.6). To take up slight misalignment or lack of parallelism a large-radius spherical seating should be inserted under the reaction beam. Under load, spherical seatings lock and will not correct any eccentricity of load during the test. Small-diameter ball seatings tend to lead to instability of the loading trains and increase the likelihood of violent collapse of the train under load. If an angular misalignment of more than a few degrees is noted, the reaction system should be repacked to eliminate the error.

In general the pressure gauges fitted to the hydraulic jack are not suitable for measuring the load applied to the test pile. Test-quality gauges may be fitted and calibrated with the jack in a compression loading frame. Such calibrated jacks may provide acceptable accuracy of load measurement for increasing load; however, the unloading cycle is usually non-linear owing to friction within the hydraulic jack. A typical calibration is shown in Figure 9.7; note the jack and gauge must be clearly marked and used together.

Load measurement is preferably carried out using a load cell, and several types are available. Commonly adopted load measuring devices, in order of accuracy, are:

1   hydraulic load capsule, maximum capacity 450 tonne;
2   load columns, maximum capacity 1000 tonne;
3   proving rings, maximum capacity 200 tonne; and
4   strain gauged load cells of various types, maximum capacity 200 tonne.

For very large loads an array (preferably at least three) of load columns or strain-gauged load cells may be used. It is important that the load cell is calibrated regularly and that, for those devices sensitive to temperature, suitable corrections are determined.

Hydraulic load cell capsules are basically short, large-diameter hydraulic cylinders (Figure 9.8). The piston is bonded to the casing with a rubber element, sealing the oil-filled chamber. The oil pressure is measured with an accurate pressure gauge,

Figure 9.6 Details of load train. (Photograph courtesy of Sir Robert McAlpine & Sons Ltd, London.)

or a pressure transducer may be substituted. The pressure gauge may be fitted with a capillary tube for reading clear of the test system. The accuracy of these devices is probably about ±1%.

Load columns are short columns of steel or aluminium which have a hole machined in the central section to increase the deformation of the column (Figure 9.9). Deformation is measured using a sensitive dial indicator. The devices were developed by the National Physical Laboratory, but are now manufactured commercially. The devices are sensitive to changes in temperature; however, it should be possible to measure load to an accuracy of ±1%.

Proving rings are machined from forgings of mild or high tensile steel (Figure 9.10). Deformations are measured with a dial indicator or a displacement transducer. The deformation of the device under load is quite large, hence the sensitivity to temperature changes is reduced. Loads may be measured to an accuracy of 0.5%. However, the slender proportions of the ring can lead to instability of the load train.

Strain gauged load cells are basically machined steel columns, fitted with bonded electrical resistance strain gauges or vibrating wire strain gauges (Figure 9.11). The load

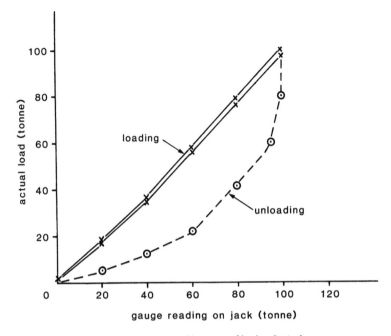

Figure 9.7 Typical calibration of hydraulic jack.

cells, particularly the electrical resistance type, should be environmentally protected. The load cell developed by the Mining Research Development Establishment has been found to be particularly satisfactory. Special readout units are required for the load cells, although battery powered devices suitable for site observations are made by several firms. The overall accuracy of the system should be better than 0.5%. The load cells are usually temperature compensated although extreme changes of temperature should be avoided.

9.2.2.3 Measurement of settlement

The measurement of deformation is perhaps the most difficult observation to make reliably. The use of dial gauges calibrated in 0.01 mm may give a false sense of accuracy to the observations. In practice, measurement to this precision is not necessary for pile testing and it is extremely difficult to maintain this degree of accuracy over a long period of time. The most commonly used systems of settlement measurement are now electronic displacement transducers, although dial test indicators or optical levelling may also be used as a supplement.

Electronic displacement transducers, either of a potentiometer type with a sliding voltage divider, or a linear variable differential transformer (LVDT), are the modern replacement of dial gauges, and have the advantage of being able to be logged continuously throughout a test, minimizing manual readings. The resolution of the displacement transducer should be around 0.01 mm.

Figure 9.8 Hydraulic load capsule. (Photograph courtesy of A. Macklow-Smith Ltd., Camberley.)

Figure 9.9 Load column. (Photograph courtesy of A. Macklow-Smith Ltd., Camberley.)

Figure 9.10 Proving ring.

Figure 9.11 Load cell.

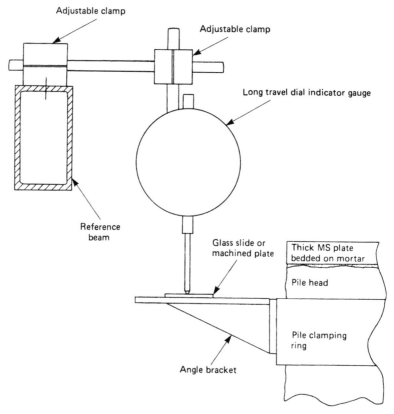

Adjustable clamp

Adjustable clamp

Long travel dial indicator gauge

Reference beam

Glass slide or machined plate

Thick MS plate bedded on mortar

Pile head

Pile clamping ring

Angle bracket

Figure 9.12 Measuring pile deformation using dial gauges. (Illustration courtesy of CIRIA, London.)

Dial gauges used to be invariably employed to measure pile deflection under load. The precision was of the order of $+/- 0.01$ mm but these days the difficulty of reading them remotely for safety reasons and the preference for automated testing means that displacement transducers are generally employed. Where dial gauges are used they should be mounted as shown in Figure 9.12. The same general principles are applied to the mounting of displacement transducers.

For short-term tests it is sometimes convenient to use magnetic clamps to mount transducers. Normally four are used for a pile test, so that any bending of the pile head can be detected.

The measuring points should be on the pile head or on brackets mounted on the side of the pile. Small machined steel or brass plates should be used as a datum for the transducer stem. On exposed sites, polythene bags provide a convenient method of protecting the devices. Care must be taken to ensure that the reference plates are perpendicular to the pile axis, so that any sway of the pile has no vertical component.

Stable datum beams are quite difficult to set up, and the basic requirements are set out below. Firm foundations for the reference beams should be constructed, away from any loaded areas. If possible, adjacent piles should be used, or a separate concrete block supported by steel stakes constructed. If practicable, the datum beam foundations should be at least four pile diameters from the test pile. It is noted that a compromise must frequently be made between a long beam prone to vibration and temperature-induced deflections, and a shorter beam with foundations in the zone of influence of the test pile or kentledge.

The reference beam should be a heavy steel section or a rectangular hollow section tube clamped at one end. The other end is free to slide, but should be restrained against lateral and vertical movement (see Figure 9.13). Such a system will allow longitudinal movement of the beam caused by changes in temperature. The beam should be shaded from direct sunlight, and in hot climates it may be necessary to carry out the test at night when temperature fluctuations are less marked.

To check for gross errors in the observations or disturbance to the datum beams, it is good practice to carry out precise level checks of the test pile and datum beams at intervals throughout the test. If tension piles are being used to provide the reaction, these should also be levelled to check for undue movements.

The accuracy of any *optical levelling system* is poorer than that of displacement transducers or dial gauges by a factor of at least 10. However the absolute accuracy of the system may be of a similar order, particularly in situations where it is difficult to establish a stable reference beam. Using optical methods it is difficult to arrange a constant rate of penetration test or to check the cessation of movement under a load

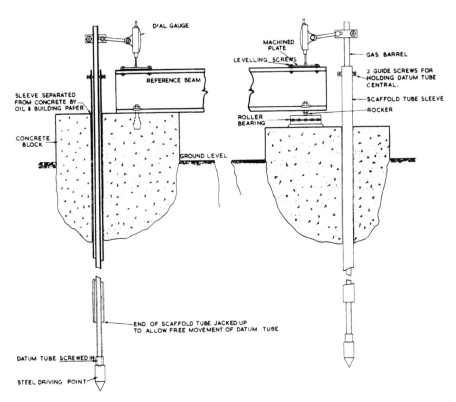

Figure 9.13 Details of datum beam support. (Building Research Establishment: reproduced by permission of the Controller, HMSO. Crown copyright.)

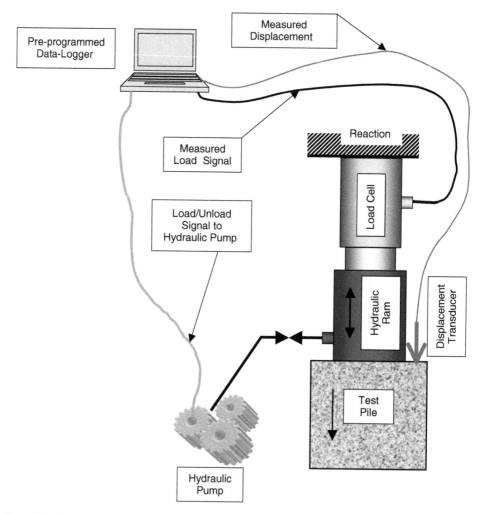

Figure 9.14 Schematic diagram of basic method of computer control of a maintained load pile test. (Illustration courtesy of PMC Limited, Durham.)

increment owing to the coarseness of the least count. However, the optical system does permit the instrument and reference point to be well outside the zone of disturbance. A precise levelling system is of suitable accuracy with a scale such that an overall accuracy of ± 0.2 mm is attainable.

Electrical load and displacement transducers allow automatic recording of the pile test data. However, care is required in setting up the equipment, particularly in respect of shielding them from adverse weather. Modern notebook computers incorporating suitable data acquisition software may be used to record all data, including load and displacement; indeed it is practicable to automate the whole of the test procedure, recognizing perhaps the need for human intervention in specific circumstances.

The degree of sophistication of the data-logging equipment provided depends to some extent on the volume of data to be recorded and on the number of piles to be tested. Some piling contractors and specialist pile testing companies such as PMC Limited have developed equipment, and the quality of the output can be significantly better than that achieved with conventional test methods. Specialized pile testing companies will use pre-programmed pile tests with automatic load maintenance and stage increases together with settlement recording and detection of cessation of pile movement. The basic elements of such a system are shown schematically in Figure 9.14.

To provide more information on the behaviour of the test pile, additional instrumentation is occasionally deemed necessary. This type of instrumentation is described in section 9.2.5.

9.2.3 Pile load test procedures

The pile loading test that is most commonly carried out is the maintained load test but alternatively, the constant rate of penetration test is sometimes employed. The maintained load test is convenient for testing end-bearing piles and for determining the load-settlement characteristics in clay soils. Usually, however, it is not as suitable for determining the true ultimate capacity of a pile.

The constant rate of penetration test procedure is best suited to determining the ultimate bearing capacity of a pile, as the method of testing is closely related to the test procedures used to obtain the shear strength of the soil. In clay soils, rapid pile testing may approximate to undrained loading conditions, whilst a constant load for several days may be necessary to allow full dissipation of pore pressures.

9.2.3.1 Maintained load tests

The maintained load test requires careful specification of loading increments and periods for which these increments are held constant. The ICE *Specification for Piling* sets out a suitable minimum scheme (Table 9.2), and the ASTM Test Designation D-1143 specifies a similar procedure. In addition, limits are placed on the rate of movement before the next load stage is added. The ICE *Specification for Piling* limits the rate of movement to 0.24 mm/h, provided the rate is decreasing, for pile head displacements greater than 24 mm, but less than 0.1 mm/h for head displacements of less than 10 mm. Between these pile head displacement values, a rate of less than 0.1% × pile head displacement/h is recommended. Some engineers prefer a limit of 0.1 mm/h, and a frequent requirement is to hold the load constant for 24 h at design load. The ISO/DIS 22477-1 Draft document (2006) recommends a cessation of movement criterion of 0.1 mm per 20 min and a time interval of at least 60 min per loading step. It also permits a pre-load of $0.05Q_{max}$ to check the loading train and instrumentation, unloaded to a small residual value of the order of 10 kN prior to starting the test proper. The particular details of the loading path followed are not of great significance, but it is important that the holding time at each load increment is the same so as to lead to the same degree of soil consolidation.

Table 9.2 Suggested load increments and holding times

Load*	Minimum time of holding load
25% DVL	30 min
50% DVL	30 min
75% DVL	30 min
100% DVL	1 h
75% DVL	10 min
50% DVL	10 min
25% DVL	10 min
0	1 h
100% DVL	6 h
100% DVL + 25% SWL	1 h
100% DVL + 50% SWL	6 h
100% DVL + 25% SWL	10 min
100% DVL	10 min
75% DVL	10 min
50% DVL	10 min
25% DVL	10 min
0	1 h

*SWL denotes specified working load; DVL denotes design verification load.

To determine these slow rates of movement a sensitive settlement measuring system is required. In granular soils or soft rocks the cessation of movement is easy to establish. However, in clay soils consolidation settlement will occur over an extended period. Plotting the results as the test proceeds is a useful way of resolving any ambiguities in the results (see Figure 9.15).

The maintained load test may be used to test contract piles to check acceptance criteria. Such piles are usually loaded to 100% of the design verification load (DVL) plus 50% of the specified working load to avoid overstressing the soil whilst still proving an adequate reserve of strength in the pile-soil system. The use of the DVL was introduced to take into account items such as downdrag friction load, which will be in the reverse direction during a test to that which may exist under long-term loading, and the fact that piles on a site may not be tested from the same level as may pertain for the piles in normal use. The ICE Specification also contains an extended proof load procedure which may be useful, particularly for piles bearing on cohesionless soil or on rock, and where further information may be desirable to verify design. On special test piles that will not form part of the permanent works, loading can be increased to failure, enabling a check to be made on the design parameters (usually by the CRP method).

For particular projects special loading paths may be called for, for example, to simulate repeated loading. Such loading paths may easily be arranged but careful supervision is a necessity. Short-period, cyclic or sinusoidal loading requires the use of sophisticated servo-controlled equipment.

9.2.3.2 Constant rate of penetration test

The constant rate of penetration (CRP) form of testing is confined to special test piles, and is usually carried out at the pre-contract stage (Whittaker and Cooke, 1961). As the name implies, the loading is strain controlled and a set rate of penetration of the

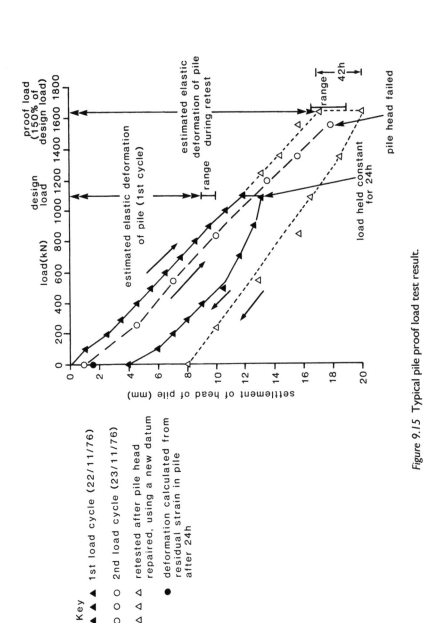

Figure 9.15 Typical pile proof load test result.

Key

▲ ▲ ▲   1st load cycle (22/11/76)

○ ○ ○   2nd load cycle (23/11/76)

△ △ △   retested after pile head
        repaired, using a new datum

● deformation calculated from
  residual strain in pile
  after 24h

pile load is specified. Rates of penetration between 0.5 and 2 mm/mm are commonly used, the lower rate being used in clay soils.

Great care is required to maintain a steady rate of penetration, but this is difficult to achieve with manual pumping without some practice. If manual control is adopted time-penetration plots should be drawn as the test proceeds, so that deviations from the required path are easily identified. For this type of testing it is much more satisfactory to use a hydraulic power pact, and control the rate of penetration through software linked to a displacement transducer. Control of the jack movement is achieved by manual operation of the rate of oil flow. Fully automatic systems have been devised but are complex, and present difficulties in operation on site.

Loading of the CRP test should be continued to a deflection that goes to or beyond failure of the pile. This may be, typically, until a constant load has been recorded or until the penetration is at least 10% of the pile diameter (see Figure 9.15).

For this test, large travel of the jack and deflection transducers should be allowed for. It may be necessary to reset the transducers during the test, or alternatively, machined spacing blocks can be inserted under the gauge stems.

Once the equipment has been set up the test may be completed within an hour or so, and a rapid assessment of the ultimate capacity of the pile made. However, an experienced team of three or four people is required to carry out the test efficiently.

Tension piles may be tested using a constant rate of uplift procedure, which is similar to the CRP test. The full tensile capacity of a pile is normally mobilized at a displacement of a few millimetres and the imposed rate of uplift should be correspondingly reduced to say 0.1 to 0.3-mm per minute.

9.2.3.3 Recording the results

It is essential that all the relevant data are recorded and to this end it is useful to use a standard proforma (see Table 9.3). Similarly the load, settlement and time data should be recorded clearly throughout the test. It is important that any problems or unexpected occurrences are reported in the test records, as this information is most useful when interpretation of the results is made.

It is customary to present load-time, settlement-time and load-settlement graphs of the results. It is important that these last two graphs are plotted during a maintained loading procedure so that modifications to the test can be made if required.

Pile tests are relatively expensive and the results of such tests can have a considerable influence on the construction of the permanent works. It is therefore imperative that an adequate number of qualified staff are available to set up and carry out the test with engineering supervision throughout the test period. In addition at least one technician should attend the tests, to control the hydraulic jack, etc. Similarly, interpretation of the results and preparation of a report on the test programme should not be stinted, and sufficient staff time should be allocated.

Special pile tests (including CRP tests), involving a higher level of instrumentation generally involve a larger team to carry them out successfully. Setting up the instrumentation should be carried out by experienced technicians or engineers. Staff time involved in carrying such special pile tests can be considerable, and represents a significant cost in the overall test programme.

Table 9.3 Essential information to accompany pile test sheets

Item	Information needed
Dates	Date pile was installed. Date of test.
Location	Sufficient detail to permit the site to be located and also the position of the pile relative to the works. The pile number should always be stated.
Pile type	Various categories of pile type are set out in Weltman and Little (1977) (undergoing revision).
Pile installation details	If a bored pile (i.e. non-displacement), the depth of temporary casing, how much concrete, mix details, how placed (e.g. trémie or chute into dry bore), and any special circumstances, particularly regarding groundwater. If driven (i.e. displacement), type of hammer, weight, drop, final set. The driving record should accompany pile test report.
Pile dimensions	Nominal diameter (or section) weight per metre run as applicable. Size of underream, bulb, 'wings', etc. Length of the pile – include the entire pile length.
Installed level (driven) Concreted level (bored) Trimmed level and toe level	Give full data so that there is no doubt regarding the reduced level of pile toe, the ground level *at the time of the test* and the level of the top of the pile either at the end of installation, after trimming or both (above or below ground level).
Orientation	State whether vertical or raking (with degree of rake if applicable).
Design Load of pile	State Design Load or indicate on plot.
Type of test set-up and settlement measuring system	Whether kentledge or tension pile test. Leading plan dimensions of the set-up including reference system. Method of measuring settlement, subsidiary levelling of reference beams, temperature corrections.
Weather	Brief comment on weather conditions and extremes of temperature during test.
Soil information	It is not generally feasible to provide full soil information. If a report exists, the name of the company who produced the report should be given, together with a reference number. The position of relevant site investigation boreholes should be given on the pile location plan. Summary logs of nearby boreholes with SPT N values and cohesion values alongside are useful. In some cases, a Bored Piling Contractor will log the boring – such logs should be given with an indication of their source. The *reduced levels* of boreholes must be given.

9.2.4 Interpretation of the results

A considerable amount of data may be computed from a pile test, and with more sophisticated instrumentation a greater understanding of the pile-soil interaction may be achieved. As discussed in section 9.1, it is important that the designer is satisfied that the behaviour of an isolated test pile under relatively short-term loading stresses the bearing stratum in a similar manner to the piles in the foundation. Interpretation of the pile test data may be carried out on several levels:

1 qualitative inspection of the load settlement curves;
2 a check for compliance with load and deformation specifications; and

3   a back-analysis of the data to provide information on the soil stiffness and strength criteria.

9.2.4.1 Load settlement curves

Typical shapes of load settlement curves are given by Tomlinson (1977). In particular unusual load settlement curves should be thoroughly investigated as they may indicate unexpected geological problems, such as soft sensitive clay, defects in the pile shaft, or poor construction techniques.

From straightforward pile tests in which the load settlement characteristics of the head of a pile are measured, there is as yet no satisfactory published method which allows separation of the components of shaft friction or adhesion and end bearing with a satisfactory degree of confidence. Randolph (1994) has discussed how the use of software based on load transfer curves for the shaft and base, and allowing for the compressibility of the pile, may be used to fit the measured load-settlement response. In this way, estimates of the limiting shaft and base resistance may be made, together with shear modulus values of the soil.

In the past it has generally been found that in order to study the division of load between a pile shaft and base, it is necessary to install a system of strain gauges. This can have the advantage of also providing guidance on the distribution of load along a pile shaft and contributing to design knowledge, subject to adequate provision of gauges. However, this requires considerable skill and care (see section 9.2.5).

9.2.4.2 Estimation of ultimate load

The ultimate load of a pile is usually not well defined; typical load-deformation curves are shown in Figures 9.15 and 9.16. In practice an exact definition of the ultimate load

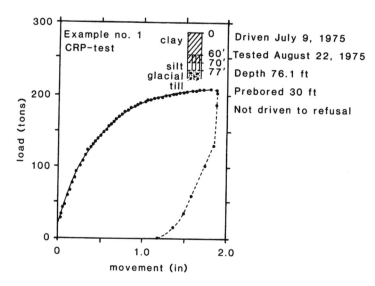

Figure 9.16 Results of a CRP test. (After Fellenius, 1980.)

is not all that important, provided an adequate factor of safety is clearly demonstrated. Two simple criteria which cover most situations are:

1 the load at which settlement continues to increase without further increase in load; and
2 the load causing a settlement of 10% of the pile diameter (base diameter). The latter limit is likely to give a low estimate of the ultimate load as it is unlikely that general yielding of the soil around the pile will have been initiated.

Other definitions of ultimate load are summarized by Tomlinson (1977), while Fellenius (1980) compares a number of methods of estimating the ultimate load. Unfortunately many of the proposed techniques are either empirical or are based on set deformation criteria. Several of the methods are also sensitive to the shape of the load-deformation curve, and it is preferable to use a considerable number of load increments to define the shape clearly.

Two methods of estimating the ultimate pile load are shown in Figures 9.17 and 9.18 after Chin (1970) and Brinch Hansen (1963). Chin's method assumes the form of the load-deformation curve is hyperbolic and is an empirical method. However, it is useful in that if the results are plotted at the time of the test, deviations from the expected curve are easily seen and the test procedure can be checked. It may in some cases be used to obtain an estimate of shaft or base load separately. Brinch Hansen's method is very simple to use, and in practice has been found to give consistent results.

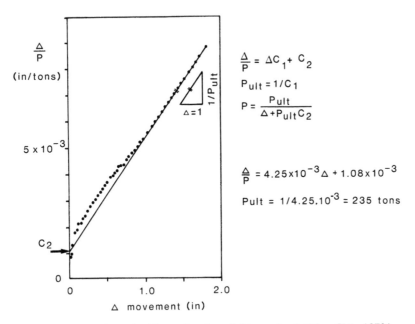

Figure 9.17 Chin's method for estimation of ultimate load. (After Chin, 1970.)

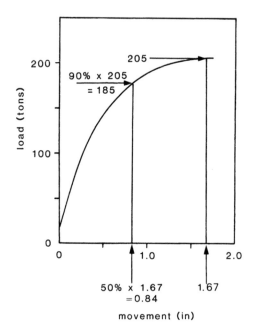

Figure 9.18 Brinch Hansen's 90% method. (After Brinch Hansen, 1963.)

9.2.4.3 Settlement criteria

Settlement limits at design load or 1.5 times the design load are frequently specified, on the basis of previous experience (e.g. a maximum of 10 mm settlement at design load). Such specified settlement limits should be realistic and take into account the likely elastic compression of the pile shaft. It is sometimes difficult to discern how the specified criteria for a single pile will relate to the behaviour of the pile group. A much more satisfactory approach is to use the pile test data to assess the stiffness of the soil mass using the solutions outlined in Chapter 4. This data may then be applied to the pile group to estimate the overall performance of the foundation under the design loads. This approach is particularly useful in the case of piled structures carrying high lateral loads, as the stiffness of the soil is difficult to establish and may well be affected by the piling operation. However, a single pile will not stress the ground to a depth comparable to the depth stressed by a pile group and some extrapolation of the results will generally be required.

Contract test piles are frequently loaded to 1.5 times the design load as a performance test. Such a load test should avoid significant permanent deformations in the surrounding soil and the alteration of the load-deformation characteristics of the pile so tested. As these tests are not taken to failure it is sometimes difficult to assess the acceptability of the pile. In such situations it is useful to estimate the likely elastic settlement of the pile and then to draw up a realistic and unambiguous specification for test pile performance. Such an assessment can quickly be made by summing the compression of the shaft and settlement of the toe. For such calculations it is necessary to estimate the shaft adhesion from the site investigation data. In practice, by making appropriate assumptions, upper and lower limits to the settlement may be established.

9.2.5 Special pile instrumentation

To obtain a greater understanding of the pile-soil behaviour, it is sometimes desirable to install further instrumentation in the test piles. It should be stressed that such instrumentation demands significant investment in cost and manpower, if reliable results are to be obtained. In general, for site use simple robust equipment should be used to obtain the necessary information, albeit with some loss of accuracy. There are several case histories reported in the literature (Reese *et al.*, 1975) for fully instrumented piles; these are really research projects and are not considered in detail here.

9.2.5.1 Distribution of load

Special high-capacity load cells were originally developed by the BRE (Whitaker, 1963). The cells are steel multi-column devices, the strain being measured by electric resistance strain gauges (Figure 9.19). Simpler modular load cell systems have subsequently been developed by the BRE (Price and Wardle, 1983; Price *et al.*, 2004). The load transducer comprises a steel tube fitted with an internal vibrating wire gauge. Load is transferred to the transducer, across a discontinuity in the pile formed by a soft rubber sheet, by steel bars bonded into the concrete. Load cells of the required

Figure 9.19 BRE pile load cells. (Building Research Establishment: reproduced by permission of the Controller, HMSO. Crown copyright.)

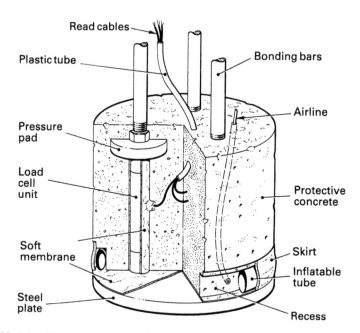

Figure 9.20 Modular pile load cell. (Building Research Establishment: reproduced by permission of the Controller, HMSO. Crown copyright.)

desired capacity may be made using an array of transducers (Figure 9.20). These cells are robust and have the advantage that the structural collapse of the pile cannot occur, following corrosion of the transducer elements.

The measurement of base load in a driven pile is not as easy so easily accomplished as for cast-in-place piles. Normally special load cells occupying the full cross-section of the pile are unsuitable, although with some segmental piles the load cell can be inserted at the joints. A more satisfactory technique is to cast vibrating wire strain gauges into the pile (Figure 9.21). The gauge installation may be calibrated by load-testing a short length of pile. Stability of the system is reasonable for short-term loading, and the gauges should not be damaged by driving. Long-term load measurement requires correction for creep strains in the concrete, and a consequent uncertainty in the estimation of the pile load. Several gauges should be installed at any level to allow for inevitable losses. The cables should be continuous and may be led out through a duct below the head of the pile. If a segmental pile system is being used, forming waterproof cable joints at each segment junction can be difficult to achieve.

A similar system may be used for bored piles (O'Riordan, 1982), but unfortunately the sensitivity of the system is reduced owing to lower stresses in the concrete. Variation of the modulus of the concrete will also add to the uncertainty in relating strain to load. Protecting gauges during concreting is difficult, and a considerable number of gauges may be damaged. Successful installations of vibrating wire strain gauges are much easier to achieve at the head of bored piles, where concreting can be carefully monitored.

Figure 9.21 Vibrating wire strain gauge installation. (Photograph courtesy of PMC Limited, Durham.)

An alternative, lower cost method to separate shaft and base resistance of a cast-in-place pile is to form a compressible base to the shaft. Although possessing a small but finite compressive strength, high-density polystyrene has been successfully used, and the pile is deformed sufficiently to compress the packing in the later stages of the test to mobilize the base resistance. Water-filled rubber bags have been used to similar effect. A similar technique is to test two piles, one constructed with a void at the base.

Steel piles may be instrumented using vibrating wire or wire resistance strain gauges. With the latter system, the sealed gauges which can be welded to the steel are to be preferred, and waterproof cabling needs to be used. The strain gauges effectively turn the steel pile into a load measuring column and computation of the pile load is straightforward. With tubular piles the gauges may be fixed to the inside and sealed in. Further mechanical protection is not necessary except to restrain the cables during driving. With H-section piles the strain gauges and cables need to be protected with steel ducts (a difficult operation which can seriously affect the stiffness of the pile and cause stability problems during driving).

In general, wire resistance strain gauges should be used only in situations where they can be sealed into a dry compartment. With this system, small electrical voltages are measured and it is therefore necessary that cable earth leakage is small. Similarly, the readout unit needs to be set up in a dry atmosphere, not an easy thing to arrange on a civil engineering site. If the dynamic response of piles to driving is to be investigated, this type of gauge is one of the few suitable systems.

9.2.5.2 Deformation measurement

Axial deformation of large-diameter piles may be measured using a simple rod extensometer (Figure 9.22). Such extensometers should be free to move in the enclosing tubes, and care should be taken to exclude debris. Using the information provided by these devices a crude estimate of shaft adhesion and base resistance may be made. Similarly, magnetic ring extensometers may be cast into the pile.

Gross lateral deformation of piles may be measured by incorporating inclinometer tubes into the pile and surveying these with a suitable instrument. The computed

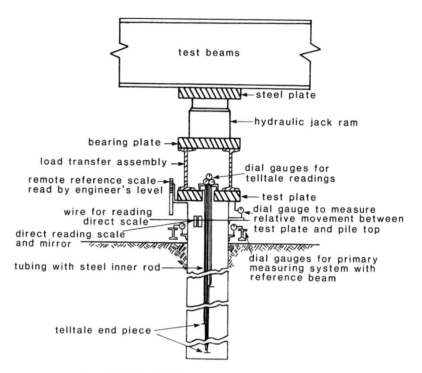

Figure 9.22 Typical pile test using rod extensometers.

profile will be accurate to a few tens of millimetres, unfortunately not sufficiently accurate for bending stresses to be estimated. Strain gauges may be used in steel piling to determine the variation of bending strain with depth and hence to compute stresses and deformations. Many sets of gauges are required to perform this reliably, and such installations are used only for research purposes (Reese *et al.*, 1975).

In lateral pile load tests it is desirable to measure rotation of the pile head. This may be accomplished by using a sensitive spirit level mounted on the pile, or by using an electro-level. Such observations are also valuable if groups of piles are tested to determine the global stiffness of the group.

Lateral deformations may be measured using a datum beam and dial gauges. These measurements suffer the same problems as for vertical measurements (section 9.2.2.3). For long-term observations on pile groups, optical methods are to be preferred. Suitable targets are shown in Figure 9.23. A single second theodolite fitted with a parallel plate micrometer may be used for the observations. An accuracy of about ±0.2 mm may be achieved using this system over a long period of time. However, reliable, stable datum marks are a necessity and these may be difficult to establish, especially in areas underlain by thick alluvial deposits. Geodetic levelling techniques may be used to observe vertical deformation to a similar order of accuracy. The absolute accuracy of such level surveys is again dependent on the establishment of a reliable, stable datum (Cheney, 1973).

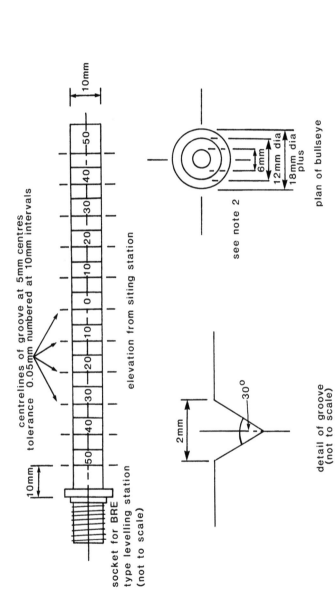

10mm

centrelines of groove at 5mm centres
tolerance 0.05mm numbered at 10mm intervals

| -50 | -40 | -30 | -20 | -10 | 0 | 10 | 20 | 30 | 40 | 50 |

elevation from siting station

10mm

socket for BRE
type levelling station
(not to scale)

2mm

30°

detail of groove
(not to scale)

see note 2

6mm
12mm dia
18mm dia
plus

plan of bullseye

Figure 9.23 Survey targets.

Notes
1 material stainless steel;
2 bullseye to be target of concentric grooves 1 mm wide around 1 mm drilled centre point. The target to be
   the end of standard socket for BRE type station.

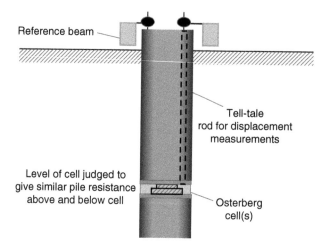

Figure 9.24 Schematic diagram of Osterberg cell method for pile load testing.

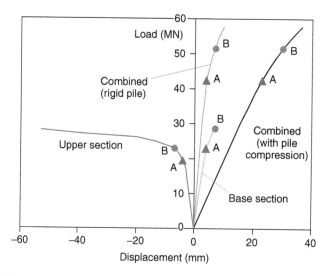

Figure 9.25 Example load–displacement responses from single-cell Osterberg test.

9.2.6 Other forms of pile test

9.2.6.1 Osterberg cell

A novel method for testing large diameter bored piles was introduced in the 1980s by Dr Osterberg (Osterberg, 1989). The method comprises hydraulically operated flat-jacks, referred to as 'Osterberg cells', which are built into the pile during construction, and are used to load the top and bottom section of the pile against each other (see Figure 9.24). In this way, the pile resistance may be proven to (approximately) twice the highest load applied by the jacks, which in turn is limited by the lower of the capacities of pile sections above and below the jack. The jack is positioned at a level where the shaft capacity of the pile above the jacks is approximately equal to the combined shaft and base capacity below the jacks.

The expansion of the jack is measured through the volume of oil pumped and also by independent transducers that span across the jack. Tell-tale rods built into the pile allow determination of the upward movement of the top of the jack, and hence also the downward movement of the bottom. A more sophisticated arrangement of jacks is possible, for example, by installing two levels, which allows an intermediate length of pile shaft to be failed independently, thus giving a direct measurement of shaft friction. An application using such a system of two levels of jacks has been described by Randolph (2003).

For a single level of jacks, as shown in Figure 9.23, either the upper (shaft) or lower (base) section of the pile will reach failure first, and typical results may be as indicated in Figure 9.26. In the standard method for interpreting an Osterberg cell pile test (assuming a single level of jacks), the load–displacement plots of the separate upper and lower sections of the pile are first plotted. Ignoring any compression of the pile, the total load–displacement response at the pile head is then estimated by summing the separate loads mobilized in each section of the pile at equal displacements. This is illustrated above for typical points A (4 mm displacement) and B (7 mm displacement), where the total pile resistance is summed and plotted at the give displacement to give the combined (rigid pile) curve.

An improved method makes approximate allowance for the compressibility of the upper section of the pile. For each displacement, the average load that would be mobilized in the upper section is first estimated (the simplest approach being to take the load from the lower section plus half the load from the upper section), and hence the compression of the upper section of the pile is estimated. This compression is added to the initial displacement for which the loads were evaluated. Considering point A, at a displacement of 4 mm, the upper and lower section loads are 19.5 and 23 MN, respectively. The average load that would then occur in the upper section for those conditions is therefore:

$$P_{\text{average}} \sim 0.5(P_{\text{lower}} + (P_{\text{upper}} + P_{\text{lower}}) = P_{\text{lower}} + 0.5P_{\text{upper}} = 32.75\,\text{MN}$$

The estimated compression of the upper section is then $\Delta w = P_{\text{average}}L/EA = 18.9\,\text{mm}$ (taking $L = 85\,\text{m}$, $EA = 147\,\text{MN}$). The true pile head displacement is therefore the sum of the original displacement, 4 mm, and the pile compression, 18.9 mm, giving a total of 22.9 mm. The corrected curve is shown for comparison with the uncorrected (rigid pile) case. The limitation of this approach is that the actual displacements in the upper section of the pile are much larger than the original displacement, and thus the load mobilized will be higher. A better approach is to undertake a full load transfer model of each section separately, and then analyze the complete pile subsequently.

While the Osterberg cell method of pile testing has some limitations, in particular due to the different direction of loading of the upper part of the pile, and problems associated with creating a gap in the vicinity of the loading jacks, the system has the potential for testing cast in situ piles of extremely high capacity.

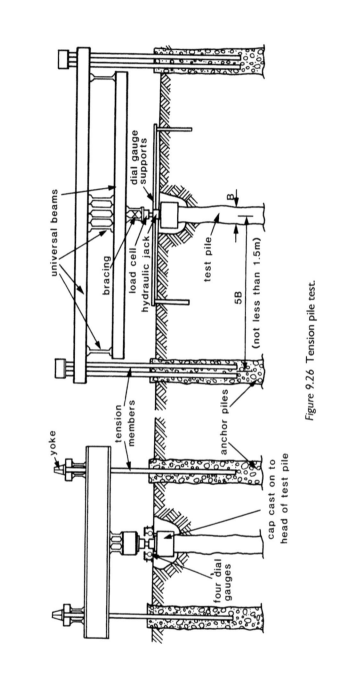

Figure 9.26 Tension pile test.

9.2.6.2 Tension pile tests

This test method is similar to compression testing of piles, and maintained load or constant rate of uplift procedures may be used. The reaction system may be provided by a strong-back resting on adjacent piles or spread foundations. Such a test arrangement is shown diagrammatically in Figure 9.26. Two jacking points on either side of the pile are used to avoid bending the test pile, and care is required to load the jacks evenly. Some cracking of reinforced concrete tension piles is to be expected if loads are high unless the reinforcement is debonded except in the bottom section of pile.

9.2.6.3 Lateral load tests

For structures carrying major lateral loads it is desirable to carry out lateral load tests to confirm the design assumptions. The applied loads are relatively small and it is convenient to use an adjoining pile as a reaction (Figure 9.27). Provision must be made for a stable datum for deformation measurement as for vertical tests. If two identical piles are being tested it may be more convenient to measure the relative movement between them using an extensometer. To increase the reliability of the back-analysis it is worthwhile measuring the rotation of the pile head. Foundations are frequently subjected to lateral, vertical and moment loading. It is usually not possible to arrange for the pile test to mimic such combined loads. Instead, application of a lateral load near to the ground-level existing at the time is often chosen for such tests. Similarly it is difficult to simulate fixed head conditions. Data from such a simplified load test may be analyzed using the principles outlined in Chapter 4, and the computed soil parameters may then be used to predict the response of the pile or pile group to a more complicated loading regime. Bearing in mind the process of analysis that has to be used in such circumstances, where possible the testing should be

Figure 9.27 Lateral pile load test. (Photograph courtesy of Westpile Ltd., Uxbridge.)

planned and loading levels arranged to give the minimum degree of extrapolation in the analysis.

9.3 Dynamic testing of piles

The use of dynamic testing methods for estimating the ultimate capacity and integrity of piles is now widespread. Assessment of pile capacity during driving has been under-taken routinely by the use of empirical driving formulae such as those discussed in section 4.1.4. There are many limitations to such formulae, particularly in respect of uncertainties in the proportion of the available driving energy that is transmit-ted to the pile. The accuracy of dynamic methods of estimating pile capacity has been improved significantly using modern instrumentation and fast data acquisition systems that allow measurement of the dynamic force and velocity waves in the pile during driving. The data may be analyzed at different levels of sophistication, as described in the following sections, to give estimates of the static pile capacity at time of driving. Similar techniques, with lower energy input, may be used to assess the integrity of piles (particularly cast-in-situ piles), as discussed in Chapter 8.

9.3.1 Wave equation analysis

The propagation of driving energy along a pile, allowing for interaction with the surrounding soil, may be analyzed with sufficient accuracy using a 'one-dimensional' idealization. In this idealization, only vertical (strictly speaking 'axial'), displacement of the pile is considered, and the governing differential equation is

$$(AE)_p \frac{\partial^2 w}{\partial z^2} = (A\rho)_p \frac{\partial^2 w}{\partial t^2} - f \tag{9.1}$$

or

$$\frac{\partial^2 w}{\partial z^2} = \frac{1}{c^2} \frac{\partial^2 w}{\partial t^2} - \frac{f}{(AE)_p} \tag{9.2}$$

where $(AE)_p$ is the cross-sectional stiffness of the pile.
        $(A\rho)_p$ is the mass per unit length of the pile.
        c is the wave propagation speed in the pile $(= \sqrt{E/\rho})$.
        w is the vertical displacement of the pile.
        z is the distance down the pile.
        t is the time variable.
        f is the mobilized soil resistance per unit length of pile.

Historically, this equation has been implemented using finite difference or finite ele-ment techniques, with the pile being modelled as a discrete assembly of mass points interconnected by springs. This model, originating in the work of Smith (1960), forms the basis of a range of computer programs for studying pile drivability.

Equation (9.2) may also be solved numerically by using the characteristic solutions, which are of the form

$$w = g(z - ct) + h(x + ct) \tag{9.3}$$

where g and h are unspecified functions which represent downward (increasing z) and upward travelling waves respectively. Taking downward displacement and compressive strain and stress as positive, the force, F, and particle velocity, v, in the pile are given by

$$F = -(AE)_p \frac{\partial w}{\partial z} = -(AE)_p \left(g' + h' \right) \tag{9.4}$$

$$v = \frac{\partial w}{\partial t} = -c \left(g' - h' \right) \tag{9.5}$$

where the prime denotes the derivative of the function with respect to its argument.

The velocity and force can each be considered as made up of two components, one due to the downward travelling wave (represented by the function g) and one due to the upward travelling wave (represented by the function h).Using subscripts d and u for these two components, the velocity is

$$v = v_d + v_u = -cg' + ch' \tag{9.6}$$

The force F is similarly expressed as:

$$F = F_d + F_a = -(AE)_p g' - (AE)_p h' = Z(v_d - v_u) \tag{9.7}$$

where $Z = (AE)_p / c$ and is referred to as the pile impedance. [Note, some authors have referred to the pile impedance as $Z = E/c$, relating axial stress and velocity rather than force and velocity. The more common definition of pile impedance as $Z = (AE)_p / c$ will be adopted here.]

The relationships given above may be used to model the passage of waves down and up piles of varying cross-section, allowing for interaction with the surrounding soil. It is helpful to consider the pile as being made up of a number of elements, each of length Δz, with the soil resistance acting at nodes at the mid-point of each element (see Figure 9.28). Numerical implementation of the characteristic solutions involves tracing the passage of the downward and upward travelling waves from one element interface to the next. The time increment, Δt, is chosen such that each wave travels across one element in the time increment (giving $\Delta t = \Delta z/c$).

Between nodes i and $i+1$, the soil resistance may be taken as T_i, the value of which will depend on the local soil displacement and velocity (see later). Taking T_i as positive when acting upwards on the pile (that is, with the soil resisting downward motion of the pile), the soil resistance will lead to upward and downward waves of magnitude

$$\Delta F_u = -\Delta F_d = T_i/2 \tag{9.8}$$

These waves will lead to modification of the waves propagating up and down the pile.

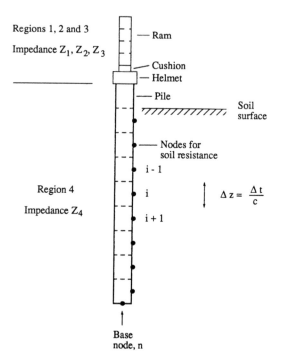

Figure 9.28 Idealization of pile as elastic rod with soil interaction at discrete nodes.

The procedure for calculating new values of wave velocities at each node is shown schematically in Figure 9.29. Thus, consider the downward and upward waves at nodes $i-1$ and $i+1$, at time t. The new downward travelling wave at node i at time $t + \Delta t$ is given by

$$(v_d)_i[t + \Delta t] = (v_d)_{i-1}[t] - T_{i-1}[t + \Delta t]/(2Z) \tag{9.9}$$

while the new upward travelling wave fractionally above node i is

$$(v_u)_i[t + \Delta t] = (v_u)_{i-1}[t] - T_{i-1}[t + \Delta t]/(2Z) \tag{9.10}$$

At the base of the pile, the downward travelling wave will be reflected, with the magnitude of the reflected wave dependent on the base resistance, Q_b, offered by the soil. The axial force in the pile must balance the base resistance, which leads to an expression for the reflected (upward travelling) wave velocity of

$$(v_u)_n[t + \Delta t] = (v_u)_n[t + \Delta t] - Q_b[t + \Delta t]/Z \tag{9.11}$$

The base velocity (the nth node) is

$$v_n = 2v_u + Q_b/Z = 2v_d - Q_b/Z \tag{9.12}$$

where all quantities refer to time $t + \Delta t$.

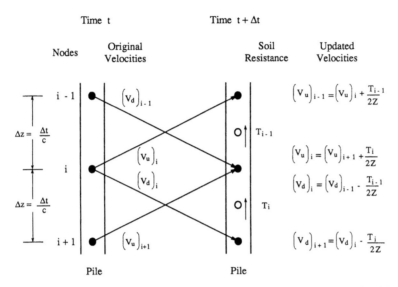

Figure 9.29 Modification of downward and upward waves due to soil interaction. (After Middendorp and van Weele, 1986.)

For a force F_d arriving at the pile tip, equation (9.11) implies a reflected force of

$$F_u = -Zv_u = Q_b - F_d \tag{9.13}$$

The magnitude of the reflected wave thus varies from $-F_d$, where the tip resistance is zero, to F_d, where the base velocity is zero and the base resistance is twice the magnitude of the incident force (see equation (9.12).

It should be noted that the characteristic solutions of the wave equation may be used to model the pile hammer assembly as well as the pile, in order to perform drivability studies. The geometry and mass density of each component (ram, cushion, helmet) is matched, and the ram is given an initial velocity to model the impact. As discussed by Middendorp and van Weele (1986), a relatively crude model of the hammer will generally suffice to give adequate results. Analytical solutions for hammer impact are also available (Take *et al.*, 1999).

9.3.1.1 Dynamic soil model along pile shaft

Accurate prediction of the performance of piles during driving requires modelling of the dynamic response of the soil around (and, for open-ended pipe piles, inside) the pile, both along the shaft and at the base. Following traditional approaches for the analysis of machine foundations, the soil response can generally be represented by a combination of a spring and dashpot. However, it is also necessary to consider limiting values of soil resistance where, along the shaft, the pile will slip past the soil and, at the tip, the pile will penetrate the soil plastically.

In the original work of Smith (1960), which still forms the basis of many commercially available pile driving programs, the soil response was modelled conceptually as a

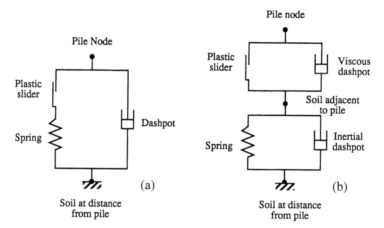

Figure 9.30 Soil models for dynamic response: (a) original soil model of Smith, 1960; (b) soil model of Randolph and Simons, 1986.

spring and plastic slider, in parallel with a dashpot (Figure 9.30(a)). For such a model, the soil resistance (per unit length of pile) may be written as

$$T = T_s + T_d = Kw + Cv \tag{9.14}$$

where the subscripts s and d refer to static and dynamic resistances, respectively, and v is the local velocity. Smith (1960) simplified this response to

$$T = \frac{w}{Q} T_{max} (1 + Jv) \tag{9.15}$$

subject to the term outside the bracket being limited to T_{max}, the limiting static resistance. The parameters Q and J are referred to as the 'quake' (the pile displacement for full mobilization of the static resistance T_{max}) and 'damping constant' (with units of s/m or equivalent). Typical values for Q and J recommended by Smith (1960) are 2 to 3 mm, and 0.1 to 0.2 s/m respectively.

A more sophisticated model, based on elastodynamic theory, has been introduced by Randolph and Simons (1986) (see also Randolph (1990)). As shown in Figure 9.30, the model separates out inertial or radiation damping due to the surrounding soil (which operates prior to slip between pile and soil), from viscous damping that may lead to an increase in the limiting resistance, T_{max}. The spring constant, K, and inertial dashpot constant, C, are related to the shear modulus, G, and mass density, p_s of the soil, by

$$K = 2.75\, G \tag{9.16}$$

$$C = \frac{\pi dG}{V_s} = \pi d \sqrt{G\rho_s} \tag{9.17}$$

where V_s is the shear wave velocity in the soil and d is the pile diameter.

Any viscous enhancement of the static resistance is modelled by taking the limiting dynamic resistance as

$$T_d = T_{max}\left[1 + \alpha(\Delta v/v_0)^\beta\right] \qquad (9.18)$$

where $v_0 = 1$ m/s and Δv is the *relative* velocity between the pile and the soil. It is more correct to use the relative velocity in this equation, rather than the absolute pile velocity, since the main viscous effects will be confined to the zone of high shear strain rate immediately adjacent to the pile. Typical values for the viscous parameters may be taken as $\beta = 0.2$ (following Gibson and Coyle (1968) and Litkouhi and Poskitt (1980)) and α in the range 0 for dry sand, up to 1 or possibly higher for clay coils.

9.3.1.2 Dynamic soil model at pile base

Suitable models for the dynamic response at the pile base have been discussed in detail by Deeks and Randolph (1995), building on the analogue model first proposed by Lysmer and Richart (1966). The basic model is very similar to that in Figure 9.30(b), but with the addition of a lumped mass at the pile node. For undrained conditions ($v = 0.5$), as appropriate for pile driving, the stiffness, damping and mass parameters are given by (Deeks and Randolph, 1995)

$$K = \frac{2Gd}{1-v} \qquad (9.19)$$

$$C = \frac{0.85d^2}{1-v}\frac{G}{V_s} \qquad (9.20)$$

$$m = 0.16d^3\rho_s \qquad (9.21)$$

where ρ_s is the density of the soil beneath the pile tip.

In practice, the lumped mass makes little difference, as it represents only 60% of the volume of a hemisphere (of diameter, d) of soil at the pile tip. Comparing the magnitudes of the stiffness and inertial dashpot terms for the pile shaft and base, it may be seen that the relative contribution from the dashpot at the pile base is only about 40% of that along the shaft. However, this still represents a significant increase in stiffness relative to static loading.

Further discussion and comparison of the original Smith formulations for dynamic pile-soil interaction, and those described above, may be found in Randolph (2000), where application to open-ended driven piles is also considered.

9.3.2 Dynamic stress wave data

Wave equation analysis, as described in the previous section, is used in determining the drivability of piles, using commercially available software such as the WEAP family (Goble and Rausche, 1976, 1986; Rausche et al., 1988). However, a potentially more significant application lies in the interpretation of dynamic stress wave data measured during pile driving. The data form the basis of a dynamic pile test, which may obviate the need for more costly static loading tests.

Measurement of stress wave data is achieved by the use of light-weight strain cells and accelerometers that are attached to the pile a few diameters below the hammer. A description of the instrumentation, and one of the most widely used monitoring systems (the Pile Driving Analyzer, or PDA) has been given by Likins (1984).

The first stage of interpretation of stress wave data is performed in real-time in the field by equipment such as the Pile Driving Analyzer. Strain and acceleration data are processed, generally through electronic hardware, to obtain force and velocity data. From these data, various parameters may be derived. Thus, integration with time of the product of force and velocity up to the time at which the product becomes negative leads to a figure for the maximum energy transmitted to the pile. This allows the overall operating efficiency of the hammer to be assessed in terms of its rated energy. If additional information is available on the ram velocity at impact, then the energy losses may be subdivided into mechanical losses in the hammer, and losses in the impact process due to inelasticity and bounce of the components.

In traditional pile driving formulae, one of the largest sources of error in estimating the overall pile capacity is uncertainty in the energy transmitted to the pile. Measurement of the actually transmitted energy allows use of simple pile driving formulae with increased confidence. Such formulae provide a means whereby information obtained on instrumented piles may be extrapolated in order to assess the quality of uninstrumented piles driven on the same site. Of course, for piles where stress wave data are obtained, more sophisticated techniques may be used to assess the pile capacity.

The relationships presented in section 9.3.1 allow the dynamic and static soil resistance to be estimated from the stress wave data. Equation (9.8) implies that, as the stress wave travels down the pile, the magnitude of the force will decrease by half the total (dynamic plus static) shaft resistance, T. Thus, at the bottom of the pile, the downward travelling force is

$$F_d = F_{d0} - T/2 \tag{9.22}$$

where F_{d0} is the original value at the pile head. Similarly, equation (9.13) may be used to obtain the upward travelling force, after reflection at the pile tip as

$$F_u = Q_b - F_d = Q_b + T/2 - F_{d0} \tag{9.23}$$

On the way back up the pile, *provided the particle velocity at each position is still downwards, implying upward forces from the soil on the pile,* the upward travelling wave will be augmented by half the shaft resistance, to give a final return wave of

$$F_{ur} = Q_b + T - F_{d0} \tag{9.24}$$

where the subscript r refers to the return (upward travelling) wave at a time $2l/c$ later than the time at which the value of F_{d0} was obtained (l being the length of pile below the instrumentation point). The total dynamic pile capacity is then

$$R = Q_b + T = F_{d0} + F_{ur} \tag{9.25}$$

Equations (9.6) and (9.7) may be used to derive the upward and downward components of force from the net force and particle velocity at the instrumentation level, so that equation (9.25) may be re-written:

$$R = 0.5(F_0 + Zv_0) + 0.5(F_r - Zv_r) \tag{9.26}$$

where subscripts 0 and r refer to times t_0 (generally close to the peak transmitted force) and $t_r = t_0 + 2l/c$, respectively. This equation is the basis for estimating the total dynamic pile capacity directly from the stress wave measurements. A search may be made for the value of to which gives the largest value of capacity.

9.3.2.1 Case method of analysis

Since the dynamic capacity will be greater than the current static capacity, a simple method is needed to estimate the static capacity in the field, without the need for a full numerical analysis of the pile. In the Case approach, which has gained widespread acceptance, this estimate is made on the basis that all the dynamic enhancement of capacity occurs at the pile tip, with a dynamic component of resistance that is proportional to the pile tip velocity, v_b. Thus the dynamic tip resistance is taken as

$$Q_d = j_c Z v_b \tag{9.27}$$

where j_c is the Case damping coefficient (Table 9.4). These simplifying assumptions lead to an expression for the static pile capacity, R_s

$$R_s = 0.5(1 - j_c)(F_0 + Zv_0) + 0.5(1 + j_c)(F_r - Zv_r) \tag{9.28}$$

The assumptions regarding the dynamic soil resistance are clearly an oversimplification, and the deduced static pile capacity can be very sensitive to the value adopted for the damping parameter, j_c. However, the above expression can provide useful guidance on the static pile capacity where it is possible to calibrate the parameter j_c for a particular site. Where no static load tests are carried out, guidelines for j_c as given below may be adopted (Rausche et al., 1985).

Figure 9.31 shows example stress wave data, for a 300-mm square concrete pile driven with a 6-tonne hammer. In the early part of the blow, before any reflections from the soil resistance arrive at the instrumentation point, the force and velocity are directly proportional ($F = Zv$) since only a downward travelling wave component is present. As reflections from the soil arrive, the force curve separates from the velocity

Table 9.4 Suggested values for case damping coefficient

Soil type in bearing strata	Suggested range of j_c	Correlation value of j_c
Sand	0.05–0.20	0.05
Silty sand/sandy silt	0.15–0.30	0.15
Silt	0.20–0.45	0.30
Silty clay/clayey silt	0.40–0.70	0.55
Clay	0.60–1.10	1.10

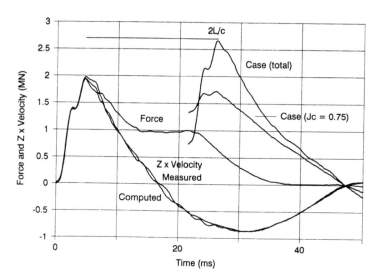

Figure 9.31 Examples of stress wave data.

curve, with the separation showing the magnitude of the shaft resistance. At a time $2l/c$ after impact (where l is the length of pile below the instrumentation point), reflection from the pile tip occurs, and the forms of the force and velocity curves change. In this particular example (taken from a re-drive), little of the blow reaches the pile tip, and the reflection from the base is minimal.

Implementation of equation (9.28) to give the Case estimate of pile capacity is shown in the latter part of the blow (after the return wave arrives back at the instrumentation point). While the total dynamic resistance (using $j_c = 0$) reaches a peak of 2.67 MN, the use of $j_c = 0.75$ gives an estimate of the maximum 'static' resistance of about 1.73 MN.

9.3.2.2 Detailed matching of stress wove data

The Case formula (equation (9.28)) offers only an approximate estimate of pile capacity, and a more reliable estimate requires detailed matching of the stress wave data. This process is an iterative one, where the soil parameters for each element down the pile are varied until an acceptable fit is obtained between measurements and computed results. In order to avoid uncertainties in modelling the hammer, either the measured force signal or the measured velocity signal is used as an upper boundary condition in the computer model. The fit is then obtained in terms of the other variable. The iterative process, with examples of the effect of varying different parameters, has been described by Goble *et al.* (1980).

It is possible to automate the matching process, with the computer optimizing the soil parameters in order to minimize some measure of the difference between measured and computed response (Dolwin and Poskitt, 1982). However, it has been found that computation time can become excessive, particularly for long piles, unless the search zone for each parameter is restricted by operator intervention. It is rather more straightforward to carry out the matching process manually. Experience soon enables assessment of where values of soil resistance, damping or stiffness need to be adjusted in order to achieve an improved fit. A satisfactory fit may generally be achieved after 5–10 iterations of adjusting the parameters and re-computing the response.

Limitations in the soil models used for pile driving analysis entail that the computer simulation will not exactly match the real situation. A consequence is that the final distribution of soil parameters should not be considered as unique, but rather as a best fit obtained by one particular operator. Generally, the total static resistance computed will show little variation provided a reasonable fit is obtained. However, the distribution of resistance down the pile, and the proportion of the resistance at the pile base, may show considerable variation (Middendorp and van Weele, 1986).

An interesting investigation of operator dependence in the analysis of stress wave measurements has been reported by Fellenius (1988). Eighteen operators were given four sets of stress wave data to analyze, covering a range of pile types and soil conditions. One of the sets of data was from a re-drive of a pile that was subjected to a static load test the following day. All the operators were using the same computer program, CAPWAP, which is one of the most widely utilized programs for such analyses, originating from the work of Goble and his co-workers (Goble and Rausche, 1979).

The study by Fellenius shows a good measure of agreement among the participants in predicting the static pile resistance, with the coefficient of variation being under 10% for three out of the four cases considered. For the one pile that was subjected to a static load test, the predictions spanned the measured static capacity.

An example of matching stress wave data is shown in Figure 9.31. The measured force has been used as input data, and measured and computed velocities are compared. The agreement is reasonably good. The fit has been obtained for a static shaft capacity of 1.78 MN, and assumed base capacity of 2.24 MN (giving 4.02 MN total capacity). However, the blow was not sufficient to mobilize the full base resistance. Analysis of the data showed a residual force at the pile base of 1.17 MN, and a maximum base resistance during the blow of 1.50 kN. The minimum static pile capacity then lies between $1.78 + 1.17 = 2.95$ MN, and $1.78 + 1.50 = 3.28$ MN. In this case, without the use of a heavier hammer, it is not possible to be more specific concerning the ultimate capacity of the pile. Note, however, that the minimum pile capacity exceeds the simple Case estimate, due to partial rebound of the pile during the return passage of the stress wave.

9.3.3 General comments

There are a number of different systems available commercially to undertake dynamic pile monitoring and interpretation. Probably the two most widely used are (1) the Pile Driving Analyzer and CAPWAP program from Pile Dynamics Inc., USA, and (2) the TNO equipment (Figure 9.32) and TNO-WAVF program, from the Institute TNO for Building Materials and Building Structures, Holland.

Regardless of which system is used, there are a number of intrinsic limitations in assessing the static pile capacity from dynamic data:

1   The capacity of a driven pile generally increases with time following installation. This phenomenon, referred to as 'set-up', is generally attributed to dissipation

Figure 9.32 The TNO dynamic pile test equipment. (Photograph courtesy of Institute TNO, Delft.)

of excess pore pressures generated during installation. Case studies have shown that, particularly in soft cohesive soil, the pile capacity may increase by a factor of 4 to 5 following installation, over a time period of several weeks or months, depending on the consolidation characteristics of the soil and the pile diameter (Randolph *et al.*, 1979). Where the pile capacity is to be estimated from dynamic measurements, it is necessary to allow for such set-up. This may be achieved by 're-striking' the pile after an appropriate time delay. It is important to ensure that the full pile capacity is mobilized in such a re-strike within the first few blows, so as not to reduce the long-term performance of the pile.

2   One of the major pitfalls to be avoided is trying to estimate the capacity from hammer blows of insufficient energy to fail the pile. This is quite common during re-drive tests and, in such cases, analysis of the data will only yield an estimate of the maximum static resistance mobilized during that blow.

3   Use of the Case formula to estimate the static capacity of the pile is not recommended without a full numerical matching of the stress wave data as corroboration. Even where this has been done, and a reasonable estimate of damping parameter, j_c, is available, care should be taken to ensure that the conditions assumed in deriving the relationship are met in practice. Two particular conditions are (a) that there are no major changes in cross-section (or impedance) of the pile along its length, and (b) that the pile velocity remains positive (downwards) over the major part of the return time of the stress wave. In many case, the latter condition is not fulfilled, and the Case formula will underpredict the pile capacity. A correction may be made, taking account of the proportion of pile that was rebounding during the return passage of the stress wave, and using the earlier part of the blow to estimate the soil resistance in that region.

4   The two main causes of differences between static and dynamic performance of the pile are (a) viscous damping and (b) inertial damping. The dynamic shaft capacity of a pile may exceed the static capacity by a factor of 2 or more, due to

viscous effects. Similarly, the dynamic base capacity may exceed the static capacity significantly due to viscous effects. In both cases, differences between dynamic and static capacity are greater in soft cohesive soil, than in stiffer or coarser material. Allowance for damping is made by appropriate choice of damping parameters in the dynamic analysis. However, the deduced static capacity is relatively sensitive to the choice of damping constant, and it is clear that further research is needed in order to provide better guidance on damping parameters for different soil types. It is strongly recommended that whenever possible at least one static load test be performed on a given site in order to calibrate the dynamic analyses.

5   One area which has received insufficient attention is the different response of H-section piles and open-ended pipe piles under dynamic and static conditions. It may be shown that both types of pile tend to drive in an 'unplugged' condition, with soil moving up the inside of the pipe, or filling the space between the flanges of the H-pile. However, during a static load test the reverse is true. The frictional resistance of the soil plug is such that both types of pile will tend to fail as a solid body. Thus, during a dynamic test, these piles will show relatively high shaft friction but low end bearing. By contrast, during a static test, the shaft capacity will be just that on the outside of the pile, while the end-bearing resistance will act over the gross area of the pile. It is essential that estimates of the static capacity of such piles take account of differences in the failure modes during dynamic and static penetration.

6   Residual stresses acting down the length of the pile can have a significant effect on the calculated pile response under dynamic conditions. These stresses will generally build up along the pile, with some locked in end-bearing stress, balanced by negative shear stresses acting along the pile shaft (particularly the lower part). Accurate assessment of pile capacity and, in particular, the relative shaft and base capacity, necessitates modelling of the residual forces locked into the pile between each blow.

Dynamic methods for estimating the static resistance of piles have evolved considerably over the last 20 years. However, it is still recommended that such methods are not relied upon as the sole means of assessing the acceptability of piles, except in areas where the geology is similar and a large body of experience has been accumulated. For general applications, it is suggested that the dynamic methods should be calibrated against static load tests, before extensive testing of contract piles is undertaken. Used in this way, dynamic monitoring of piles during driving can enable economies to be made in the static pile testing programme, and allow a larger proportion of contract piles to be tested, giving increased confidence in the foundation performance.

9.4 Statnamic testing

The Statnamic test was was developed by the Berminghammer Foundation Equipment in Canada in the late 1980s as a means of testing high capacity piles without the need for an expensive reaction system, and avoiding some of the problems associated with conventional dynamic pile testing (Bermingham and Janes, 1989). The test uses the principle of equivalence of action and reaction, with the action in this case comprising

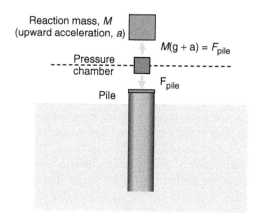

Figure 9.33 Schematic diagram of Statnamic pile test.

upward acceleration of a mass, M, by an acceleration, a (see Figure 9.33). The reaction (provided by the pile) is then

$$F_{\text{pile}} = M(g+a) \qquad (9.29)$$

where g is the acceleration due to gravity. The motive force is provided by a combustion chamber burning solid propellant fuel, leading to expanding high pressure gases that accelerate the mass upward by 10 to 20 g. In this way, a reaction mass with weight (Mg) of only 5 to 10% of the required maximum pile load is needed.

The duration of loading, loading rate and maximum applied load may be controlled by the piston and cylinder size of the combustion chamber, mass of fuel and total reaction mass. During the test, the load transferred to the pile is monitored by a load cell situated immediately below the combustion chamber, and the pile head displacement is measured by a photo-voltaic laser sensor and a remote reference laser source. Velocity and acceleration at the pile head may be obtained by differentiating the displacement response. Test loads of up to 30 MN may be applied to the pile, and the technique may also be used to conduct lateral pile load tests.

Figure 9.34 shows a comparison of typical load, displacement and velocity profiles down a pile during dynamic, Statnamic and static load tests. Whereas in a dynamic test the stress-waves propagate up and down the pile within the time-scale of the applied impulse, for the Statnamic test the pile moves more as a rigid body. Although the velocities are significantly higher than in a static test, so that corrections for viscous damping are necessary, the in-phase motion of the pile obviates the need for a full dynamic analysis.

The typical time-scale for the Statnamic impulse is 100 to 200 ms, compared with a typical travel time for a shock wave propagating down the pile of perhaps 6 to 10 ms. The ratio is referred to as the Wave Number, N_w (Middendorp and Bielefeld, 1995), and is typically 10 to 15 for a Statnamic test, compared with 1 to 2 for a dynamic test. For wave numbers greater than about 10, a dynamic analysis that treats the pile as a single degree of freedom system is sufficient since the pile is moving essentially in phase.

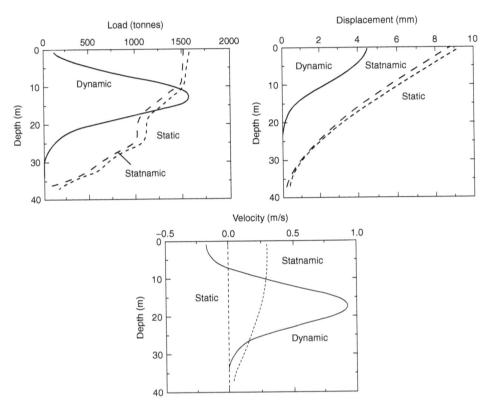

Figure 9.34 Profiles of load, displacement and velocity during dynamic, Statnamic and static testing.

The force and displacement signals from a Statnamic test will always show a peak force reached prior to the peak displacement, as shown in Figure 9.35. The first step is to plot the total force against the displacement, and then to subtract the inertia term, Ma, where M is the pile mass and a is the acceleration (obtained by differentiating the displacement signal twice). The resulting response represents the combined static and dynamic soil resistance.

The next step is to adjust the soil response curve to allow for viscous damping effects. The simplest approach is to take the 'unloading point' (zero velocity) as the best estimate of the maximum mobilized force during the test, and to evaluate the damping coefficient, C, by comparing the force at that point ($F_{v=0}$) with the maximum soil force, $F_{soilmax}$. If the pile head velocity at the point of maximum soil force is v^*, then the damping coefficient is

$$C = \frac{F_{soilmax} - F_{v=0}}{v^*} \tag{9.30}$$

Once C is evaluated, the 'static' load–displacement response may be deduced from

$$F_{static} = F_{soil} - Cv \tag{9.31}$$

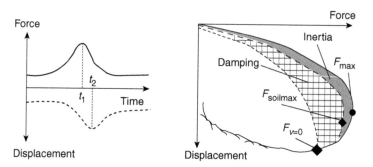

Figure 9.35 Force and displacement response during Statnamic test.

Alternative, more sophisticated, approaches have also been proposed by Matsumoto *et al.* (1994) and by Kusakabe and Matsumoto (1995).

9.5 Safety aspects of pile testing

This section summarizes some of the discussion of static load testing equipment from section 9.2.2, focusing on safety issues. Substantial loads are mobilized during a pile test and a large amount of strain energy is stored in the equipment. It is therefore imperative that all parts of the pile testing apparatus are carefully designed and that short-cuts or poor workmanship are not entertained for what is after all a temporary structure. Failures usually result in the explosive release of energy, for example the head of a concrete pile may fly apart under the action of eccentric loads that could then lead to catastrophic failure of the loading system.

Major parts of the equipment should be clearly marked with the design load and care should be taken not to overload the equipment. Reaction beams, in particular, should be used with the spreader plates, cross-heads, or grillage to support kentledge for which they were designed, and grillage beams should be at the correct spacing. It is good practice for working drawings of the test layout to be sent to site together with the testing equipment.

Eccentricities of load can cause unexpected high bending moments in the pile and loading frame. They may, in extreme cases, lead to twisting and collapse of the load column, a very dangerous situation. Thus care should be taken to align the elements of the load column and ensure all surfaces mate properly. The reaction beam should be carefully levelled and set with the web vertical. To increase the stability of the system it is better to use double beams.

Kentledge constitutes a heavy structure and stability of the stack is important. Cribbages should be soundly constructed and the timbers clipped together with dogs if necessary. Cribs should be of sufficient plan area to prevent bearing failure of the soil. On soft ground timber mattresses may be used to spread the load. The kentledge should be carefully stacked and interlocked if possible. Stability of the outer blocks can be increased by suitably placed lacing. The stack should be symmetrical about the pile and the mass of kentledge should exceed the test load by a reasonable margin.

Tension piles are reasonably straightforward to design, and should be reinforced for the full length of the shaft. High stresses in the concrete will occur at the heads of the pile around tie bar anchors so that the reinforcement for these areas should be carefully designed. Any splices in the reinforcement should be designed with a generous lap. If extended main steel is to be used as anchor straps, these should be in mild steel and any field welds should be designed properly and not left to ad hoc decisions by the welder on site. During the test the heads of the tension piles should be levelled from time to time to check that soil failure is not occurring.

Ground anchors have little inherent stiffness against lateral movement, and tests should be carried out with multiple anchors. Good alignment of the equipment is particularly important for this type of reaction. When stressing the tendons care should be taken to ensure that each strand carries an equal load.

High pressure jacks present a special problem, and checks should be made to ensure that pressure lines are not frayed and that couplings do not leak. The hydraulic fluid should be clean, and the various filters in the pumping system should be in place.

During the test all unnecessary personnel should keep clear of the test area, especially when load is being increased. The hydraulic pump should be sited clear of the loaded area, and as little time as necessary should be taken to read the dial gauges but remotely read transducers are preferred. It is important that regular visual inspections are made of the equipment to check on its satisfactory performance. Obvious signs of distress are:

1   excessive deflection of beams
2   yielding of tension bars of piles
3   horizontal deflection of frames
4   movement of kentledge blocks
5   difficulty in maintaining load on the pile
6   uplift movement of any tensile reaction.

The test should not be left unattended whilst under load. In some circumstances it may be necessary to fence the area. Also, lighting should be adequate at night so that operatives and equipment can be moved safely, and the area should be kept tidy and free from rubbish.

Chapter 10

Choice of piling method and economics of design

10.1 General

It is pertinent, before beginning to design a piled foundation, to contemplate the underlying reasons why piles may be chosen as the supporting elements in that particular case. Some of the reasons advanced are related to engineering requirements, but others are concerned more with convenience and speed of action. These issues should not be confused and sheer expediency should not be allowed to override valid engineering requirements.

Engineering reasons are often expressed as a need to place foundations into firm ground, below soils of great variability or doubtful long-term competence, or to place foundations at such a level that settlements, and more particularly differential settlements, will be kept to acceptable values. In addition there is value in being able to provide a foundation system that can be fairly readily tailored so as to match each load concentration with a local support concentration, perhaps avoiding the use of large ground-bearing slabs that may be difficult to accommodate. It is also economically advantageous, and from an engineering point of view desirable, to avoid where possible having to carry out deep excavations simply for the purpose of installing footings of a conventional kind, perhaps below water or in circumstances where adjacent buildings might be put at increased risk.

From the expediency point of view there are of course advantages in using a competent specialist service for work below ground, since the trade skills necessary at the foundation stage do not match very well those necessary for structural work. Piling can often be used therefore to save time and to simplify the work of following trades and there are in certain cases advantages to be gained where one specialist foundation contractor can provide a combined service of piling, retaining wall construction and perhaps ground anchoring or other related works within a single contractual package.

All engineering design involves the choice of a solution to a particular problem that is effective, practical and economic, and these objectives apply to piling as to any other item.

To be effective the piles must carry the loads that the supported structure imparts to them, together with any additional forces that may result from deformations of the soil mass in which they are embedded. They must also be sound, durable and free from significant defects, and their design must recognize fully the properties of the ground and the implications of groundwater movements so that deformations or settlements will not cause unacceptable strains in the supported or adjacent structures.

To be practical, the piles must be of a type that will permit access for piling equipment to the locations where they are required, the design must recognize the limits of what is possible in current practice with regard to the equipment available, and the method of construction must recognize and seek to minimize difficulties related to ground conditions that could impede proper construction.

To be economic, the design should maximize the bearing capacity of each pile while at the same time providing for an adequate margin of safety against failure or excessive deformation of either individual piles or pile groups. The materials of the pile need also to be reasonably stressed and not used wastefully.

It is sometimes thought that, for any particular ground conditions and structural application, there is but one best choice of pile type that will result in sound and durable results. However it is more often than not the case that any one of a number of methods could be employed to yield a satisfactory end product and the real choice may rest on such items as access for equipment, environmental disturbance, or on the important consideration of overall cost. So many variables exist in practice, not least being the detail of ground conditions, that it is not possible to produce a set of rules that will invariably lead to easy identification of a best solution.

It is clear that the factors involved in making a best choice require knowledge of ground conditions, equipment installation procedures, and materials, and that several skills are needed on the part of both the designer and the contractor carrying out the work. In so far as the usual practice of carrying out pile design, selection and the implementation of the work is concerned, several procedures are adopted. Some larger firms of consulting engineers with geotechnical skills carry out pile design within their own companies and frequently discuss their requirements with specialist contractors in the process of formulating a solution. Other structural engineers set out guidelines for pile design, perhaps stipulating settlement or deformation requirements and occasionally specifying the type of pile that they consider appropriate. They then leave detailed design in the hands of specialist contractors with a request for competitive quotations. Yet others provide a pile design for the works, but leave specialist contractors the option of submitting an alternative scheme based on any other proposal that will achieve the same ends. Each of these approaches can work satisfactorily in practice, but there is a need to recognize that certain skills and experience are necessary on the part of those involved, and full information to allow both proper design and construction must be made available to all those involved. Many contractual disputes centre around inappropriate site investigation, unfortunate pile type selection, bad specification or ill informed supervision and the scene is set for difficulty at the very start of the work.

10.2 Limitations on method choice imposed by ground conditions

10.2.1 Water and rocks

It is clear that a proper choice of piling method can only be based on adequate site investigation and on a knowledge of the geology and history of use of the site in question. Without such an appreciation serious difficulties may be encountered at the construction stage or subsequently, perhaps entailing a change of method during the work, a requirement for a different type of plant, and in certain conditions even remedial work. Some examples of items that can give rise to difficulties are outlined in the following paragraphs.

The presence of groundwater and its possible fluctuations can be significant in certain construction procedures in relation to most bored piles and care should be taken to obtain as much relevant information as possible. It is particularly relevant to pile head casting levels, lengths of temporary casing necessary to maintain bore stability, the use of stabilizing fluids such as bentonite or polymer suspensions, concrete quality and the method of placing concrete.

Hard strata, such as dense granular soils and rocks, must be identified and their properties quantified. If for example a driven pile is chosen in circumstances where considerable penetration of dense upper soils is required, then time and energy will be consumed in achieving this and piles may be damaged during the driving process. It is not in general sensible to seek to achieve significant penetrations into hard rock-bearing strata by any straightforward driving process and the specification of very small sets and prolonged hard driving is often counterproductive and very expensive. In so far as bored piles are concerned, the costs of hard rock penetration are also high, and lack of adequate information may add considerably to the expense of the work because larger penetrations are called for than are necessary.

Boulder clays pose a particularly intractable problem because conventional site investigation cannot easily lead to identification of either true boulder dimensions or frequency, nor are Standard Penetration Tests capable of defining the true nature of the soil. It has sometimes been recommended that, where practicable in such ground, large-diameter bore holes and direct inspection should be used rather than the small-diameter standard investigation holes. Boulders can severely hinder many pile installation procedures and can exert a very important influence on the cost of piling. Piles, particularly preformed types, can be damaged, and for all types of pile, delays in construction can occur. In the presence of large quantities of boulders it is worth considering how the boring length in the difficult strata may be minimized.

It is not without reason that it has sometimes been said that most claims for extra payment in piling result either from battles with adverse groundwater conditions or in seeking to penetrate rock or other hard strata.

10.2.2 Other ground-related difficulties

Other ground-related difficulties that can present a problem include down drag on piles which may be due to consolidation of upper soil layers after pile installation, or heave which may result from swelling of clays around piles when trees have been removed from a site prior to construction. A similar condition may arise because of a deep excavation following pile installation and both these effects are often underestimated in terms of the forces that are caused in the piles. It is frequently also not realized that if attempts are made to improve pile performance by base grouting, and subsequently there is substantial removal of overburden around the pile heads, the effects of the base grouting can be lost.

The forces resulting on piles in either conditions of down drag or heave may be reduced by providing a slip layer around the pile over the affected length, or the pile may be reinforced and lengthened if necessary to cater for the forces as they occur. In such circumstances, it is usually better to choose the simple solution of making piles longer and stronger than to experiment with more fragile slip-coating solutions. Where the heave is related to deep basement excavation only, these measures may not be necessary, depending on analysis of the amount of movement expected, since some small degree of cracking in a finished pile may be acceptable.

The general stability of excavations or sloping ground surrounding a piling site can lead to slip failures and movements which produce large forces on the piles. These forces are sometimes difficult to quantify, but must be taken into consideration when preparing a pile design. Piles have little ability to resist the forces that result from major slip failures in the soil around them, since these forces are often very large.

Large diameter piles may have their bases enlarged in stiff stable water-free ground by underreaming, but if there are severe water inflows or bands of cohesionless material at the level chosen for the pile bases, then it may be impossible to ensure completely stable base excavations or a clean bearing surface. The conditions where instability is likely are often indicated by silt or sand seams in a clay or by significant reductions in moisture content as compared with those of the general clay mass.

In the re-use of sites that have previously been occupied by chemical works or gas works, it is frequently found that the ground contains residual chemicals which could adversely affect the durability of one or several types of pile. Protective coatings may be required in the case of preformed piles, or in the case of cast-in-place piles a protective lining or sleeve may be needed.

It is clear from these few but important examples of difficulties which result from particular soil conditions that the means of avoiding or minimizing undesirable effects can differ between one pile type and another and that in certain cases particular types of pile maybe ruled out as unsuitable. In practice there are many ground-condition influences that bear on the economy and proper choice of piles for any job and one ignores or underestimates the importance of carrying out a proper and appropriate site investigation at one's peril. In general terms, the larger the job in question, the more dramatic and expensive the end consequences of insufficient information become.

10.2.3 Pre-boring for driven piles

Pre-boring is a commonly referenced method for easing the passage of some driven piles into the ground. However, its use can also be misunderstood or misguided. It is not a satisfactory way of overcoming significant obstructions to enable piles to be driven because that which impedes the driven pile will also in general impede progress of the pre-boring tool.

Pre-boring in sand and gravel presents a problem because of the inherent instability of the soil through which the pre-bore passes. When such soil is dense, pre-bores may stand open temporarily because of arching and the influence of temporary pore water suction. However, as soon as a piling tube or pile enters the bore and the hammer begins striking, the upper granular soil collapses into the lower part of the bore. The lower section of the bore will possibly not collapse in this circumstance at the initial driving strokes because the soil is relatively more dense and the hammer influence more remote. The result is frequently that because of re-compacted debris in the lower bore, piles will not drive back to the same depth as originally bored. Only if the bore is temporarily cased to prevent collapse, and if the casing is of large enough diameter to allow access for the final pile, can a satisfactory load bearing unit be inserted, albeit with loss of potential friction resulting from loss of displacement effects and the need for in-filling around the pile.

As an alternative to trying to form an open hole in sand soils, the pre-boring tool is sometimes used simply to stir up the ground, leaving disturbed soil in position. This may be sufficient to deal with dense soil near ground level. However, if deep bores are attempted after this manner, again when a piling tube or pile is entered and driving begins, the loosened material is compacted down into the lower part of the bore and becomes virtually indistinguishable from the original natural soil. Piles will frequently not drive back to the depth of the pre-bore or may behave inconsistently under applied load.

It is therefore not generally satisfactory to use deep pre-bore methods in sands, for example, for the purpose of ensuring that piles reach a deeper stratum such as rock unless special temporary casing methods are adopted.

Pre-boring sockets into rock or very hard soils for the supposed purpose of enhancing end bearing or reaching strong soil, where there are overlying fill, sand or clay layers, is also generally futile. For the same reasons as stated above, it will be found that without guaranteed bore stability and measures to prevent soil from collapsing into the socket, a satisfactory load bearing and consistent unit cannot be formed because of debris falling before the pile arrives.

Pre-bores are satisfactory only under specific circumstances:

1   To loosen dense upper crust soils and enable long piles to be driven without breakage. Long piles struck at the head are really slender columns and so the possibilities of buckling failure can be very real.
2   To make an open hole in stiff clays or similar cohesive soils into which a pile is pre-entered. The purpose in this instance is to avoid or diminish soil heave. If using the method for the purpose of eliminating ground heave, it is generally legitimate to choose the area of the bore so that the pile cross-sectional area is just slightly larger.

Jobs with pre-boring are frequently associated with claims and cost overruns, partly because it is difficult to synchronize the activities of boring and driving machines with consequent delay, and partly because, where the motivation is to achieve stringent 'sets' this may be a major source of damage to equipment.

10.3 Structural consideration of pile use

Piling is used for the support of a wide range of structures and the choice of pile must be appropriate to the requirements of each type of structure. A major and prestigious structure, for example, will warrant perhaps the use of large and deep piles merely to keep the possibilities of structural movements in the completed work to an absolute minimum, whereas for some purely functional industrial structure, settlement may be of less consequence and simple bearing capacity may become the overriding feature.

It is often not apparent at first sight how important the control of settlement and differential settlement may be. For example, fine tolerances are sometimes specified for items such as machinery bases and automatic stacking systems in warehouses. In structures such as bridges, the sensitivity of the system may be very dependent on the assumptions of deck design, whether it is a case of a simply supported span, spans in a continuous beam arrangement or integral spans with continuity to a supporting abutment.

It should be recognized that structural design and foundation design should not be carried out in isolation. The loads from the structure will lead to deformations of the ground which depend on soil types, and these deformations will in turn lead to internal stress changes within the structure, thus changing the loads on the piles. Many cases exist where such considerations may be unimportant, perhaps where the structure rests on piles bearing onto hard rock, and it is therefore unnecessary to become too involved in detailed analysis. However, cases also exist in plenty where soil/structure interaction effects are important in terms of both the finished structure and adjacent existing structures. These considerations may have a determining influence on the selection of pile types.

10.3.1 The use of piles for vertical load bearing

The majority of piles are designed to carry vertical loads and any secondary lateral loads to which they may be subjected may be small. Secondary loads derive in the main from structures with a certain degree of redundancy and cannot by their nature be calculated exactly. Derived loads usually represent combinations of dead loads, live loads and wind loads, and the final structure is usually designed on the worst loading condition, say for each column. It is important in considering real cases, and in particular probable structural settlements, to take into account realistic conditions, although each pile must of course still be capable of carrying safely the required maximum structural load.

The starting point of most structural engineering designs is an assumption of a non-deforming structure, although it is possible for simpler types of structure to incorporate the concept of a settling foundation in order to assess how loads may change in such a circumstance. This step in the procedure is then often followed by the acceptance of some differential settlement of the foundations on a purely empirical basis. Such empirical deformation criteria in relation to structural damage are discussed by a wide variety of authors, e.g. Sowers (1962), Skempton and McDonald (1956), Meyerhof (1947), Polshin and Tokar (1957), Bjerrum (1963) and Burland and Wroth (1975). Summary information is given on these methods by Lamb and Whitman (1969) and in Padfield and Sharrock (1983).

In order to satisfy the designer of a piled foundation who seeks the nearest possible approximation to a non-deforming foundation, a great deal of attention is devoted to ensuring that the deformation of each individual pile, and of all the supporting piles acting as a group, is minimized. The choice of depth required for piles, particularly where load is carried mainly by friction, is often based on this requirement, and in general the deeper the piles are, the more likely is settlement restriction likely to be achieved. This means that in many cases factors of safety and costs are raised in order to achieve this result.

10.3.2 Pile raft and structure interactions

In the past it was common practice, in the case of piles which carried most of their loads by friction or adhesion, to make all the piles on a site of the same length, but this practice has largely been abandoned, provided piles in the same group are kept to similar lengths and that very large variations are avoided between adjacent piles at close centre-to-centre spacing. It is necessary in such cases still to found all piles in the same basic stratum, or at least in strata which have broadly similar deformation characteristics. Most driven piles carry a high proportion of load by end bearing on dense strata or rock, and the possibility of substantial length variations does not there-fore arise except where, perhaps because of underlying geological faults, the bearing stratum changes radically in level across a site.

The interaction between piles and rafts has been discussed in Chapter 5 (sections 5.2 and 5.3) and it is clear that where ground conditions permit, advantage can be taken of load sharing effects to improve the economy of design. The piles may also be used as settlement reducers and substantial control is theoretically possible over the deforma-tion of the raft and the supported structure. The practice of using a raft and piles as a combined system has been used on a somewhat empirical basis on a small scale in the past, but the possibilities of using this type of approach in a much more satisfactory way are now worth consideration when the soils underlying the raft are competent. Naturally, if the ground beneath the raft is fill or is soft or otherwise inconsistent, such ideas are precluded.

An alternative approach to pile/structure interaction has been suggested by Fleming (1984) in which the load/settlement characteristics of piles are regarded only as boundary conditions within which alterations to performance may be made by placing a pile settlement control element on each pile head.

The principle involved is indicated in Figures 10.1 and 10.2. Figure 10.1 indi-cates the result of a load test on a single pile. If say a suitable rubber or otherwise deformable block is placed on the pile head and load and settlement are then measured at the upper surface of the block, the load/settlement relationship can be modified. By selection of the thickness of the block, or its deformation behaviour, the

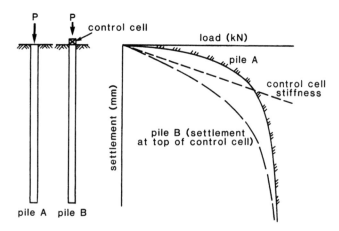

Figure 10.1 Structural settlement control cell.

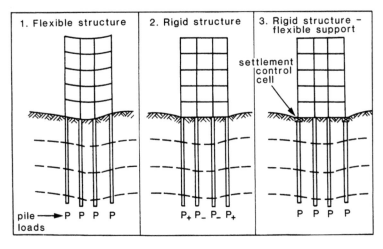

Figure 10.2 Deformation of piled structure on clay soil.

characteristic of the pile may be tuned to any suitable but similar form within the boundary.

In Figure 10.2 three representations of a structure are shown. The first diagram describes a 'flexible' building resting on a clay soil with equal load acting on each pile from a column. As the soil deforms or 'dishes', the structure will deform without alteration of pile loads. The second diagram shows a 'rigid' building in the same situation, but in this case, as the soil layers 'dish', the edges of the structure and supporting piles are pushed more deeply into the soil mass. Hence the load on external piles will go up and on internal piles will go down. The structure is now called upon to accept internal shear stresses, and it is these additional stresses which place the design of the structure at greater risk. The third diagram shows how the desirable features of constancy of support which occur in the flexible case can be combined with the presence of a rigid structure, minimizing the internal stresses to which the structure may be subject as a result of forced ground distortion. The piles deform more or less as in the flexible case but the rigid structure experiences the correct support loads corresponding to the design assumptions.

Since for many structures settlement (particularly in clays) is time-dependent, it may not be possible to choose the stiffness of the blocks or control cells to place on the pile heads to cope exactly with the longer-term consolidation movements. A major part, perhaps of the order of 60% of the deformation of the soil mass around piles in stiff clay, is settlement that occurs during construction and the deformation of each block will be more or less concurrent with the development of settlement, so that there should be little problem in choosing stiffnesses up to this stage. Even in the later consolidation stage when settlement increases but load stays constant, much smaller load shifts between piles would be necessary to ensure equal settlements at pile heads than would occur in any conventional structural support arrangement. Some compromise choice of stiffness might prove best – say basing deformation calculations on immediate settlement plus half of the remaining consolidation settlement.

It seems generally to be true, making reference to several papers in the 'Settlement of Structures' Conference (British Geotechnical Society, 1975) that settlement of structures on piles in clay soils occurs more rapidly than normal one-dimensional consolidation theory would imply and it has been suggested that this may at least in part be due to the piles also acting as a vertical drainage system. This deformation control system has not been exploited but it could have clear advantages since it does not seek the absolute minimization of pile settlements, could be advantageous for many particular types of structure and could be adapted to situations such as where tall buildings adjoin low level surrounding structures. Two factors have inhibited use of the system, namely the precision with which calculation of settlement at each pile position could be carried out and the development of simple, durable and low-cost materials for the control cell medium. However, computer calculation methods are now available that enable the former problem to be solved to an adequate degree of accuracy, and effective basic cell materials or composites may give the life and durability required. Such a system, properly used, would more nearly justify the structural engineer's assumption of a non-differentially-deforming structure, would place the problem of ensuring that this condition is met in the hands of the geotechnical engineer, and would remove to a large extent the need for present empirical differential-settlement limiting criteria which relate to observation of historical events rather than informed calculation of load shifts within structures.

It is clear that there is scope for the improvement of pile support systems in some of the ways discussed above, and that the engineer, albeit at some additional cost, may be able to promote more economical structures.

10.3.3 The use of piles for lateral load bearing

The design of piles subject to lateral loading has been discussed in Chapters 4 and 5. In practice one has to make a choice between the use of vertical piles used singly or in groups to carry such loads or of groups incorporating at least some piles installed to an angle of rake. The capacity of a pile as a structural unit to carry shear loads at its head depends on the strength of the section, and when the forces become high, one is impelled to find some structurally acceptable solution which keeps stresses within reasonable limits.

However, in choosing the possible option of raking piles one should be aware of the problems and limitations that may be involved. Some of the factors involved are as follows:

1   Raking piles are usually more expensive than vertical piles. This is partly involved with extra time taken to set up and maintain the equipment in position, the less efficient use of hammers in the case of driven piles, and the difficulties of concrete placing in bored piles.
2   The standards of tolerance that can be maintained in the installation of raking piles are not as good as for vertical piles. Most analyses of pile groups of this kind ignore the effect of tolerances, but if tolerances are properly taken into account they can have a significant effect on calculated pile loads, depending on pile grouping and numbers, with small groups being usually most sensitive.

3   Where the upper part of a raking pile is embedded in a soil that is likely to suffer time-dependent settlement, the pile will in due course be subject to bending stresses unrelated to the structural design load conditions. This may require increase of strength of the section, which is in turn reflected in costs.

4   Many machines used for pile installation carry the pile driving or forming equipment on a long mast, so that they become intrinsically less stable, particularly as the line of the pile gets further from the vertical position. In certain cases, when working close to river banks or railway lines, for example, there is a major limitation on how machinery can be positioned to produce the desired end result.

5   Design of groups involving raking and vertical piles and with loads that are both vertical and horizontal should have regard to the constancy of the relationship between these. If, for example, the vertical load is near constant, but the horizontal force varies greatly, then it is better to employ groupings with rakers balanced in two opposed directions rather than to have an arrangement of vertical piles plus piles raking in one direction only. This is simply to minimize the shears in the heads of the piles when horizontal load falls to a minimum value (see Chapter 5, section 5.4.2).

6   The use of raking piles to 'spread' load under vertically loaded foundations, where the piles are fully embedded in the soil mass and where the whole foundation is expected to undergo significant consolidation/creep settlement, must lead to large bending stresses being developed in the piles. In certain cases this can lead to such stress levels in the piles that the section will suffer damage, which may in turn lead to severe problems in the supported structure.

It should, however, be said that where groups of raking piles derive their axial capacity from strata that are hard and relatively non-deformable, they provide a stiffness in terms of laterally applied forces which can be very desirable. The main issue in design is to avoid large and unquantifiable secondary stresses, and provided this can be achieved all will be well.

Where there are very heavy lateral loads to be carried and neither raking piles nor single piles other than perhaps those of very large diameter are suitable, then diaphragm piers or 'barrettes' have a useful potential application. They can be given high stiffness in the direction of applied horizontal loading without fear of the problem of major secondary stresses.

10.3.4 Tolerances in the installation of piles and their structural effects

It is not possible to install piles consistently in the right position nor precisely to the alignment, which is required by a design. The achievable tolerances are strongly related to such factors as accurate setting up of the equipment, fixity in machine parts, presence of obstructions in the ground and wide variations in the properties of the soil especially near the point of entry of the pile into the surface, inclination of strata, and operator error. Tolerances that can normally and reasonably be achieved are quoted for example in the publication *Specification for Piling and Embedded Retaining Walls* (ICE, 2007). This publication is quite detailed, having sections on piles (and tolerances where appropriate) in the following categories:

(i)   piling in general
(ii)   precast reinforced and prestressed concrete
(iii)   bored
(iv)   bored CFA (Augercast)
(v)   driven cast-in-place
(vi)   steel bearing
(vii)   timber
(viii) friction reduction on piles
(ix)   non-destructive methods of testing
(x)   static load tests
(xi)   diaphragm walls
(xii)   secant and hard/soft secant
(xiii) contiguous wall, etc.

It also contains sections on many other items including, support fluid, contract documentation and method of measurement.

It may be noted that a useful general commentary on this specification, prepared by the Federation of Piling Specialists, is also available from the same publisher.

The proper positioning of the machine used to drive or bore the piles is a primary requirement, but once a pile or boring tool has penetrated some short way into the soil, correction of an initial error becomes more and more difficult and eventually is virtually impossible in the case of a driven pile and very costly and difficult in the case of a bored pile.

If a single pile is installed to carry a single load, positional errors mean that the load is applied eccentrically, and if an additive alignment error also occurs and upper soils are very weak, then the resulting bending moment applied at the pile head will increase for some short distance below the pile head until soil constraints come into effect and prevent further increases. In the case of a rigidly capped pair of piles the same will be true in the direction at right angles to the line joining pile centres, and for positional and alignment errors in the same direction. However, for rigidly capped groups of 3 or more piles or where the error is in the direction of the long axis of the cap in the case of a pair of piles, the main effect is not to produce significant moments in the piles but rather to adjust the vertical loads in the individual members of the group.

Whereas in the case of driven precast concrete piles, reinforcement is often fairly substantial to cater for handling stresses, for bored cast-in-place piles it is often nominal only and tolerance errors may require additional reinforcement, as will laterally applied loads, heave, etc. One of the potential advantages of cast-in-place piles is that the reinforcement may be tailored to fit, though not on a pile-by-pile basis except where some deviation or oddity is discovered before concrete placing. The pile also starts its life in an unstressed state. Reinforcement specified is therefore often based on the achievement of standard tolerances.

It should be noted that the British Standard BS 8004: Foundations has been superceded by European CEN execution codes in regard to bored piles and diaphragm walls and that other codes for piles and ground engineering activities from the same source will shortly follow. The Bored Pile code is now BS 1536 and the Diaphragm Wall code is BS 1537. Because these new codes are related to construction only, the BS 8004 is being revised and is to be re-issued when the necessary adjustments to bring about conformity have been made. One of the changes among others that this brings about is in terms of pile positional tolerances where there is a general increase from 75 mm to 100 mm. This may affect pile reinforcement unless the particular specifications are adjusted differently on a job-by-job basis.

It is evident, since tolerances are generally the same for piles of large or small diameter, that by far the most severe conditions are associated with piles of small cross-section. If steel reinforcement becomes large in quantity, the cost of piling will be increased and the spacing of reinforcement bars within the section could be reduced to unacceptable values. It is generally held that the clear distance between bars or bundles of bars should be maintained at 100 mm in cast-in-place piles if possible and it should never be less than about 80 mm where normal high slump concrete is used with 20 mm aggregate.

Where lateral loads are applied to piles, the bending effects will add to the tolerance effects, and reinforcement has to be provided for the combined moments and any consequent changes in axial load.

From a theoretical point of view the case of an axially loaded bent pile has been considered in order to determine the maximum load that can be taken by the pile section, the maximum bending moment in the pile, and the maximum lateral soils reaction induced along the pile in relation to permissible values (Broms, 1963). It is apparent from this type of analysis that the possibility of overstressing really only appears with very slender piles in soft ground and with fairly severe deviations of alignment. Work by Francis *et al.* (1965) bears out that a pile with initial alignment imperfections has the same buckling load as a straight pile, though it may deflect more.

The methods described in Chapter 4 (section 4.3) for the calculation of bending moments in laterally loaded piles may equally be used in the consideration of tolerance effects, taking the eccentricity of load as applying a moment and a direct stress at the pile head, and resolving the forces in the axial and lateral directions for the alignment error.

The analysis of positional and alignment errors by a simple method has also been described by Fleming and Lane (1971) and it has been used extensively in routine pile design conditions. The method is based on the work of Broms (1964). Reinforcement for the section is easily calculated using standard concrete design procedures.

The use of the methods demonstrates several noteable principles, as follows:

1   Maximum moments occur just below the top of the restraining stratum if the effects of alignment deviation and eccentricity are additive. If the effects are in opposed direction, the maximum moment occurs nearer to the pile head.

2   Maximum effort should be devoted to ensuring that a pile that is not fixed in position at the head is located accurately.

3   A higher standard of alignment accuracy is desirable in that part of the piles which passes through soft upper soils than at greater depths. Deviations near the pile base are unimportant where the piles are reasonably long.

4   Checking of pile positions prior to placing concrete in cast-in-place piles is highly desirable where the piles are heavily reinforced. If an error is then detected, steps may still be taken to cater for extra bending stresses through the provision of reinforcing steel.

In general tie beams between pile heads to minimize lateral deflections are only necessary where the upper ground is soft and the piles are slender, and if tolerance effects have not been taken into account adequately at the pile design stage.

For large-diameter piles or for most conventional piles where the upper ground is firm or dense, calculation indicates and experience shows that tie beams are generally unnecessary. The fixing of pile heads into a stiff cap reduces moments to about half of what they would be in a free-headed case.

Reinforcement may also be required in driven piles to cater for uplift forces where piles are driven into soils that heave as a result of volume displacement. This includes all clays and most other soils occurring in moderately dense to dense states.

The above considerations demonstrate how tolerances can have significance in relation to the selection of pile type on any particular contract, and how they influence subsequent structural work at ground level and the amount of the steel reinforcement that is necessary.

10.4 Constructional consideration of pile use

10.4.1 Site access

The problems encountered in gaining access to sites are numerous, and many invitations to tender for piling work are accompanied by the sensible requirement that before tendering the specialist contractor must visit the site and be satisfied as to the means of getting in and out and that there is adequate working space.

Naturally if a small and inaccessible site is considered, perhaps in the basement of an existing building, a machine that can work in those conditions will be required, and this often means the selection of bored piling equipment rather than driven, because small boring machines are more readily available. Some driven steel or iron piles are constructed in this way or under such conditions, for example in Sweden, but the system does not appear to be used much elsewhere.

When working in enclosed areas it has to be remembered that if standard diesel engines are used the exhaust fumes must be ducted away to open air. Otherwise the equipment for such work can be powered either by remotely provided compressed air, electricity or hydraulic pumps.

Whatever the size of a piling site, the physical dimensions of entrances and of the slopes and stability of access ramps require consideration. Ramps should not in general slope more steeply than about 1:10. It is rarely sensible to have to undertake a partial dismantling job on a piling machine just to get it into a site, but there are times when such action becomes necessary. Occasionally piling equipment is picked up bodily and lowered into deep basements by heavy cranes.

Several questions must be asked in relation to the disposition of piles within a site:

1   Is there enough space for all the equipment needed? This may include extra craneage, various materials that must be stored in readiness for use, and items

of ancillary plant. It is not just a matter of fitting the equipment into the site, since manoeuvring space is needed, and double-handling materials as work proceeds in order to make space is not economical. Inadequate working space constitutes a safety hazard.

2   Is the headroom adequate to allow the intended equipment to function? Roofs, overhead electrical cables and pipes, and overhangs on buildings all impose constraints. Occasionally cut-down equipment is made for special low headroom circumstances, but this nearly always results in some other functional penalty which is reflected in a lower rate of output than normal and a higher price.

3   Are the clearances from existing structures sufficient to allow pile construction? There are many types of equipment, each with its own dimensional characteristics, but as a generalization the centres of piles must lie no closer to adjacent walls than about 0.75 m, although a few machines can work closer than this. It is particularly difficult to work into the corners of a site, where buildings surround the corner on each side, and minimum distance from each wall in such a case may have to be 1.0 m to the pile centre or perhaps a little more. This kind of information is readily available from specialist contractors.

4   Is the site surface sufficient to bear the weight of the piling equipment? Many piling machines have self-weights in the range 35 to 100 tonnes or more, including the self-weight of the base unit upon which equipment is mounted. With constant and heavy usage, undrained and badly prepared site working platforms can quickly become quagmires. This in turn delays pile installation and can endanger the stability of the machines. Particular care should be exercised to ensure that old excavations are properly backfilled for the same reason. To attempt to work on poorly prepared, waterlogged or unsafe ground is false economy and is dangerous. In general a carpet of hardcore is laid to enable piling to be carried out without hindrance and more and more this is becoming a designed item rather than an expression of opinion by a site supervisor. Machine loadings under all working conditions, previous excavation history, sub-surface soils and carpet materials are all required to be assessed and calculations presented before work is permitted to start. The design of such granular platforms is covered in detail by the BRE document 'Working platforms for tracked plant', (BRE 2004) prepared under a contract let by the Federation of Piling Specialists (FPS) in the United Kingdom.

Sloping site surfaces are not generally satisfactory to work on except for the smallest and lightest types of equipment. Traditional tripod percussive boring equipment is light and adaptable and may be used in cases where no alternative may be possible. Small diameter rotary drills of the kind used for rock drilling may also be appropriate, working from designed platforms.

10.4.2 Contract size

In the case of cast-in-place piles, when a site is large enough and where the number of piles to be installed justifies it, several machines may be used to speed the work. The order of working has always to be thought about because it may be unwise to construct immediately adjacent piles in sequence. The risk is that the installation of a pile nearby may damage an already-cast pile, either by displacement in the case of a driven pile or by allowing concrete to flow between piles in the case of a bored pile, up until the time when the concrete in the first pile has achieved an initial set. Attempts to use too many machines on a site that is really too small for them will lead to logistical difficulties, problems with spoil removal and concrete supply, and the complications may grow, bearing in mind that on large jobs a main contractor may be seeking to follow up piling as closely as possible with cap and beam construction.

The selection of a piling method for a particular site is to some extent governed by the size of the job to be carried out. Transportation cost is a function of piling machine size and the distance to be travelled to reach the site. Small machines are far less costly to transport than large machines, and if there are not many piles to be constructed, they begin with a distinct advantage in terms of overall cost. Economy implies that small equipment and small numbers of piles go together, though such a convenient arrangement is not always possible. In the United Kingdom regulations govern the movement of heavy loads by road and in certain cases up to seven days' notice must be given of impending travel. Rapid changes of intention for whatever reason generally involve delays and, as would be expected, cost is increased.

Large contracts do, however, offer a piling specialist contractor continuity of use of expensive equipment and some freedom from plant time loss in moves. They are therefore in general competitively priced and every effort is made to keep output high. Hence, just as small machines are best for small jobs, so large machines are most appropriate for large jobs.

10.4.3 Limitations on various types of piling equipment

Every piling machine is designed to operate within certain limits of pile length and cross-sectional area, and the power of machines varies to cater for the heaviest work that may correspondingly be expected. Table 10.1 indicates approximately the normal range of lengths available for a selection of types of pile often used for work on land-based sites. Wise clients will always seek to establish clearly the ranges of operation of machines provided to carry out work, particularly with regard to depth, before placing

Table 10.1 Normal range of pile lengths typically available

Pile type	Normal maximum length (m)
Steel 'H'	30–50*
Steel pipe	30–40*
Precast concrete	30 m without join*
Precast concrete—joined sections	30–40 m**
Driven cast-in-place—permanent concrete shell	25–30 m*
Driven cast-in-place—withdrawn tube	20–30 m
Bored—tripod small-diameter	20–25 m*
Bored—rotary small-diameter	25–30 m*
Bored—rotary large-diameter	50–60 m
Bored—continuous-flight auger	25–30 m

Note
*exceptionally longer.
**very exceptionally much longer.

Table 10.2 Typical design load ranges for different pile types and sizes

Pile type	Size (mm)	Typical design load (kN)
Steel 'H'	up to 350 × 400	600–4400
Steel pipe	up to 900	400–7300
Precast concrete	200–600 sq	600–2000
Precast concrete—	up to 300 sq	550–1500
jointed sections	up to 390 hex.	700–2000
Driven cast-in-place—permanent concrete shell or light steel casing	400–750	500–2000 shell 600–2500 light steel casing
Driven cast-in-place, withdrawn tube	275–550	350–2000
Bored small-diameter (tripod or rotary)	150–600	100–1500
Bored—large-diameter	750–2100	2000–30000+
Bored—continuous-flight auger	300–750	400–2600

orders, just in case ground conditions are found to require piles that are longer than the equipment can provide. This is an embarrassing condition, which arises all too often. A prudent client will also have the pile design checked, irrespective of where responsibility for it may theoretically lie.

Similarly, maximum load bearing capacity varies with the pile cross-sectional area, subject to the ground being of sufficient competence. Some examples of approximate maximum capacities are summarized in Table 10.2, and fuller details of pile dimensions and appropriate loadings will be found in Chapter 3. Precast concrete piles, which are formed of lengths jointed together by proprietary jointing systems, usually employ high strength concrete and have joints that are designed to provide similar or greater strength in all respects to the main body of the pile. It should however be noted that it is more difficult to accommodate bending stresses in piles when the concrete is stressed to a level near the maximum permissible values than in the case of piles with lower axial stresses.

In normal usage, the term 'small-diameter' as applied to bored piles includes piles up to 600 mm diameter, and the term 'large-diameter' refers to piles of larger diameter. In the large-diameter range, the sizes generally on offer range from 750 mm diameter rising in stages of 150 mm to 1500 mm diameter and thereafter rising in stages of 300 mm.

10.4.4 Pile base enlargement and equipment

In stiff clay soils, or in soft rock such as chalk, marl or soft mudstones in dry conditions, bases may be enlarged by underreaming to approximately three times the shaft diameter when required, but in the United Kingdom bases of more than 5400 mm diameter have rarely been used. Underreaming is not generally permitted in any water-bearing soil. Large underreams have been known to collapse in stiff clays with sand seams or where there are sand/silt seams of low natural moisture contents.

For large-diameter piles, the larger sizes in the available range require powerful and substantial machines, fitted with appropriate robust augers or drilling buckets. The changing of these tools when required is relatively simple and quick, so that piles of several diameters may be selected to match the loading variations of any one site. However, if every available size should be chosen, many boring tools and temporary casings might be required so that not only would on-site establishment costs rise, but a great deal of storage space would be needed.

Advancement of bores into hard rock, for example to provide frictional sockets, may call for special equipment. Aggressive augers with special teeth are most commonly used for this purpose, but core barrels, down-the-hole hammers, heavy chisels and, unusually, rock roller bits may be employed. There is only a limited range of sizes available for down-the-hole hammers and rock-roller bits, and for both rock roller bits and core barrel use, a direct or reverse circulation technique is usual. The enlargement of pile bases in medium hard to hard rocks can be done, but requires very special equipment and is very expensive and would be relevant only in very special cases.

Penetration of rotary bores into boulder-laden clays may be obstructed by large boulders which will not pass into the drilling tool, and these are usually dealt with by chiseling; in general terms the heavier the chisel the more effective it will be. Large and powerful boring equipment, which may carry a range of versatile ancillary tools, is expensive to transport, maintain and operate but it often has limitations for small sites which can be quite important. Enlargement of pile bases in glacial boulder laden clays is not advised.

The bases of certain driven cast-in-place piles may also be enlarged within limits in certain systems, notably in the case of the traditional based Franki pile. This pile can frequently be used to provide piles of required capacity at a higher level on a cohesionless bearing stratum, for example, that would otherwise be possible with any comparable straight shafted solution. It may be noted that, although the Franki based pile is very effective as a load bearing unit, the enlargement is mainly restricted to a limited range of cohesionless ground conditions, the reasons being that the maximum capacity is mostly governed by the shaft stress and that its production rates are of the order of one half of those available from a conventional straight shafted pile, even when using hydraulic hammers for the shaft drive of both.

It should be pointed out that although the maximum lengths given in Table 10.1 are possible, ground conditions often mean that such lengths are neither possible nor necessary.

10.4.5 Production and efficiency

The time taken to complete a contract is an important consideration in most instances, and this in turn depends on the type of pile selected and the equipment used in installation. On large contracts driven piling can be very rapid where upper soils are soft down to a good bearing stratum, with output from efficient equipment being several hundreds of metres per day. In general hydraulic hammers, being heavier, having greater efficiency and a higher striking rate, can produce about twice the daily output of simple drop hammers of the same general weight. However, care should be taken not to 'overdrive' precast concrete piles, particularly when passing through a strong crust or upper layer into weaker ground, since large tensile reflected waves can crack such piles or lead to breakage.

A new and rapid feature of pile driving that has become very useful is the development of automatic logging introduced by Cementation Skanska. It permits a full automatic driving log to be produced for each and every pile installed. The record then allows direct comparison of all piles installed, and also direct comparison of the logs with Dutch Cone penetration records; this in turn allows the results to be associated effectively with well conducted static load tests. Using the TIMESET, CEMSOLVE, CEMSET suite of programs for static load test analysis the system can be highly effective as a control system.

Likewise, in ideal conditions and where very little temporary casing is needed to stabilize soft soils near the ground surface, rotary boring of small diameter piles can be very rapid.

However, driving piles through relatively stiff cohesive strata, or boring piles to substantial depths through soft and unstable strata where long temporary casings are needed, leads to reduced output and therefore increased costs. Fortunately the conditions that slow down bored piling are often those in which driven piling thrives and vice versa, so that economic choice becomes fairly obvious.

In the case of large bored piles, there is a slowing down of the processes as the diameter increases, but the use of long temporary casings is probably the factor of greatest influence in production. These casings may be installed relatively quickly if they are short, and a 'mudding in' process, which involves creating a slurry column of soil in bentonite suspension, may be used to aid insertion down to a depth of about 10 m. To place much longer casings through cohesionless soils, a large vibrator may be attached to the casing and by this method casings of the order of 25 m have been inserted. However, the effectiveness of such vibrators diminishes rapidly in the presence of clay soils and their use for situations where interbedded layers of cohesive and cohesionless soils occur is strictly limited. Therefore, if large bored piles are required to go to great depths in unstable soils, bentonite or polymer suspensions remain the most practical way to sustain an open bore hole throughout the process of pile formation. It is clear that the more elaborate the method becomes, the more plant will be required and in general the slower the process will become.

The continuous-flight auger method (CFA), with grout or concrete injection through the stem of the auger as it is withdrawn, represents a neat way of using bored piles to avoid casing problems and in circumstances where, but for environmental considerations, a driven pile would be used. It provides a rapid installation rate, perhaps just a little slower than for a comparable solution using driven piles.

The use of on-board computers to monitor the construction of continuous-flight auger (Augercast) piles is now common among major specialist contractors in the United Kingdom following their introduction by Cementation Skanska. Full records of construction can be printed on-site and analyzed off-site by a central computer using a program designed to scan for such items as under-supply of concrete in relation to auger lift and other possible untoward events. It is believed that this procedure can aid considerably in the production of reliable piles. More recently it has proved possible to carry out the concreting of piles automatically and this is general practice on Cementation contracts. Further, automatic pile excavation is also possible under well defined ground conditions in such a way that rotations of the auger are minimized and risk of draw-in of surrounding soils is minimized (Fleming, 1995 + discussion). An additional development in this field has been the wider use of automatic auger cleaners for safety reasons, though this is still an area of further development.

Rotary displacement, or screw, piles have been available in a limited way for many years, the Atlas system of Franki in Belgium being a good example. In the last few years there has been renewed development in this field and many displacement tools have been developed, all working on the same basic soil displacement principles but aiming for greater construction speed. Some of the difficulties encountered are related to knowing exactly how far piles may be driven in given circumstances and also because quite severe wear is experienced on the boring/displacement heads. On the other hand, the piles appear to perform well as load bearing units. This type of pile is likely to develop its own area of the piling market corresponding to certain defined ground conditions but it requires reasonably powerful machines to provide the necessary torque and the heavy wear problems remain to be addressed more fully.

The main advantages of this latter type of pile lies in relatively low noise emission and the reduced volume of cart-away soil from the site. This may be particularly useful where the ground is contaminated and where there are high disposal costs.

Advice on production rates for all types of pile is always best sought from an experienced specialist contractor who is in possession of the full facts of the job in question.

10.4.6 Environmental factors

Increasing attention has been directed to this subject in recent years and some discussion has been included in Chapter 7 (section 7.1.4) with regard to driven piles. Although the duration of a piling contract may be short in comparison with the whole contract period, noise and vibration perception may be more acute during the piling phase.

In the United Kingdom, the Control of Pollution Act (1974) provides a legislative framework for, amongst other things, the control of construction site noise. The Act defines noise as including vibration and it provides for the publication and approval of Codes of Practice, the approved code being British Standard 5228. A section of the Code (Part 4) deals specifically with piling noise.

Because some of the traditionally noisy methods of piling have become unacceptable in many urban and suburban locations, plant manufactures have responded by seeking quieter alternative techniques.

The noise from pile driving displays high peaks associated with each hammer blow, and some manufactures concerned with sheet pile driving have for example enclosed the hammer in an acoustic box. This leads to some appreciable noise reduction but can affect the efficiency of the process of installation. Another development is that quieter hydraulic hammers have been devised and tested and are likely to appear in the market shortly.

In the case of bored piling equipment, much of the noise emanates from the engine providing the power, and it is possible to reduce the steady noise level by improving the soundproofing qualities of the engine enclosure, but this does increase the difficulties of routine maintenance. Some of the noise is associated with the driving and extraction of temporary casings and this is not easy to reduce. However, continuous flight auger equipment does not involve a casing process and therefore this equipment is currently the quietest method available for pile installation.

Driven cast-in-place piling can be made less noisy by the use of an internal drop hammer acting on a plug at the bottom of the driving tube and by attention to the sound-proofing of the machine's engine enclosure, but the process is generally slower than using a top hammer configuration and therefore is more costly.

Excessive noise presents a health hazard by inducing premature deafness, although this is more a problem for the site operatives than for neighbouring residents (Noise at Work Regulation, 1989). Operatives now wear ear defenders as a matter of routine on sites.

It is widely recognized that noise and vibration, although related, are not amenable to similar treatment. In the main, noise from site is airborne and consequently the prediction of noise levels is relatively straightforward, given the noise characteristics and pattern of use of the equipment. On the other hand, the transmission of vibration is largely determined by site soil conditions and the particular nature of the structures involved. There are cases cited in which vibrations have not materially affected immediately adjacent structures, but have affected structures at a greater distance from the source. Some general guidance may be derived from the study of case histories of similar situations (Construction Research and Information Association, in preparation; Attewell and Farmer, 1973). Useful references on the whole subject of ground vibrations are provided also by Skipp (1984) and by Building Research Establishment Digest No. 353 (1990).

Piling processes produce vibrations of two basic types, namely free and forced vibrations. Free vibrations are normally associated with impact loading, such as when a hammer strikes a pile or a steel casing, and the frequency is then determined by the pile, soil and structure characteristics. Forced vibrations occur when a vibratory tool, such as a heavy casing vibrator, is used to produce continuous vibrations with well defined frequency characteristics. Continuous forced vibrations are more likely to set up resonant conditions in structural members of certain dimensions, which in consequence may vibrate at much greater amplitudes than other non-resonant members.

In general human perception of vibrations occurs at levels that are low in comparison with thresholds of risk for structural damage. The British Standards Institution has issued a 'Guide to the evaluation of human exposure to vibration in buildings (1 Hz to 80 Hz)' (BS 6472 : BSI (1992)) which sets out acceptable criteria for vibration levels in various different types of accommodation. The vast majority of piling processes currently in use would give rise to vibrational energy within the stated 1 to 80 Hz frequency range.

The International Standards Organization is working on criteria for structural damage caused by vibration, measured at foundation level in terms of peak particle velocities, for various categories of structure, above which there is initially a risk of architectural damage in the form of minor cracks in plaster, etc. Compliance with the proposed criteria can normally be achieved, and where structural damage occurs below the relevant thresholds, it is usually because some other cause is also present. It is frequently feared by adjacent building occupiers that vibrations caused by piling will adversely affect computer operation on their premises. This matter is dealt with in an article by Boyle (1990).

Apart from the direct effect of transmitted vibrations, it is also possible to initiate settlement of the ground by reason of the compaction effect on loose soils such as fine sands or silty sands. However other materials also deform as a result of repeated loading.

Various methods of reducing vibration transmission from piling have been advocated with conflicting claims as to their success. It may for example help to use low-displacement piles rather than piles of large displacement, but in the last resort it may be necessary to change the piling equipment and adopt a different piling technique. This is a costly exercise when the problems of the particular site are discovered too late.

The incidence of damage due to piling vibrations in the United Kingdom is very low, but it is clear that because of human sensitivity, many people confronted with the problem for the first time become understandably fearful for the safety of their property. Confidence will often be restored by undertaking a programme of monitoring and programming work so as to minimize the periods of annoyance.

One of the main considerations in recent years particularly in the United Kingdom has been the disposal of arisings associated with non-displacement piling and costs associated with its removal to landfill sites. If pile bore arisings are in any way contaminated the costs dramatically escalate. This has led to soil displacement piling methods becoming more popular and development of piling plant to install piles by displacement means (other than simple pile driving). These generally involve screwing an auger into the ground rather than using the auger to remove soil in forming a pile bore. There may be a slight increase in environmental damage from the plant generally necessary to install a screw type of pile of comparable capacity as it will use more power and produce more emissions but this is not particularly significant and in balance is probably less if the transport of spoil from an augered pile and its subsequent handling is taken into account. As an aside, the disposal of bentonite slurries used in temporary support of pile bores is also becoming more problematic and the use of polymer muds has advantages in this respect.

10.5 Cost considerations

It must be obvious that there are so many variations of pile type, ground conditions, structural requirements and other factors involved, that there is no simple way of determining cost effectiveness in a particular case, apart from competitive price comparisons and perhaps several trial designs.

In the United Kingdom the piling market is fairly evenly divided between driven or displacement piles and bored or replacement types, although the ground conditions of some areas favour one type more than another. Elsewhere in the world much depends on areas with predominating ground conditions and even with traditional piling processes.

Where soft strata overlie a definite granular or rock stratum and there is a well defined depth at which substantial end bearing capacities can be achieved, a driven pile will usually prove most attractive for conventional uses, provided environmental or other unusual circumstances do not intervene. If the bearing stratum is within say 25 m of ground surface, then a simple unlined concrete cast-in-place type of pile is often the least expensive choice, and although it increases cost, some systems of this kind make provision for possible base enlargement. Expansion of the pile foot is often not necessary, but where it is effective, it can sometimes lead to significant saving in pile length and hence overall economy may be improved if there is a stratum within a suitable depth range.

If the bearing stratum is at a greater depth, say normally down to 40 m, then a segmental jointed precast pile has much to commend it. A balance has to be struck between the segment length, joint cost, and driving time in order to make best use of the system.

Where the depth of pile required becomes much greater, or where harder driving can be expected on the way to a deeper founding stratum, driven steel 'H' sections may prove economic. They can easily be raked and have high tensile as well as compressive bearing capacity. In work in ports and harbours steel box piles are frequently used, since they are compatible with sheet pile driving and have similar durability. The 'H' pile, because of its low displacement volume, can be used to minimize heave problems which might otherwise result from the use of larger-section piles, but to be used efficiently they need to be stressed to high levels.

It is of interest to note that engineers are now beginning to think in environmental terms in particular with a view to the recovery and/or re-use of piles which have served one life and which may now be extracted and either re-used or re-smelted. In this sense steel piles have an advantage in that they are easy to withdraw in most circumstances. By comparison with most concrete piles, which are difficult and costly to extract, it may be that H piles and tubes will find new uses in city centre areas but this may take a long time to happen. The recovery of the reinforced concrete piles and indeed of rendering the ground in which they were originally formed re-usable, is unlikely to be far from simple but of course the value of city sites may still be a powerful factor influencing such considerations as will the costs of trying to obtain re-insurance on piles installed by others under a different contract.

Tubular piles are used mainly for port and harbour works. They may be driven open-ended or closed-ended depending on whether there is a suitable stratum from which a closed-ended system can take benefit. They have good characteristics with regard to multi-directional lateral loading conditions. Open-ended tubes can plug during driving subject to their diameter and the specific ground conditions but great care is required if it is decided to clean out upper soil from the plug and replace it with concrete. If, for example, a clay is grabbed out and then underlying sand is allowed to blow into the tube, even in the slightest degree, it can change the bearing capacity of the pile dramatically so that considerable re-driving will be necessary.

Pile plug inducers are occasionally used, offering a partially plugged section that will readily drive through clays but are sufficient to cause a full blocking mechanism when the inducer enters medium dense to dense sand. The reason such devices are not well known is probably because they are appropriate for some conditions; but not all. For example, clays may plug open ended piles relatively easily; closed ended piles will stop generally close to the top of a cohesionless stratum, but piles with plug inducers have the objective of trying to secure a pre-set penetration. Such inducers are generally aimed at stopping the pile in a medium dense to dense cohesionless soil but at the same time giving a total penetration which may, for example, mainly be aimed at resisting lateral load.

Intrinsically steel is an expensive material and therefore steel piles are mainly used for the special purposes as described.

Permanently lined cast-in-place piles formed in steel tubes are invariably more expensive than piles of similar size without such linings, but piles formed within concrete shells can sometimes be competitive with the simpler forms of driven cast-in-place pile. The use of such shells is now infrequent but they can be treated appropriately in aggressive ground or where down-drag is anticipated, and they may be manufactured in short lengths, thus permitting low headroom conditions.

In so far as conventional bored piles are concerned, in general they find their best use in clay soils where open bore holes can be formed without the use of long temporary casings. In terms of cost per metre, they can seldom achieve the low prices of standard driven cast-in-place piles of similar diameter, when the driven pile is used in its ideal conditions. They are less efficient than driven piles as load transfer units but they can be installed rapidly, they are efficient in terms of installation energy use, they are relatively vibration-free particularly when a rotary system is used, and they have the advantage that the soil is at least seen and its properties can be appreciated during the process of installation.

As the length of casing needed to stabilize upper soft soils increases, so also does the cost of any given diameter of pile, and where there is a need for very deep casings or in those circumstances where casing has to be abandoned in favour of bentonite or polymer suspension, the cost per metre run of pile will rise rapidly to perhaps twice the cost of the simplest unlined similar pile.

Piles bored by percussion methods cannot generally be constructed as fast as when rotary boring equipment is used, and are therefore more expensive per metre of comparable size, but nevertheless these piles still find substantial use on small sites. The large-diameter version of the percussion bored pile had merit in the construction of secant walls when required but equipment for this purpose has now become rare. It had good characteristics for producing truly vertical piles in particular.

In the United Kingdom continuous flight auger (CFA) piling has become a dominant force in the market for normal land based piling. Its use is generally regulated by the ICE Specification for Piling and Embedded Retaining Walls (1996) which is now widely applied. Specifically this calls for electronic monitoring in a detailed way and although certain of its requirements may need minor amendment in the light of the commentary provided by the Federation of Piling Specialists, it has been a great force for good in the production of better quality work. In this field of work it should be recognized that small diameter piles and machines of generous power involve much less risk than machines of meagre power attempting to construct piles on which they struggle. A key issue is to restrict the auger turns per metre of depth advanced to as near a minimum as possible, particularly if the auger has passed through upper sandy soils, avoiding rotating augers unnecessarily and drawing sand laterally onto the auger with associated loosening effects. Well constructed continuous flight auger piles can still be constructed rapidly, are sensibly predictable in performance and are relatively inexpensive.

There is therefore a wide range of pile prices that cannot be generalized because of the different circumstances of use. Any variation from standard procedures in the construction of piles in general increases the cost. Raking piles are notably more expensive than vertical piles, permanently lined piles than unlined piles, and heavily reinforced than nominally reinforced piles.

As stated previously, the costs of establishing large equipment on site are much greater than for small equipment, and the choice of plant depends very much on the volume of work to be carried out. However, having chosen the best equipment and method there are still several factors that are of strong influence on the price of the contract.

Load tests on piles are expensive, especially if required to have elaborate instrumentation; but they are sometimes a requirement of a local controlling authority. As an alternative to such testing, engineers from time to time negotiate with local authorities on the basis of avoiding a test by providing a higher factor of safety in design calculations where the ground is well known. This is particularly sensible on small contracts and where there is an extensive history of similar work in similar ground. In such cases, not only does the cost of a load test become an inordinately large fraction of total cost, but the physical act of carrying out a pile test may disrupt and delay work by obstructing the site for perhaps several weeks, depending on acceptable curing times for concrete. On sites where large-diameter piles are used it is often simply impractical and unreasonably costly to apply a load test to the larger piles, and subject to previous history of use, a similar procedure to avoid testing may be adopted or a pile may be tested to half scale (not scaling the depth).

On the other hand where sites are large enough or circumstances otherwise justify load tests that are perhaps rather more than ordinary, then it is better to carry out the tests to a high standard rather than to substitute some inferior but uncertain alternative. This is not however to advocate fully instrumented tests at every opportunity, though there may be appropriate occasions. A well conducted static test, carried out under full and accurate computer load control and with all data fully processed, is capable of providing good quality information about ultimate shaft friction, ultimate end bearing, soil stiffnesses and a good deal more besides without instrumentation within the pile body. While good testing has some noticeble cost implications, the kind of information available has a quality and reliability that can never be matched by any rapid testing system and with current computer control systems there are potentially significant cost savings to be made.

Integrity testing is inexpensive, and is now sufficiently well established in the hands of experienced companies to provide good indications of major defects in piles without the great cost of load testing. The information is not in any way equivalent to results from proper static load testing. To take most advantage of its use on large and potentially troublesome sites, it is best to start testing early in a contract rather than to leave all testing to the end. The statistical chance of finding a faulty pile by a random load test on any site is usually very small indeed, because of the small percentage of piles that can be tested at any reasonable and acceptable cost. On the other hand it has always to be realized that what is measured in an integrity test is only an acoustic property and that a load bearing problem in a pile may only be inferred from it in any case, if present. The system works but is subject to mis-interpretations from time to time.

Another cost to be considered directly with piling is the cost of cutting back piles to a final trimmed level, and it is here that a great deal of damage can be done by the use of crude methods. Various improvements in this process are currently appearing and these include the use of hydraulic fracturing equipment placed in holes drilled in each pile head at appropriate positions or the use of non-explosive demolition agents such as 'Bristar'. In the case where the latter process is used, holes are formed in each pile head at suitably chosen positions and subsequently filled or part-filled with a mixture of an inorganic silicate compound and an organic compound. This sets in the hole after a short time and then expands to shatter surrounding concrete. These methods may appear to be more costly than other simpler procedures, but when the damage that can be caused to pile/structure connections from using unrefined methods is fully appreciated, they may in the end provide a good and economic solution to this problem. The specific detail of any proposal along these lines needs to be considered and proof of success provided.

The cost of pile caps and beams is closely related to the early design decisions in a contract. A large-diameter pile may be used instead of numbers of smaller piles, and it is then necessary to consider the combined cost of the piles and any capping arrangements in assessing the cost effectiveness of one solution as opposed to the other. Large-diameter piles rarely need substantial caps nor ground restraint arrangements at the pile head. Some useful notes on the design of pile caps are provided by Whittle and Beattie (1972).

Outside the normal limits of site costs, there are certain risk factors that must be evaluated and against which insurances will normally afford some protection. These include the problems of damage to services and to adjacent structures. Although a contractor may be insured against such unfortunate occurrences, it should not be forgotten that when the unforeseen event occurs, it absorbs a great deal of time and effort, which is costly to the contractor and the other parties to the contract, and it is unlikely that these costs can ever fully be recovered.

Unforeseen ground conditions represent perhaps the most common source of claim from specialist contractors on piling contracts and these are sometimes inevitable and invariably unwelcome. It is very difficult to choose a piling method in the first place that would minimize such claims, but when something does go wrong, the most important factor operating in the case is often delay. Promptness in isolating the cause of a problem and providing a remedy should be a primary objective.

10.6 Legal disputes

However one may seek to avoid legal disputes, they do from time to time occur, often originating from delays and disputed ground conditions. Many disputes are however centred in the conditions of contract and misunderstandings of what is really specified or required. Those administering contracts are often without detailed technical knowledge and if and when they seek advice may not recognize if the advice is good or bad. Legal disputes are time consuming, costly and frequently are resolved before they come to court, with financial loss on both sides. Costs often exceed anything that is likely to be obtained in damages and a sudden understanding of the situation at a stage when the law case is well developed, does not promote happiness.

A first rule when difficulties arise is to seek expert help from perhaps more than one source so that there can be some confidence in the real causes of the event. At this stage if it is clear that there is wrong information supplied, bad specification, unsuitable pile design choice or some problem within the operation of the contract, such as having supervision in the hands of the wrong people, then steps should be taken to rectify the error without delay. If a contractor has made a mistake, then the sooner it is recognized and corrected the better.

There is an unfortunate, but natural, tendency to imagine that once a job is under way, the problems are commonly of the making of the other side. The thought of taking radical evasive action is then often seen as a display of weakness, which is not easy to accept.

Such contract disputes frequently come to rely on correspondence, meeting minutes and construction records. For this reason alone, if a contract begins to develop serious problems, it is essential to record every action and have it agreed at the time when it is carried out.

The costs of delay escalate rapidly. A primary objective must be to return the job quickly to providing the solution required and as near to its original target production as possible.

Legal costs can rapidly develop to the point where they rival the cost of the job as a whole. More often than not some early common sense would settle the issues.

References

Abdrabbo, C.M. (1976) *A Model Scale Study of Axially Loaded Pile Foundations*, Ph.D. Thesis, University of Southampton.

Addenbrooke, T.I. (1994) A flexibility number for the displacement controlled design of multi propped retaining walls, *Ground Engineering*, September: 41–45.

American Petroleum Institute (1997) Recommended practice for planning, designing and constructing fixed offshore platforms – Load and resistance factor design, API-RP-2A, 21st edn.

API (1993) *RP-2A: Recommended Practice for Planning, Designing and Constructing Fixed Offshore Platforms*, Washington, DC: American Petroleum Institute.

Attewell, P.B. and I.W. Farmer (1973) Attenuation of ground vibrations from pile driving, *Ground Engineering*, 6(4): 26–29.

Baguelin F., R. Frank and Y.H. Said (1977) Theoretical study of lateral reaction mechanism of piles, *Géotechnique*, 27(3): 405–434.

Baguelin, F. and R. Frank (1979) Theoretical studies of piles using the finite element method, *Int. Conf. on Numerical Methods in Offshore Piling*, London: ICE, pp. 33–91.

Baldi, G., R. Bellotti, V.N. Ghionna, M. Jamiolkowski and D.C.F. Lo Presti (1989) Modulus of sands from CPTs and DMTs, *Proc. 12th Int. Conf. on Soil Mech. and Found. Eng.*, Rio de Janeiro, Brazil, Vol. 1, pp. 165–170.

Baligh, M.M. (1986) Undrained deep penetration, *Géotechnique*, 36(4): 471–485; 487–501.

Banerjee, P.K. and R. Butterfield (1981) *Boundary Element Method in Engineering Science*, Maidenhead: McGraw-Hill.

Banerjee, P.K. and T.G. Davies (1978) The behaviour of axially and laterally loaded single piles embedded in nonhomogeneous soils, *Géotechnique*, 28(3): 309–326.

Banerjee, P.K. and R.M.C. Driscoll (1976) Program for the analysis of pile groups of any geometry subjected to horizontal and vertical loads and moments, PGROUP, (2.1). HECB/B/7, Department of Transport, HECB, London.

Barry, D.L. (1983) Material durability in aggressive ground. CIRIA Report RO98M, London.

Bartholomew, R.F. (1979) The protection of concrete piles in aggressive ground conditions: an international appreciation, *Proc. Conf. Recent Piling Developments in the Design and Construction of Piles*, pp. 131–141.

Barton, Y.O. (1982) *Laterally Loaded Model Piles in Sand: Centrifuge Tests and Finite Element Analyses*, Ph.D. Thesis, University of Cambridge.

Berezantzev, V.C., V. Khristoforov and V. Golubkov (1961) Load bearing capacity and deformation of piled foundations, *Proc. 5th Int. Conf. on Soil Mech. and Found. Eng.*, Paris, Vol. 2, pp. 11–15.

Bermingham, P. and M. Janes (1989) An innovative approach to load testing of high capacity piles, *Proc. Int. Conf. on Piling and Deep Foundations*, London, pp. 409–413.

Bica, A.V.D. and C.R.I. Clayton (1989) Limit equilibrium design methods for free embedded cantilever walls in granular materials. *Proc. Instn. Civ. Engrs.*, London, Part 1, October, pp. 879–898.

Bishop, A.W. (1971) Shear strength parameters for undisturbed and remoulded soil specimens, in R.H.G. Parry and G.T. Foulis (eds), *Stress Strain Behaviour of Soils* (Proc. Roscoc Mem. Symp.), Blackie Publishing Group, Cambridge, London. pp. 3–58.

Bishop, A.W., D.L. Webb and P. I. Lewin (1965) Undisturbed samples of London Clay from the Ashford Common Shaft: Strength – effective stress relationships, *Géotechnique*, 15(1): 1–31.

Bjerrum, L. (1963) Allowable settlement of structures (discussion), *Proc. 3rd Euro Conf. on Soil Mech. and Foundn. Engng.*, Wiesbaden. Vol. 2, pp. 135.

Bjerrum, L. and O. Eide (1956) Stability of strutted excavations in clay, *Géotechnique*, 6(1): 32–47.

Bolton, M.D. (1986) The strength and dilatancy of sands, *Géotechnique*, 36(1): 65–78.

Boyle, S. (1990) The effect of piling operations in the vicinity of computing systems, *Ground Engineering*, June: 23–27.

Bransby, P.L. and G.W.E. Milligan (1975) Sail deformations near cantilever sheet pile walls, *Géotechnique*, 25(2): 175–195.

BRE (2004) *Working Platforms for Tracked Plant*, Watford: BRE Bookshop.

Brinch Hansen, J. (1961) The ultimate resistance of rigid piles against transversal forces, Copenhagen: Geoteknisk Institut. Bull. No. 12.

Brinch Hansen, J. (1963) Discussion: hyperbolic stress-strain response cohesive soils, *ASCE J. Soil Mech. and Found. Div.*, 89(SM4): 241–242.

British Geotechnical Society (1975) *Settlement of Structures*. London: Pentech Press.

Broms, B.B. (1963) Allowable bearing capacity of initially bent piles, *ASCE Journal of Soil Mechanics*, 89(SM5): 73.

Broms, B.B. (1964) Lateral resistance of piles in cohesionless soils, *ASCE J. Soil Mech. and Found. Div.*, 90(SM3): 123–156.

Broms, B.B. (1964) Lateral resistance of piles in cohesive soils, *ASCE J. Soil Mech. and Found. Div.*, 90(SM2): 27–63.

Broms, B.B. (1965) Discussion to paper by Y. Yoshimi, *ASCE J. Soil Mech. and Found. Div.*, 91(SM4): 199–205.

Broms, B.B. (1978) *Precast Piling Practice*. Balken Piling Ltd.

Brooker, E.W. and H.O. Ireland (1965) Earth pressures at rest related to stress history, *Can. Geotech. J.*, 2(1): 1–15.

Brown, P.T. (1969) Numerical analyses of uniformly loaded circular rafts on elastic layers of finite depth, *Géotechnique*, 19(2): 301–306.

Bruno D. (1999) *Dynamic and Static Load Testing of Driven Piles in Sand*, Ph.D. Thesis, The University of Western Australia.

BSI (1986) *Code of Practice for Foundations*. BS 8004:1986. London: British Standards Institution.

BSI (1992) *Guide to Evaluation of Human Exposure to Vibration in Buildings* (1 Hz to 80 Hz). BS 6472:1992. London: British Standards Institution.

BSI (1994) *Code of Practice for Earth Retaining Structures*. BS 8002:1994. London: British Standards Institution.

BSI (1999) *Copper/Chromium/Arsenic Preparations for Wood Preservation*. BS 4072:1999. London: British Standards Institution.

BSI (2002) *Concrete Complimentary British Standard to BS EN 206-1: 2000 Concrete Specification, Performance, Production and Conformity*. BS 8500:2002. London: British Standards Institution.

BSI (2003) *Geotechnical Investigation and Testing. Identification and Classification of Rock*. BS EN ISO:14689-1. London: British Standards Institution.

BSI (2004) *Geotechnical Investigation and Testing – Identification and Classification of Soil. Principles for a Classification*. BS EN ISO: 14688-2:2004. London: British Standards Institution.

BSI (2004) *Hot Rolled Products of Structural Steels. General Technical Delivery Conditions*. BS EN 10025:2005. London: British Standards Institution.

BSI (2005) *Execution of Special Geotechnical Works. Micropiles*. BS EN 14199:2005. London: British Standards Institution.

Building Research Establishment Digest 353 (1990) Building Research Establishment, Watford.

Burgess, I.W. (1976) The stability of slender piles during driving, *Géotechnique*, 26(2): 281–292.

Burgess, I.W. (1980) Reply to discussion by Ly B.L., *Géotechnique*, 30(3): 322–323.

Burland, J.B. (1973) Shaft friction of piles in clay – a simple fundamental approach, *Ground Engineering*, 6(3): 30–42.

Burland, J.B., B.B. Broms and V.F.B. de Mello (1977) Behaviour of foundations and structures, *Proc. 9th Int. Conf. on Soil Mech. and Found. Eng.*, Tokyo, Vol. 2, pp. 495–546.

Burland, J.E. and R.W. Cooke (1974) The design of bored piles in stiff clays, *Ground Engineering*, 7(4): 28–30; 33–35.

Burland, J.B. and D. Twine (1988) The shaft friction of bored piles in terms of effective strength, *Proc. 1st Int. Geot. Sem. on Deep Foundations on Bored and Auger Piles*, Ghent, pp. 411–420.

Burland, J.B. and C.P. Wroth (1975) Settlement of buildings and associated damage, *Proc Conf. Settlement of Structures*, Cambridge: Pentech Press, pp. 611.

Butterfield, R. and P.K. Banerjee (1971a) The elastic analysis of compressible piles and pile groups, *Géotechnique*, 21(1): 43–60.

Butterfield, R. and P.K. Banerjee (1971b) The problem of pile group – pile cap interaction, *Géotechnique*, 21(2): 135–142.

Butterfield, R. and R.A. Douglas (1981) Flexibility coefficients for the design of piles and pile groups. *CIRIA Technical Note 108*.

Caputo, V. and C. Viggiani (1984) Pile foundation analysis: a simple approach to non linearity effects, *Rivista Italiana di Geotecnica*, 18(2): 32–51.

Carriglio, F., V.N. Ghionna, M. Jamiolkowski and R. Lancellotta (1990) Stiffness and penetration resistance of sands versus state parameter, *ASCE Journal of Geotechnical Engineering*, 116(6): 1015–1020.

Carter, J.P. and F.H. Kulhawy (1988) Analysis and design of drilled shaft foundations socketed into rock. Report to Electric Power Research Institute, California, Project RP1493–4, Report EL-5918.

Chandler, B.C. (1998) My Thuan bridge: update on bored pile foundations. *Proc. Australasian Bridge Conference*, Sydney.

Chandler, R.J. (1968) The shaft friction of piles in cohesive soils in terms of effective stresses, *Civ. Eng. Public Wks. Rev.* 63: 48–51.

Chandler, R.J. (1969) The effect of weathering on the shear strength properties of Keuper Marl, *Géotechnique*, 19(3): 321–334.

Chandler, R.J. (1972) Lias clay: weathering processes and their effect on shear strength, *Géotechnique*, 22(3): 403–431.

Chandler, R.J. and J.P. Martins (1982) An experimental study of skin friction around piles in clay, *Géotechnique*, 32(2): 119–132.

Cheney, J.E. (1973) Techniques and equipment using the surveyors level for accurate measurement of building movement, *Proc. Symp. Field Instrumentation in Geotech. Eng.*, Butterworth, London.

Chin, F.K. (1970) Estimation of the ultimate load of piles not carried to failure, *Proc. 2nd S.E. Asia Conf. on Soil Eng.* pp. 81–90.

Chow, F.C. (1997) *Investigations in the Behaviour of Displacement Piles for Offshore Foundations*. Ph.D Thesis, Imperial College, UK.

Chow, F.C., R.J. Jardine, F. Brucy and J.F. Nauroy (1998) Effects of time on capacity of pipe piles in dense marine sand, *ASCE J. Geotech. and Geoenv. Eng. Div*, 124(3): 254–264.

Claessen, A.I.M. and E. Horvat (1974) Reducing negative friction with bitumen slip layers, *ASCE J. Geotech. Eng. Div.*, 100(GT8): 925–944.

Clancy, P. and M.F. Randolph (1993) An approximate analysis procedure for piled raft foundations, *Int J. Num. and Anal. Methods in Geomechanics*, 17(12): 849–869.

Clancy, P. and M.F. Randolph (1996) Simple design tools for piled raft foundations, *Géotechnique*, 46(2): 313–328.

Clough, G.W. and T.D. O'Rourke (1990) Construction induced movements of in situ walls. *Proc. ASCE Conf. on Design and Performance of Earth Retaining Structures*, Geotech Special Publication No 25, ASCE, New York, pp. 439–470.

Cole, K.W. (1972) Uplift of piles due to driving displacement, *Civ. Eng. Public Wks.* March, pp. 263–269.

Cole, K.W. (1980) The South abutment of Kessock Bridge. Scotland, *Proc. IABSE Conf.*, Vienna.

Combarieu, O. and A. Morbois (1982) Fondations mixtes semelle-pieux. *Annales de lInst. Tech. du Bat. et des Trav. Pub.*, No. 410, December.

Cooke, R.W. (1974) Settlement of friction pile foundations, *Proc. Conf. on Toll Buildings*, Kuala Lumpur, pp. 7–19.

Cooke, R.W. (1986) Piled raft foundations on stiff clays: a contribution to design philosophy, *Géotechnique*, 36(2): 169–203.

Cooke, R.W., D.W. Bryden Smith, M.N. Gooch and D.F. Sillet (1981) Some observations of the foundation loading and settlement of a multi-storey building on a piled raft foundation in London clay, *Proc. ICE*, 107(1): 433–460.

Cooke, R.W. and G. Price (1978) Strains and displacements around friction piles. Building Research Station CP 28/78, October.

Cooke, R.W., G. Price and K.W. Tarr (1979) Friction piles under vertical working load conditions – load transfer and settlement, *Géotechnique*, 29(2): 113–147.

Cooke, R.W., G. Price and K.W. Tarr (1980) Jacked piles in London clay: interaction and group behaviour under working conditions, *Géotechnique*, 30(2): 449–471.

Coyle, H.M. and R.R. Castello (1981) New design correlations for piles in sand, *ASCE J. Geotech. Eng. Div.*, 107(GT7): 965–986.

Crammond, N.J. and P.J. Nixon (1993) Deterioration of concrete foundation piles as a result of thaumasite formation. *Sixth International Conference on the Durability of Building Materials*, Japan, E & F SPM, Vol. 1: 295–305.

Cundall, P.A. (1976) Explicit finite difference methods in geomechanics, in Numerical Methods in Engineering, *Proc. EF Conference on Numerical Methods in Geomechanics*, Blacksburg, VA, Vol. 1: 132–150.

Davisson, M.T. (1963) Estimating buckling loads for piles, *Proc. 2nd Pan-Amer. Conf. on Soil Mech. and Found. Eng.*, Brazil, Vol. 1: 351–371.

Davisson, M.T. and H.L. Gill (1963) Laterally loaded piles in a layered soil system, *ASCE J. Soil Mech. and Found. Div.*, 89(SM3): 63–94.

Davisson, M.T. and K.E. Robinson (1965) Bending and buckling of partially embedded piles, *Proc. 6th Int. Conf. on Soil Mech. and Found. Eng.*, Vol. 2: 243–246.

De Beer, E.E. and M. Wallays (1972) Forces induced in piles by unsymmetrical surcharges on the soil around the piles, *Proc. 5th Eur. Conf. on Soil Mech. and Found. Eng.*, Madrid, Vol. 1: 325–332.

De Mello, V.F.B. (1969) Foundations of buildings on clay, State of the Art Report, *Proc. 7th Int. Conf. on Soil Mech. and Found. Eng.*, Vol. 1: 49–136.

De Nicola, A. and M.F. Randolph (1993) Tensile and compressive shaft capacity of piles in sand, *ASCE J. Geot. Eng. Div.*, 119(12): 1952–1973.

De Nicola, A. and M.F. Randolph (1997) The plugging behaviour of driven and jacked piles in sand, *Géotechnique*, 47(4): 841–856.

Deeks, A.J. and M.F. Randolph (1995) A simple model for inelastic footing response to transient loading, *Int. J. Num. and Anal. Meth. in Geom.*, 19(5): 307–329.

Dolwin, J.D. and T. Poskitt (1982) An optimisation method for pile driving analysis, *Proc. 2nd Int. Conf. on Num. Methods in Offshore Piling*, Austin, 91–106.

Dumbleton, M.J. and G. West (1976) Preliminary sources of information for site investigation in Britain. Department of the Environment, Transport and Road Research Laboratory Report LR403, revised edition.

Dunnavant, T.W. and M.W. O'Neill (1989) Experimental p-y model for submerged, stiff clay, *ASCE Journal of Geotechnical Engineering*, 115(1): 95–114.

Elson, W.K. (1984) Design of laterally loaded piles. CIRIA Report 103, London.

Evangelista, A. and C. Viggiani (1976) Accuracy of numerical solutions for laterally loaded piles in elastic half-space, *Proc. 2nd Int. Conf. on Numerical Methods in Geomechanics*, Blacksburg, Vol. 3: 1367–1370.

Fahey, M. (1998) Deformation and in situ stress measurement, *Proc. 1st Int. Conf. On Site Characterisation – ISC 98*, Atlanta, Vol. 1: 49–68.

Fahey, M. and M.F. Randolph (1984) Effects of disturbance on parameters derived from self-boring pressuremeter tests in sand, *Géotechnique*, 34(1): 81–97.

Fellenius, B.H. (1972) Down drag on piles in clay due to negative skin friction, *Can. Geotech. J.*, 9(4): 323–337.

Fellenius, B.H. (1980) The analysis of results from routine pile load tests, *Ground Engineering*, 6(Sept): 19–31.

Fellenius, B.H. (1988) Variation of CAPWAP results as a function of the operator, *Proc. 3rd Int. Conf. on Application of Stress-wave Theory to Piles*, Ottawa, pp. 814–825.

Fernie, R. and T. Suckling (1996) Simplified approach for estimating lateral wall movement of embedded walls in UK ground, in Mair and Taylor (eds), *Geotechnical Aspects of Underground Construction in Soft Ground*, Rotterdam, Balkema.

Fleming, W.G.K. (1984) Discussion on State of the Art paper, *Advances in Piling and Ground Treatment for Foundations Conference*, London: ICE.

Fleming, W.G.K. (1992) A new method for single pile settlement for prediction and analysis, *Géotechnique*, 42(3): 411–425.

Fleming, W.G.K. (1995) The understanding of continuous flight auger piling, its monitoring and control, *Proceedings, Institution of Civil Engineers Geotechnical Engineering*, Vol. 113, July: 157–165. Discussion by R. Smyth-Osbourne and reply, Vol. 119, Oct., 1996: 237.

Fleming, W.G.K. and P.F. Lane (1971) Tolerance requirements and construction problems in piling, *Conf. on Behaviour of Piles*, London: ICE, pp. 175–178.

Fleming, W.G.K. and S. Thorburn (1983) Recent piling advances, State of the Art Report, *Proc. Conf. on Advances in Piling and Ground Treatment for Foundations*, London: ICE.

Fleming, W.K. and Z.J. Sliwinski (1977) The use and influence of bentonite in bored pile construction. D.O.E/CIRIA Piling Development Group Report PG 3.

Fookes, P.G. and L. Collis (1975) Cracking and the Middle East, *Concrete, 1176*, 10(February): 14–19.

Foray, P., J.L. Colliat and J.F. Nauroy (1993) Bearing capacity of driven model piles in dense sands from calibration tests, *Proc. 25th Annual Offshore Technology Conference*, Houston, Paper OTC 7194, pp. 655–665.

Francescon, M. (1983) *Model Pile Tests in Clay: Stresses and Displacements Due to Installation and Axial Loading*. Ph.D. Thesis, University of Cambridge.

Francis, A.J., L.K. Stevens and P.J. Hoadley (1965) *Paper to Symposium of Soft Ground Engineering*. Brisbane: Institution of Engineers Australia.

Frank, R.A. (1974) Etude theorique du comportement des pieux sous charge verticale: Introduction de la dilatance, Dr-Ing-Thesis, Universite Pierre et Marie Curie (Paris VI), also Rapport de Recherche No 46, Laboratoire Central des Ponts et Chausses, Paris.

Franke, E., B. Lutz and Y. El-Mossallamy (1994) Measurements and numerical modelling of high rise building foundations on frankfurt clay. *Proc. Conf. on Vertical and Horizontal Deformations of Foundations an Embankments*, Texas, ASCE Geotechnical Special Publication No. 40(2):1325–1336.

Geddes, W.G.N., G.P. Martin and D.D. Land (1966) Clyde dry dock project, Greenock. *Proc. ICE.* 33: 615.

Geiger, J. (1959) Estimating the risk of damage to buildings from vibrations. *Bauingenieur*, 34, 1959.

Gens, A. and D.W. Hight (1979) The laboratory measurement of design parameters for a glacial till, *Proc. 7th European Conf. on Soil Mech. and Found. Eng.*, Vol. 2: pp. 57–65.

George, A.B., F.W. Sherrell and M.J. Tomlinson (1977) The behaviour of steel H-piles in slaty mudstone, *Proc. Conf. on Piles in weak rock*, London: ICE, pp. 95–104 (also CIRIA Technical Note 66, May 1976).

Ghosh, N. (1975) *A Model Scale Investigation of the Working Load Stiffness of Single Piles and Groups of Piles in Clay Under Centric and Eccentric Vertical Loads*, Ph.D. Thesis, University of Southampton.

Gibbs, H.J. and W.G. Holtz (1957) Research on determining the density of sands by spoon penetration testing. *Proc. 4th Int. Conf. on Soil Mech. and Found. Eng.*, London, Vol. 1, pp. 35–39.

Gibson, G.C. and H.M. Coyle (1968) Soil damping constants related to common soil properties in sands and clays, Report No.125-1, Texas Transport Institute, Texas A and M University, Houston.

Goble, G.G. and F. Rausche (1976) Wave equation analysis of pile driving – WEAP program, US Department of Transportation, Federal Highway Administration, Implementation Division, Office of Research and Development, Washington D.C. 20590.

Goble, G.G. and F. Rausche (1979) Pile drivability predictions by CAPWAP, *Proc. Int. Conf. on Numerical Methods in Offshore Piling*, London: ICE, pp. 29–36.

Goble, G.G., F. Rausche and G.E. Liking (1980) The analysis of pile driving – A state-of-the-art, *Proc. Int. Conf. on Stress-Wave Theory on Piles*, Stockholm, pp. 131–161.

Goble, G.G. and F. Rausche (1986) WEAP86 Program Documentation in 4 Vols. Federal Highway Administration, Office of Implementation, Washington DC.

Griffiths, D.V., P. Clancy and M.F. Randolph (1991), Piled raft foundation analysis by finite elements, *Proc. 7th Int. Conf. on Num. Methods in Geom.*, Cairns, Vol. 2, pp. 1153–1157.

Guo, W.D. and M.F. Randolph (1997) Vertically loaded piles in non-homogeneous media. *Int. J. Num. and Anal. Methods in Geomechanics*, 21(8): 507–532.

Hagerty, D. and R.B. Peck (1971) Heave and lateral movements due to pile driving. *ASCE J. Soil Mech. and Found. Div.*, (SM 11) (Nov.):1513–1531.

Hain, S.J. and I.K. Lee (1978) The analysis of flexible pile raft systems, *Géotechnique*, 28(1): 65–83.

Hambly, E.G. (1976) A review of current practice in the design of bridge foundations. Report for the Department of the Environment, Building Research Establishment, UK.

Hanna, A. and T.Q. Nguyen (2003) Shaft resistance of single vertical and batter piles driven in sand, *ASCE J. Geot. And Geo Envir. Eng.*, 129(7): 601–607.

Hanna, T.H. (1968) The bending of long H-section piles, *Canadian Geotech. J.*, 5(3): 150–172.

Hansbo, S. and L. Jenderby (1983) A case study of two alternative foundation principles: conventional friction piling and creep piling, *Vag-och Vattenbyggaren*, 7(8): 29–31.

Healy, P.R. and A.J. Weltman (1980) Survey of problems associated with the installation of displacement piles. CIRIA Report PG8, Storeys Gate, London.

Hepton, P. (1995) Deep rotary cored boreholes in soils using wireline drilling. *Proc. Int. Conf. on Advances in Site Investigation Practice*, London: Thomas Telford, pp. 269–280.

Hewlett, W.J. and M.F. Randolph (1988) Analysis of piled embankments, *Ground Engineering*, 22(3): 12–18.

Hiley, A. (1925) A rational pile-driving formula and its application in piling practice explained. *Engineering* (London) 119, 657, 721.

Hobbs, N.B. and P.R. Healy (1979) Piling in Chalk DOE/CIRIA Piling Development Group Report PG. 6. CIRIA, London.

Hobbs, N.B. and P. Robins (1976) Compression and tension tests on driven piles in chalk, *Géotechnique*, 26(1): 33–46.

Hodges, W.G.H. and S. Pink (1971) The use of penetrometer soundings in the estimation of pile bearing capacity and settlement for driven piles in highly weathered chalk, in Parry, R.H.G., G.T. Foulis (ed.), *Stress Strain Behaviour of Soils* (Proc. Roscoc Mem. Symp.), (Blackie Publishing Group) pp. 693–723.

Holeyman, A., J.-F. Vanden Berghe and N. Charue (2002) *Vibratory Pile Driving and Deep Soil Compaction*, Lisse: Zwets and Zeitlinger, p. 233.

Horikoshi, K. and M.F. Randolph (1996) Centrifuge modelling of piled raft foundations on clay, *Géotechnique*, 46(4): 741–752.

Horikoshi, K. and M.F. Randolph (1997) On the definition of raft-soil stiffness ratio, *Géotechnique*, 47(5): 1055–1061.

Horikoshi, K. and M.F. Randolph (1998) A contribution to optimum design of piled rafts, *Géotechnique*, 48(3): 301–317.

Horikoshi, K. and M.F. Randolph (1999) Estimation of piled raft stiffness, *Soils and Foundations*, 39(2): 59–68.

Horvath, R.G., W.A. Trow and T.C. Kenney (1980) Results of tests to determine shaft resistance of rock-socketed drilled piers. *Proc. Int. Conf. on Structural Foundations on Rock*, Sydney, Vol. 1, pp. 349–361.

Hughes, J.M.O., C.P. Wroth and D. Windle (1977) Pressuremeter tests in sand, *Géotechnique*, 27(4): 455–477.

ICE (2007) *Specification for Piling and Retraining Walls*, 2nd edn, London: Thomas Telford.

Jaeger, J.C. and A.M. Starfield (1974) *An Introduction to Applied Mathematics* (2nd Edn). Oxford: Clarendon Press.

Jardine, R.J. and F.C. Chow (1996) New design methods for offshore piles. Marine Technology Directorate, London, Publication No. 96/103.

Jardine, R.J., F.C. Chow, R.F. Overy and J.R. Standing (2005) ICP design methods for driven piles in sands and clays, UK: Thomas Telford.

Jardine, R.J., D. Potts, A.B. Fourie and J.B. Burland (1986) Studies of the influence of non-linear stress-strain characteristics in soil-structure interaction, *Géotechnique*, 36(3): 377–396.

Johanessen, I.J. and L. Bjerrum (1965) Measurement of the compression of a steel pile to rock due to settlement of the surrounding clay, *Proc. 6th Int. Conf. on Soil Mech. and Found. Eng.*, Vol. 2, pp. 261–264.

Johnston, I.W. and T.S.K. Lam (1989) Shear behaviour of regular triangular concrete/rock joints – Analysis, *ASCE J. Geot. Eng. Div.*, 115(GT5): 711–727.

Kagawa, T. (1992) Moduli and damping factors of soft marine clays, *ASCE J. Geot. Eng. Div.*, 118(9): 1360–1375.

Katzenbach, R., U. Arslan and C. Moormann (2000) *Piled Raft Foundation Projects in Germany. Design Applications of Raft Foundations*. London: Thomas Telford, pp. 323–391.

Kennet, R. (1971) Geophysical borehole logs as an aid to ground engineering, *Ground Engineering*, 4(5).

Kishida, H. and M. Uesugi (1987) Tests of interface between sand and steel in the simple shear apparatus, *Géotechnique*, 37(1): 46–52.

Kjekstad, O. and F. Stub (1978) Installation of the Elf TCP2 Condeep platform at the Frigg field. *Proc. Eur. Offshore Petroleum Conf. and Exhibition*, London, Vol. 1, pp. 121–130.

Kolk, H.J. and E. van der Velde (1996) A reliable method to determine friction capacity of piles driven into clays. *Proc. Offshore Technology Conf.*, Houston, Paper OTC 7993.

Kraft, L.M. (1990) Computing axial pile capacity in sands for offshore conditions, *Marine Geotechnology*, 9: 61–92.

Kraft, L.M. (1982) Effective stress capacity model for piles in clay, *ASCE J. Geotech. Eng. Div.*, 108(GT11): 1387–1404.

Kuhlemeyer, R.L. (1979) Static and dynamic laterally loaded floating piles, *ASCE J. Geotech. Eng. Div.*, 105(GT2): 289–304.

Kulhawy, F.H. (1984) Limiting tip and side resistance: Fact or fallacy, in Proc. of Symp. on Analysis and Design of Pile Foundations, *ASCE Spec. Conf.*, San Francisco, pp. 80–98.

Kulhawy, F.H. and K.K. Phoon (1993) Drilled shaft side resistance in clay soil to rock, Geotechnical Special Publication No. 38, Design and Performance of Deep Foundations, ASCE, New York, 172–183.

Kusakabe, O. and T. Matsumoto (1995) Statnamic tests of Shonan test program with review of signal interpretation, *Proc. 1st Int. Statnamic Seminar*, Vancouver, Canada, pp. 113–122.

Kuwabara, F. and H.G. Poulos (1989) Downdrag forces in groups of piles, *ASCE J. Geot. Eng. Div.*, 115(GT6): 806–818.

Lamb, T.W and R.V. Whitman (1969) *Soil Mechanics*. New York: John Wiley and Sons Inc., pp. 199.

Lee, C.J., M.D. Bolton and A. Al-Tabbaa (2002) Numerical modelling of group effect on the distribution of dragloads in pile foundations, *Géotechnique*, 52(5): 323–335.

Lee, J.H. and R. Salgado (1999) Determination of pile base resistance in sands, *ASCE J. Geotech. and Geoenv. Eng.*, 125(8): 673–683.

Lehane, B.M. (1992) *Experimental Investigations of Pile Behaviour Using Instrumented Field Piles*, Ph.D Thesis, Imperial College, London.

Lehane, B.M., R.J. Jardine, A.J. Bond and F.C. Chow (1994) The development of shaft friction on displacement piles in clay, *Proc. 13th Int. Conf. on Soil Mech. and Found. Eng.*, New Dehli, Vol. 2, pp. 473–476.

Lehane, B.M., R.J. Jardine, A.J. Bond and R. Frank (1993) Mechanisms of shaft friction in sand from instrumented pile tests, *ASCE J. Geot. Eng. Div.*, 119(1): 19–35.

Lehane, B.M. and M.F. Randolph (2002) Evaluation of a minimum base resistance for driven pipe piles in siliceous sand, *ASCE J. Geotech Eng. Div.*, 128(3): 198–205.

Lehane, B.M., J.A. Schneider and X. Xu (2005) A review of design methods for offshore driven piles in siliceous sand. Research Report Geo:05358, Geomechanics Group, The University of Western Australia.

Liew, S.S., S.S. Gue and Y.C. Tan (2002) Design and instrumentation results of a reinforced concrete piled raft supporting 2500 tonne oil storage tank on very soft alluvium. *Proc. 9th Int. Conf. On Piling and Deep Foundations*, Nice.

Likins, G.E. (1984) Field measurements and the pile driving analyser, *Proc. 2nd Int. Conf. on Application of Stress Wave Theory*, Stockholm, pp. 298–305.

Litkouhi, S. and T.J. Poskitt (1980) Damping constant for pile driveability calculations. *Géotechnique*, 30(1): 77–86.

Lo, M.B. (1967) Discussion to paper by Y.O. Beredugo, *Can. Geotech. J.*, 4: 353–354.

Long, M. (2001) Database for retaining wall and ground movements due to deep excavations, *ASCE Journal of Geotechnical and Geoenvironmental Engineering*, 127(3): 203.

Lord, J.A. (1976) A comparison of three types of driven cast-in-situ pile in chalk. *Géotechnique*, 26(1): 73–93.

Lord, J.A., T. Hayward and C.R.I. Clayton (2003) Shaft friction of CFA piles in chalk. CIRIA Project Report 86.

Love, J.P. and G. Milligan (2003) Design methods for basally reinforced pile-supported embankments on soft ground. *Ground Engineering*, March, pp. 39–43.

Lunne, T., P.K. Robertson and J.J.M. Powell (1997) *Cone Penetration Testing in Geotechnical Engineering*. London: Blackie Academic and Professional.

Lupini, J.F., A.E. Skinner and P.R. Vaughan (1981) The drained residual strength of cohesive soil, *Géotechnique*, 31(2): 181–213.

Ly, B.L. (1980) Discussion of Burgess, I.W., *Géotechnique*, 30(3): 321–322.

Lysmer, T. and F.E. Richart (1966) Dynamic response of footing to vertical loading, *ASCE J. Soil Mech. and Found. Eng. Div.*, 98: 85–105.

Mair, R.J. and D.M. Woods (1987) *Pressuremeter Testing, Methods and Interpretation*. London: CIRIA and Butterworth.

Mandolini, A. (2003) Design of piled raft foundations: practice and development. *Proc. 4th Int. Sem. On Deep Foundations on Bored and Auger Piles, BAP IV*, Ghent, pp. 59–80.

Mandolini, A. and C. Viggiani (1997) Settlement of piled foundations, *Géotechnique*, 47(4): 791–816.

Marcuson, W.V. and WA. Bieganousky (1977) SPT and relative density in coarse sands, *ASCE J. Geotech. Eng. Div.*, 103(GTII): 1295–1309.

Marsland, A. (1976) In-situ and laboratory tests on glacial clays at Redcar Building Research Establishment Current Paper CP 65/76 HMSO (London).

Marsland, A. and M.F. Randolph (1977) Comparison of the results from pressuremeter tests and large in situ plate tests in London clay, *Géotechnique*, 27(2): 217–243.

Marti, J. and P.A. Cundall (1982) Mixed discretization procedure for accurate solution of plasticity problems, *Int. J. Num. Methods in Eng.*, 6: 129–139.

Matlock, H.S. (1970) Correlations for design of laterally loaded piles in soft clay, *2nd Annual Offshore Technology Conf.*, Houston.

Matlock, H., W.B. Ingram, A.E. Kelley and D. Bogard (1980) Field tests of the lateral-load behaviour of pile groups in soft clay. *Proc. 12th Annual Offshore Technology Conf.*, Houston, pp. 163–174.

Matlock, H.S. and L.C. Reese (1960) Generalised solutions for laterally loaded piles, *ASCE J. Soil Mech. and Found. Div.*, 86(SM5): 63–91.

Matsumoto, T., M. Tsuzuki and Y. Michi (1994) Comparative study of static loading test and Statnamic on a steel pipe pile driven into a soft rock, *Proc. 5th Int. Conf. on Piling and Deep Foundations*, Bruges.

Mattes, N.S. and H.G. Poulos (1969) Settlement of single compressible pile, *ASCE J. Soil Mech. and Found. Div.*, 95(SM1): 198–207.

Mayne, P.W. and F.H. Kulhawy (1982) Ko-OCR relationships in soils, *ASCE J. Geotech. Eng. Div.*, 108(GT6): 851–872.

McClelland, B. (1974) Design of deep penetration piles for ocean structures, *ASCE J. Geotech. Eng. Div.*, 100(GT7): 705–747.

McClelland, B., J.A. Focht and W.J. Emrich (1969) Problems in design and installation of offshore piles, *ASCE J. Soil Mech. and Found. Div.*, 6: 1491–1513.

McKinlay, D.G., M.J. Tomlinson and W.F. Anderson (1974) Observations on the undrained strength of glacial till, *Géotechnique*, 24(4): 503–516.

Meigh, A.C. (1987) *Cone Penetration Testing, Methods and Interpretation*. London: CIRIA and Butterworth.

Meyerhof, G.G. (1947) The settlement analysis of building frames, *Structural Engineer*, 25(9): 369.

Meyerhof, G.G. (1956) Penetration tests and bearing capacity of cohesionless soils, *ASCE J. Soil Mech. Found. Div.*, 82 (SM1): 1–19.

Meyerhof; G.G. (1976) Bearing capacity and settlement of pile foundations, *ASCE J. Geot. Engng. Div.*, 102(GT3): 197–228.

Meyerhof, G.G. (1995) Behaviour of pile foundations under special loading conditions: R.M. Hardy keynote address, *Canadian Geotechnical Journal*, 32(2): 204–222.

Meyerhof, G.G. and G. Ranjan (1972) The bearing capacity of rigid piles under inclined loads in sand: I – Vertical piles, *Can. Geotech. J.* 9(4): 430–446 (see also Can. Geotech. J. 10(1):71–85 for Part II).

Middendorp, P. and M.W. Bielefeld (1995) Statnamic and the influence of stress-wave phenomena. *Proc. 1st Int. Statnamic Seminar*, Vancouver, Canada, pp. 1–30.

Middendorp, P. and A.F. van Weele (1986) Application of characteristic stress wave method in offshore practice, *Proc. 3rd Int. Conf. on Num. Methods in Offshore Piling*, Nantes, Supplement, 6–18.

Milligan, G.W.E. (1983) Soil deformations near anchored sheet pile walls, *Géotechnique*, 33(1): 41–55.

Murff, J.D. (1980) Pile capacity in a softening soil, *Int. J. Num. and Anal. Methods in Geomechanics*, 4: 185–189.

Murff, J.D. (1987) Pile capacity in calcareous sands: State of the Art, *ASCE J. Geot. Eng. Div.*, 113(GT5): 490–507.

Murff, J.D. and J.M. Hamilton (1993) P-Ultimate for undrained analysis of laterally loaded piles, *ASCE Journal of Geotechnical Engineering*, 119(1): 91–107.

Mylonakis, G. (2001) Winkler modulus for axially loaded piles, *Géotechnique*, 51(5): 455–461.

Mylonakis, G. and G. Gazetas (1998) Settlement and additional internal forces of grouped piles in layered soil, *Géotechnique*, 48(1): 55–72.

Neely, W.J. (1988) Bearing capacity of expanded-base piles in sand, *ASCE J. Geotech. Eng. Div.*, 116(GT1): 73–87.

Nixon, P.J., Longworth, T.I. and Matthews, J.D. (2003) New UK guidance on the use of concrete in aggressive ground, *Cement and Concrete Composites*, 25(8): 1177–1184.

Nordlund, R.L. (1963) Bearing capacity of piles in cohesionless soils, *ASCE J. Soil Mech and Found Eng. Div.*, 89(SM3): 1–35.

Ohsaki, Y. and R. Iwasaki (1973) On dynamic shear moduli and Poissons ratio of soil deposits, *Soils and Foundations*, 13(4): 61–73.

Omar, R.M. (1978) Discussion of Burgess IW., *Géotechnique*, 28(2): 234–239.

O'Neill, M.W., O.I. Ghazzaly and H.B. Ha (1977) Analysis of three dimensional pile groups with non-linear soil response and pile-soil-pile interaction. *Proc. 9th Offshore Tech. Conf.*, Vol. 2, pp. 245–256.

O'Neill, M.W., R.A. Hawkins and L.J. Mahar (1982) Load transfer mechanisms in piles and pile groups, *ASCE J. Geotech. Eng. Div.*, 108(GT12): 1605–1623.

O'Riordan, N.J. (1982) The mobilisation of shaft adhesion down a bored, cast-in-situ pile in the Woolwich and Reading beds, *Ground Engineering*, 15(3): 17–26.

Osterberg, J. (1989) New device for load testing driven piles and drilled shafts separates friction and end-bearing, *Proc. Int. Conf. On Piling an Deep Found.*, London, Vol. 1, pp. 421–427.

Packshaw, S. (1951) Pile driving in difficult conditions. ICE Introductory Note.

Padfield, C.J. and M.J. Sharrock (1983) Settlement of structures on clay soils, CIRIA Special Publication 27, London: CIRIA.

Pappin J.W., B. Simpson, P.J. Felton and Raison, C. (1986) Numerical analysis of flexible retaining walls. Symposium on Computer Applications in Geotechnical Engineering. The Midland Geotechnical Society, UK.

Parry, R.H.G. (1972) Some properties of heavily overconsolidated lower Oxford clay at a site near Bedford, *Géotechnique*, 22(3): 485–507.

Parry, R.H.G. (1980) A study of pile capacity for the Heather platform, *Ground Engineering*, 13(2): 26–28, 31, 37.

Parry, R.H.G. and C.W. Swain (1977) Effective stress methods of calculating skin friction of driven piles in soft clay, *Ground Engineering*, 10(3): 24–26.

Peck, R.B. (1943) Earth pressure measurements in open cuts, Chicago (ILL) Subway, *Transactions Amer. Soc. Civ. Eng.*, 108: 1008–1036.

Peck, R.B., W.E. Hanson and T.H. Thornburn (1974) *Foundation Engineering*, New York: John Wiley and Sons.

Pise, P.J. (1982) Laterally loaded piles in a two-layer soil system, *ASCE J. Geotech. Eng. Div.*, 108(GT9): 1177–1181.

Polshin, D.E. and R.A. Tokar (1957) Maximum allowable non-uniform settlement of structures, *Proc. 4th Int. Conf. on Soil Mech. and Foundn. Engng.*, London, Vol. 1, pp. 402.

Poulos, H.G. (1971) Behaviour of laterally loaded piles: I – single piles, *ASCE J. Soil Mech. and Found. Div.*, 97(SM5): 711–731.

Poulos, H.G. (1971) Behaviour of laterally loaded piles: II – pile groups, *ASCE J. Soil Mech. and Found. Div.*, 97(SM5): 733–751.

Poulos, H.G. (1980) Users guide to program DEF PIG – deformation analysis of pile groups. School of Civil Engineering, University of Sydney.

Poulos, H.G. (1994) An approximate numerical analysis of pile-raft interaction, *Int. Journal for Numerical and Analytical Methods in Geomechanics*, 18: 73–92.

Poulos, H.G. (2001) Piled-raft foundation: design and applications, *Géotechnique*, 51(2): 95–113.

Poulos, H.G. and E.H. Davis (1968) The settlement behaviour of single axially loaded incompressible piles and piers, *Géotechnique*, 18: 351–371.

Poulos, H.G. and E.H. Davis (1974) *Elastic Solutions for Soil and Rock Mechanics*. New York: John Wiley and Sons.

Poulos, H.G. and E.H. Davis (1980) *Pile Foundation Analysis and Design*. New York: John Wiley and Sons.

Poulos, H.G. and M.F. Randolph (1983) Pile group analysis: a study of two methods, *ASCE J. Geotech. Eng.*, 109(3): 355–372.

Powell, J.J.M. and I.M. Uglow (1988) The interpretation of the Marchetti dilatometer test in UK clays, *ICE Proc. Conf. Penetration Testing in the UK*.

Prakoso, W.A. and F.H. Kulhawy (2001) Contribution to piled raft optimum design, *ASCE J. of Geotech. and Geoenv. Engrg.*, 127(1): 17–24.

Preiss, K. and F. Caiserman (1975) Non-destructive integrity testing of bored piles by gamma ray scattering, *Ground Engineering*, 8(3): 44–46.

Price, G. and I.F. Wardle (1983) Recent developments in pile/soil instrumentation systems, in *Proc. Conf. Field Measurements in Geomechanics*, Zurich.

Price, G., I.F. Wardle and N.G. Price (2004) The increasing importance of monitoring the field performance of foundations to validate numerical analysis. *Proc. The Skempton Conference – Advances in Geotechnical Engineering*. London: Thomas Telford, Vol. 2, pp. 1131–1142.

Ramsey, N., R.J. Jardine, B.M. Lehane and A.M. Ridley (1998) A review of soil-steel interface testing with the ring shear apparatus. *Proc. Conf. on Offshore Site Investigation and Foundation Behaviour, Soc. for Underwater Technology*, London, pp. 237–258.

Randolph, M.F. (1977) *A Theoretical Study of the Performance of Piles*, Ph.D. Thesis, University of Cambridge.

Randolph, M.F. (1979) *Discussion in Proc. of Conf. on Recent Developments in the Design and Construction of Piles*, London: ICE, pp. 389–390.

Randolph, M.F. (1980) PIGLET: A Computer Program for the Analysis and Design of Pile groups Under General Loading Conditions. Cambridge University Engineering Department Research Report, Soils TR 91.

Randolph, M.F. (1981) Piles subjected to torsion, *ASCE J. Geotech. Eng. Div.*, 107(GT8): 1095–1111.

Randolph, M.F. (1981) The response of flexible piles to lateral loading, *Géotechnique*, 31(2): 247–259.

Randolph, M.F. (1983) Design considerations for offshore piles, *Proc. Conf. on Geotech. Practice in Offshore Engng., ASCE*, Austin, pp. 422–439.

Randolph, M.F. (1983) Design of piled raft foundations, *Proc. Int. Symp. on Recent Developments in Laboratory and Field Tests and Analysis of Geotechnical Problems*, Bangkok, pp. 525–537.

Randolph, M.F. (1983) Settlement considerations in the design of axially loaded piles, *Ground Engineering*, 16(4): 28–32.

Randolph, M.F. (1988) The axial capacity of deep foundations in calcareous soil, *Proc. Int. Conf. on Calcareous Sediments*, Perth, Vol. 2, pp. 837–857.

Randolph, M.F. (1990) Analysis of the dynamics of pile driving, in P.K. Banerjee and R. Butterfield (eds), *Developments in Soil Mechanics IV: Advanced Geotethoical Analyses*. London: Elsevier Applied Science Publishers Ltd.

Randolph, M.F. (1994) Design methods for pile groups and piled rafts, *Proc. 13th Int. Conf. on Soil Mech. and Found. Eng.*, New Delhi, 5, 61–82.

Randolph, M.F. (2000) Pile-soil interaction for dynamic and static loading, Keynote Lecture, *Proc. 6th Int. Conf. On Application of Stress–Wave Theory to Piles*, Sao Paulo, Appendix: 3–11.

Randolph, M.F. (2003) 43rd Rankine Lecture: Science and empiricism in pile foundation design, *Géotechnique*, 53(10): 847–875.

Randolph, M.F. (2003) RATZ – Load transfer analysis of axially loaded piles. User manual, University of Western Australia, Perth.

Randolph, M.F., J.P. Carter and C.P. Wroth (1979) Driven piles in clay – the effects of installation and subsequent consolidation, *Géotechnique*, 29(4): 361–393.

Randolph, M.F., J. Dolwin and R.D. Beck (1994) Design of driven piles in sand, *Géotechnique*, 44(3): 427–448.

Randolph, M.F. and G.T. Houlsby (1984) The limiting pressure on a circular pile loaded laterally in cohesive soil, *Géotechnique*, 34(4): 613–623.

Randolph, M.F., H.A. Joer and D.W. Airey (1998) Foundation design in cemented sands, *2nd Int. Seminar on Hard Soils*, Soft Rocks, Naples, 3: 1373–1387.

Randolph, M.F. and B.S. Murphy (1985) Shaft capacity of driven piles in clay, *Proc. 17th Offshore Technology Conf.*, Houston, Paper OTC 4883, pp. 371–378.

Randolph, M.F. and H.G. Poulos (1982) Estimating the flexibility of offshore pile groups, *Proc 2nd Int. Conf. on Numerical Methods in Offshore Piling*, Austin, pp. 313–328.

Randolph, M.F. and D.A. Simons (1986) An improved soil model for uni-dimensional pile driving analysis, *Proc. 3rd Int. Conf. on Num. Methods in Offshore Piling*, Nantes, pp. 1–17.

Randolph, M.F. and C.P. Wroth (1978) A simple approach to pile design and the evaluation of pile tests, in R. Lundgren (ed.), *Behaviour of Deep Foundations*, ASTM STP 670, pp. 484–499.

Randolph, M.F. and C.P. Wroth (1978) Analysis of deformation of vertically loaded piles, *ASCE J.Geotech Engng. Div.*, 104(GT12): 1465–1488.

Randolph, M.F. and C.P. Wroth (1979) An analysis of the vertical deformation of pile groups, *Géotechnique*, 29(4): 423–439.

Randolph, M.F. and C.P. Wroth (1979) An analytical solution for the consolidation around a driven pile, *Int. J. Num. and Anal. Methods in Geomechanics*, 3: 217–229.

Randolph, M.F. and C.P. Wroth (1981) Application of the failure state in undrained simple shear to the shaft capacity of driven piles, *Géotechnique*, 31(1): 143–157.

Randolph, M.F. and C.P. Wroth (1982) Recent developments in understanding the axial capacity of piles in clay, *Ground Engineering*, 15(7): 17–25, 32.

Rausche, F., G.G. Goble and G.E. Likins (1985) Dynamic determination of pile capacity, *ASCE J. Geot. Eng. Div.*, 111: 367–383.

Rausche, F., G.G. Goble and G.E. Likins (1988) Recent WEAP developments, *Proc. 3rd Int. Conf. on Application of Stress Wave Theory to Piles*, Ottawa, pp. 164–173.

Reddaway, A.L. and W.K. Elson (1982) The performance of a piled bridge abutment at Newhaven. CIRIA Technical Note 109, London.

Reddy, A.S. and A.J. Valsangkar (1970) Buckling of fully and partially embedded piles, *ASCE J. Soil Mech. and Found. Div.*, 96(SM6): 1951–1965.

Reese, L.C. (1958) Discussion of paper by B. McClelland and J.A. Focht, *Trans. ASCE*, 123: 1071–1074.

Reese, L.C. (1977) Laterally loaded piles: program documentation, *ASCE J. Geotech. Eng. Div.*, 103(GT4): 287–305.

Reese, L.C. and H.S. Matlock (1956) Non-dimensional solutions for laterally loaded piles with soil modulus proportional to depth, *Proc. 8th Texas Conf. on Soil Mech and Found. Eng.*, pp. 1–41.

Reese, L.C. and S.T. Wang (1993) LPILE 4.0, Ensoft Inc, Austin, Texas.

Reese, L.C., W.R. Cox and F.D. Koop (1974) Analysis of laterally loaded piles in sand, *6th Annual Offshore Technology Conf.*, Houston, OTC 2080, Vol. 2, pp. 473–485.

Reese, L.C., W.R. Cox and F.D. Koop (1975) Field testing and analysis of laterally loaded piles in stiff clay, *Proc. 7th Offshore Tech. Conf.*, Vol. 2, pp. 671–690.

Reese, L.C., F.T. Touma and M.W. O'Neill (1976) Behaviour of drilled piers under axial loading, *Proc. J. Geotech. Eng. Div., ASCE*, 102 (GT5): 493–510.

Reid, W.M. and N.W. Buchanan (1983) Bridge approach support piling, *Proc. Conf. on Advances in Piling and Ground Treatment for Foundations*, London: ICE.

Reul, O. and M.F. Randolph (2003) Piled rafts in overconsolidated clay – Comparison of in-situ measurements and numerical analyses, *Géotechnique*, 53(3): 301–315.

Reul, O. and M.F. Randolph (2004) Design strategies for piled rafts subjected to non-uniform vertical loading, *ASCE J. Geotech. and Geoenv. Eng. Div*, 130(1): 1–13.

Robertson, P.K. (1991) Estimation of foundation settlements in sand from CPT, *Proc. of Geotech. Eng. Congress, Geotech. Eng. Div. ASCE*, Boulder, Colorado, pp. 764–775.

Rowe, R.K. and H.H. Armitage (1987) A design method for drilled piers in soft rack, *Canadian Geotechnical Journal*, 24(1): 126–142.

Russell, D. and N. Pierpoint (1997) An assessment of design methods for piled embankments, *Ground Engineering*, 30(11): 39–44.

Russo, G. (1998) Numerical analysis of piled rafts, *Int. J. Anal. and Num. Methods in Geomechanics*, 22(6): 477–493.

Samuels, S.G. (1975) Some properties of the Gault Clay from the Ely-Ouse Essex Water Tunnel. *Géotechnique*, 25(2): 239–264.

Schofield, A.N. and C.P. Wroth (1968) *Critical State Soil Mechanics*. Maidenhead: McGraw-Hill.

Seed, H.B. and L.C. Reese (1955) The action of soft clay along friction piles, *Proc. ASCE*, 81, Paper 842.

Seidel, J. and C.M. Haberfield (1995) The axial capacity of pile sockets in rocks and hard soils, *Ground Engineering*, 28(2): 33–38.

Semple, R.M. and W.J. Rigden (1984) Shaft capacity of driven piles in clay, *Proc. ASCE National Convention*, San Francisco.

Simpson, B., G. Calabresi, H. Sommer and M. Wallays (1979) Design parameters for stiff clays. *Proc. 7th Eur. Conf. on Soil Mech. and Found. Eng.*, Brighton, Vol. 5, pp. 91–125.

Skempton, A.W. (1951) *The Bearing Capacity of Clays, in Building Research Congress*. London: ICE, pp. 180–189.

Skempton, A.W. (1959) Cast in situ bored piles in London clay, *Géotechnique*, 9: 153–173.

Skempton, A.W. (1970) First time slides in over consolidated clays, *Géotechnique*, 20(3): 320–324.

Skempton, A.W. (1974) Slope stability of cuttings in brown London Clay. Special Lecture, *9th Int. Conf. on Soil Mechanics and Found. Eng.*, Tokyo, Vol. 3.

Skempton, A.W. (1986) Standard penetration test procedures and the effects in sands of overburden pressure, relative density, particle size, ageing and overcansolidation, *Géotechnique*, 36(2): 425–447.

Skempton, A.W. and J.D. Brown (1961) A landslide in boulder clay at Selset, Yorkshire, *Géotechnique*, 11(4): 280–293.

Skempton, A.W. and D.H. MacDonald (1956) The allowable settlements of buildings, *Proc. ICE*, 5(Dec.): 727–767. (discussion *ICE*, 5(3): 724–784)

Skempton, A.W. and R.D. Northey (1952) The sensitivity of clays, *Géotechnique*, 3: 30–53.

Skipp, B.O. (1984) Dynamic ground movements – man-made vibrations, in P.B. Attewell and R.K. Taylor (eds), *Ground Movements and their Effects on Structures*, Glasgow and London: Blackie, pp. 381–434.

Sliwinski, Z.J. and W.G.K. Fleming (1975) *Practical Considerations Affecting the Construction of Diaphragm Walls, Diaphragm Walls and Anchorages*, London: ICE.

Sliwinski, Z.J. and W.G.K. Fleming (1984) *The Integrity and Performance of Bored Piles. Advances in Piling and Ground Treatment for Foundations*, London: ICE.

Smith, E.A.L. (1960) Pile driving analysis by the wave equation, *ASCE J. Soil Mech. and Found. Div.*, 86(SM4): 35–61.

Smith, I.M. and Y.K. Chow (1982) Three-dimensional analysis of pile drivability, *Proc. 2nd Int. Conf. on Numerical Methods in Offshore Pilling*, Austin, pp. 1–10.

Sokolovski, V.V. (1965) *Statics of Granular Media*, Oxford: Pergamon Press.

Sowers, G.F. (1962) Shallow foundations, G.A. Leonards (ed.), *Foundation Engineering*, New York: McGraw-Hill, pp. 525.

Stewart, D.P. (1999a) Reduction of undrained lateral pile capacity in clay due to an adjacent slope, *Australian Geomechanics*, 34(4): 17–23.

Stewart, D.P. (1999b) PYGMY user manual, version 2.1, School of Civil and Resource Engineering, The University of Western Australia.

Stewart, D.P., M.F. Randolph and R.J. Jewell (1994a) Recent developments in the design of piled bridge abutments for loading from lateral soil movements, *FHWA Conf. on Design and Construction of Deep Foundations*, Florida, Vol. 2, pp. 992–1006.

Stewart, D.P., R.J. Jewell and M.F. Randolph (1994b) Design of piled bridge abutments on soft clay for loading from lateral soil movements, *Géotechnique*, 44(2): 277–296.

Stroud, M.A. (1974) The standard penetration test in insensitive clays and soft rocks, *Proc. Eur. Symp. on Penetration Testing*, Stockholm, Vol. 2.2, pp. 367–375.

Stroud, M.A. (1989) The standard penetration test – its application and interpretation. *ICE Conf. Penetration Testing*, Birmingham: Thomas Telford, London.

Stroud, M.A. and F.G. Butler (1975) The standard penetration test and the engineering properties of glacial materials, in *Proc. Symp. on the Engineering Behaviour of Glacial Materials*, University of Birmingham.

Swain, A. (1976) *Model Ground Anchors in Clay*, Ph.D. Thesis, University of Cambridge.

Ta, L.D. and J.C. Small (1996) Analysis of piled raft systems in layered soil, *Int. J. Num. and Anal. Methods in Geomech.*, 20(1): 57–72.

Take A., A.J. Valsangkar and M.F. Randolph (1999) Analytical solution for pile drivability assessment, *Computers and Geotechnics*, 25(2): 57–74.

Teng, W.C. (1962) *Foundation Design*, New Jersey: Prentice-Hall Inc.

Terzaghi, K. (1954) Anchored bulkheads. *Trans. ASCE*, 119.

Terzaghi, K. and R.B. Peck (1948) *Soil Mechanics in Engineering Practice*, New York: John Wiley and Sons.

Terzaghi, K. and R.B. Peck (1967) *Soil Mechanics in Engineering Practice*, New York: John Wiley and Sons.

Thorburn, S., C. Laird and M.F. Randolph (1983) Storage tanks founded on soft soils reinforced with driven piles, in *Proc. Conf. on Recent Advances in Piling and Ground Treatment for Foundations*, London: ICE, pp. 157–164.

Thorburn, S. and R.S.L. MacVicar (1971) Pile load tests to failure in the Clyde alluvium (see also Discussion to Session A), *Proc. Conf. on Behaviour of Piles*, London: ICE, pp. 1–7; 53–54.

Thorburn, S. and W.J. Rigden (1980) A practical study of pile behaviour. *Proc. 12th Annual Offshore Technology Conf.*, Houston.

Thorburn, S. and J.Q. Thorburn (1977) *Review of problems associated with the construction of cast in place piles*. CIRIA Report PG2, Storeys Gate, London.

Timoshenko, S.P. and J.N. Goodier (1970) *Theory of Elasticity* (3rd edn.), New York: McGraw-Hill.

Tomlinson, M.J. (1957) The adhesion of piles driven in clay soils, *Proc. 4th Int. Conf. on Soil Mech. and Found. Eng.*, Vol. 2, pp. 66–71.

Tomlinson, M.J. (1970) Adhesion of piles in stiff clay. CIRIA Report 26, London.

Tomlinson, M.J. (1977) *Pile Design and Construction Practice*. London: Viewpoint Publications.

Tomlinson, M.J. (1986) Foundation Design and Concrete. 5th edn. London: Longman Scientific and Technical.

Toolan, F.E. and D.A. Fox (1977) Geotechnical planning of piled foundations for offshore platforms, *Proc. ICE*, 62(1): 221–244.

Toolan F.E., M.L. Lings and U.A. Mirza (1990) An appraisal of API RP2A recommendations for determining skin friction of piles in sand, *Proc. 22nd Offshore Technology Conf.*, Houston, OTC 6422, pp. 33–42.

Trenter, N.A. (1999) Engineering in glacial tills. CIRIA Publication C504, London: Storeys Gate.

Twine, D. and H. Roscoe (1996) Prop loads: guidance on design. CIRIA Research Project RP 526, CIRIA, London: Storeys Gate.

Unsworth, J.M. and W.G.K. Fleming (1990) Continuous flight auger piling instrumentation. In *Geotechnical Instrumentation in Practice*, London: Thomas Telford.

Vaughan, P.R., H.T. Lovenbury and P. Horswill (1975) The design, construction and performance of Cow Green Embankment Dam, *Géotechnique*, 25(3): 555–580.

Vesic, A.S. (1969) Experiments with instrumented pile groups in sand. ASTM STP 444, pp. 177–222.

Vesic, A.S. (1977) Design of Pile Foundations, National Co-operative Highway Research Program, Synthesis of Highway Practice No 42, Transportation Research Board, National Research Council, Washington D.C.

Viggiani, G. (1991) Dynamic measurement of small strain stiffness of fine grained soils in the triaxial apparatus. *Experimental Characterization and Modelling of Soils and Soft Rocks*, University of Naples, 75–97.

Viggiani, C. (2001) Analysis and design of piled foundations. First Arrigo Croce Lecture, *Rivista Italiana di Geotecnica*, 35(1): 47–75.

Walker, L.K., P. Darvall and P. Le (1973) Dragdown on coated and uncoated piles, *Proc. 8th Int. Conf. on Soil Mech. and Found. Eng.*, Vol. 2.1, pp. 257–262.

Weltman, A.J. (1977) Integrity testing of piles: a review. CIRIA Report PG4, Storeys Gate, London.

Weltman, A.J. and J.M. Head (1983) Site Investigation Manual. DoE/CIRIA Special Publication 25, CIRIA, London.

Weltman, A.J. and P.R. Healy (1978) Piling in boulder clay and other glacial tills. DoE/CIRIA Report PG 5, London.

Weltman, A.J. and J.A. Little (1977) A review of bearing pile types DoE/CIRIA Piling Development Group Report PG. 1.

Whitaker, T. (1963) *Load cells for measuring the base loads in bored piles and cylinder foundations.* BRS Current Paper, Engineering Series No.11.

Whitaker, T. (1975) *The Design of Piled Foundations* (2nd edn.), Oxford: Pergamon.

White, D.J. and M.D. Bolton (2002) Observing friction fatigue on a jacked pile, in S.M. Springman (ed.), *Centrifuge and Constitutive Modelling: Two extremes*, Rotterdam: Zwets and Zeitlinger, pp. 347–354.

Whittle, R.T. and D. Beattie (1972) Standard pile caps I, *Concrete*, 6(1): 34–36.

Whittle, R.T. and D. Beattie (1972) Standard pile caps II, *Concrete*, 6(2): 39–41.

Whittle, A.J. (1992) Assessment of an effective stress analysis for predicting the performance of driven piles in clays, *Proc. Conf. on Offshore Site Investigation and Foundation Behaviour, Society for Underwater Technology*, Kluwer, 28: 607–643.

Wilkins, M. (1963) Calculation of elastic–plastic flow. Lawrence Radiation Laboratory Report UCRL 7322.

Williams, A.F. and P.J.N. Pells (1981) Side resistance of rock sockets in sandstone, mudstone and shale, *Can. Geotech. J.*, 18: 502–513.

Wroth, C.P., M.F. Randolph, G.T. Houlsby and M. Fahey (1979) A review of the engineering properties of soils, with particular reference to the shear modulus. Cambridge University Research Report, CUED/D-Soils TR 75.

Youd, T.L. (1972) Compaction of sands by repeated shear straining, *ASCE Journal of Soil Mechanics*, 98(SM7): 709–725.

Zhang, L. and H.H. Einstein (1998) End bearing capacity of drilled shafts in rock, *ASCE J. of Geot. and Geoenvironmental Engineering*, 124(7): 574–584.

Index